LAW, TECHNOLOGY AND SOCIETY

This book considers the implications of the regulatory burden being borne increasingly by technological management rather than by rules of law. If crime is controlled, if human health and safety are secured, if the environment is protected, not by rules but by measures of technological management—designed into products, processes, places and so on—what should we make of this transformation?

In an era of smart regulatory technologies, how should we understand the 'regulatory environment', and the 'complexion' of its regulatory signals? How does technological management sit with the Rule of Law and with the traditional ideals of legality, legal coherence, and respect for liberty, human rights and human dignity? What is the future for the rules of criminal law, torts and contract law—are they likely to be rendered redundant? How are human informational interests to be specified and protected? Can traditional rules of law survive not only the emergent use of technological management but also a risk management mentality that pervades the collective engagement with new technologies? Even if technological management is effective, is it acceptable? Are we ready for rule by technology?

Undertaking a radical examination of the disruptive effects of technology on the law and the legal mind-set, Roger Brownsword calls for a triple act of re-imagination: first, re-imagining legal rules as one element of a larger regulatory environment of which technological management is also a part; secondly, re-imagining the Rule of Law as a constraint on the arbitrary exercise of power (whether exercised through rules or through technological measures); and, thirdly, re-imagining the future of traditional rules of criminal law, tort law, and contract law.

Roger Brownsword has professorial appointments in the Dickson Poon School of Law at King's College London and in the Department of Law at Bournemouth University, and he is an honorary Professor in Law at the University of Sheffield.

Part of the Law, Science and Society series

Series editors
John Paterson
University of Aberdeen, UK
Julian Webb
University of Melbourne, Australia

For information about the series and details of previous and forthcoming titles, see https://www.routledge.com/law/series/CAV16

A GlassHouse Book

LAW, TECHNOLOGY AND SOCIETY

Re-imagining the Regulatory Environment

Roger Brownsword

Routledge
Taylor & Francis Group
a GlassHouse Book

First published 2019
by Routledge
2 Park Square, Milton Park, Abingdon, Oxon OX14 4RN

and by Routledge
52 Vanderbilt Avenue, New York, NY 10017

a GlassHouse book

Routledge is an imprint of the Taylor & Francis Group, an informa business

British Library Cataloguing-in-Publication Data
A catalogue record for this book is available from the British Library

Library of Congress Cataloging-in-Publication Data
A catalog record for this book has been requested

ISBN: 978-0-8153-5645-5 (hbk)
ISBN: 978-0-8153-5646-2 (pbk)
ISBN: 978-1-351-12818-6 (ebk)

Typeset in Galliard
by Apex CoVantage, LLC

Printed and bound in Great Britain by
TJ International Ltd, Padstow, Cornwall

CONTENTS

PREFACE

In *Rights, Regulation and the Technological Revolution* (2008) I identified and discussed the generic challenges involved in creating the right kind of regulatory environment at a time of rapid and disruptive technological development. While it was clear that new laws were required to authorise, to support, and to limit the development and application of a raft of novel technologies, it was not clear how regulators might accommodate the deep moral differences elicited by some of these technologies (particularly by biotechnologies), how to put in place effective laws when online transactions and interactions crossed borders in the blink of an eye, and how to craft sustainable and connected legal frameworks. However, there was much unfinished business in that book and, in particular, there was more to be said about the way in which technological instruments were themselves being deployed by regulators.

While many technological applications assist regulators in monitoring compliance and in discouraging non-compliance, there is also the prospect of applying complete technological fixes—for example, replacing coin boxes with card payments, or using GPS to immobilise supermarket trolleys if someone tries to wheel them out of bounds, or automating processes so that both potential human offenders and potential human victims are taken out of the equation, thereby eliminating certain kinds of criminal activity. While technological management of crime might be effective, it changes the complexion of the regulatory environment in ways that might be corrosive of the prospects for a moral community. The fact that pervasive technological management ensures that it is impossible to act in ways that violate the personal or proprietary interests of others signifies, not a moral community, but the very antithesis of a community that strives freely to do the right thing for the right reason.

At the same time, technological management can be applied in less controversial ways, the regulatory intention being to promote human health and safety or to protect the environment. For example, while autonomous vehicles will be designed to observe road traffic laws—or, at any rate, I assume that this will be the case so long as they share highway space with driven vehicles—it would be a distortion to present the development of such vehicles as a regulatory response to road traffic violations; the purpose behind autonomous cars is not to control crime but, rather, to enhance human health and safety. Arguably, this kind of use of a technological fix is less problematic morally: it is not intended to impinge on the opportunities that regulatees have for doing the right thing; and, insofar as it reduces the opportunities for doing the wrong thing, it is regulatory crime rather than 'real' crime that is affected. However, even if the use of technological management for the general welfare is less problematic morally, it is potentially highly disruptive (impacting on the pattern of employment and the preferences of agents).

This book looks ahead to a time when technological management is a significant part of the regulatory environment, seeking to assess the implications of this kind of regulatory strategy not only in the area of criminal justice but also in the area of health and safety and environmental protection. When regulators use technological management to define what is possible and what is impossible, rather than prescribing what regulatees ought or ought not to do, what does this mean for the Rule of Law, for the ideals of legality and coherence? What does it mean for those bodies of criminal law and the law of torts that are superseded by the technological fix? And, does the law of contract have a future when the infrastructure for 'smart' transactions is technologically managed, when transactions are automated, and when 'transactors' are not human?

When we put these ideas together, we see that technological innovation impacts on the landscape of the law in three interesting ways. First, the development of new technologies means that some new laws are required but, at the same time, the use of technological management (in place of legal rules) means that some older laws are rendered redundant. In other words, technological innovation in the present century signifies a need for both more and less law. Secondly, although technological management replaces a considerable number of older duty-imposing rules, the background laws that authorise legal interventions become more important than ever in setting the social licence for the use of technological management. Thirdly, the 'risk management' and 'instrumentalist' mentality that accompanies technological management reinforces a thoroughly 'regulatory' approach to legal doctrine, an approach that jars with a traditional approach that sees law as a formalisation of some simple moral principles and that, concomitantly, understands legal reasoning as an exercise in maintaining and applying a 'coherent' body of doctrine.

If there was unfinished business in 2008, I am sure that the same is true today. In recent years, the emergence of AI, machine learning and robotics has provoked fresh concerns about the future of humanity. That future will be shaped not only by the particular tools that are developed and the ways in which they are applied but also by the way in which humans respond to and embrace new technological options. The role of lawyers in helping communities to engage in a critical and reflective way with a cascade of emerging tools is, I suggest, central to our technological futures.

The central questions and the agenda for the book, together with my developing thoughts on the concepts of the 'regulatory environment', the 'complexion' of the regulatory environment, the notion of 'regulatory coherence', the key regulatory responsibilities, and the technological disruption of the legal mind-set have been prefigured in a number of my publications, notably: 'Lost in Translation: Legality, Regulatory Margins, and Technological Management' (2011) 26 *Berkeley Technology Law Journal* 1321–1365; 'Regulatory Coherence—A European Challenge' in Kai Purnhagen and Peter Rott (eds), *Varieties of European Economic Law and Regulation: Essays in Honour of Hans Micklitz* (New York: Springer, 2014) 235–258; 'Comparatively Speaking: "Law in its Regulatory Environment"' in Maurice Adams and Dirk Heirbaut (eds), *The Method and Culture of Comparative Law* (Festschrift for Mark van Hoecke) (Oxford: Hart, 2014) 189–205; 'In the Year 2061: From Law to Technological Management' (2015) 7 *Law, Innovation and Technology* 1–51; 'Field, Frame and Focus: Methodological Issues in the New Legal World' in Rob van Gestel, Hans Micklitz, and Ed Rubin (eds), *Rethinking Legal Scholarship* (Cambridge: Cambridge University Press, 2016) 112–172; '*Law as a Moral Judgment*, the Domain of Jurisprudence, and Technological Management' in Patrick Capps and Shaun D. Pattinson (eds), *Ethical Rationalism and the Law* (Oxford: Hart, 2016) 109–130; 'Law, Liberty and Technology', in R. Brownsword, E. Scotford, and K.Yeung (eds), *The Oxford Handbook of Law, Regulation and Technology* (Oxford: Oxford University Press, 2016 [e-publication]; 2017) 41–68; 'Technological Management and the Rule of Law' (2016) 8 *Law, Innovation and Technology* 100–140; 'New Genetic Tests, New Research Findings: Do Patients and Participants Have a Right to Know—and Do They Have a Right Not to Know?' (2016) 8 *Law, Innovation and Technology* 247–267; 'From Erewhon to Alpha Go: For the Sake of Human Dignity Should We Destroy the Machines?' (2017) 9 *Law, Innovation and Technology* 117–153; 'The E-Commerce Directive, Consumer Transactions, and the Digital Single Market: Questions of Regulatory Fitness, Regulatory Disconnection and Rule Redirection' in Stefan Grundmann (ed) *European Contract Law in the Digital Age* (Cambridge: Intersentia, 2017) 165–204; 'After Brexit: Regulatory-Instrumentalism, Coherentism, and the English Law of Contract' (2018) 35 *Journal of Contract Law* 139–164; and, 'Law and Technology: Two Modes of Disruption, Three Legal Mind-Sets, and the Big Picture of Regulatory Responsibilities' (2018) 14 *Indian Journal of Law and Technology* 1–40. While there are plenty of indications of fragments of my thinking in these earlier publications, I hope that the book conveys the bigger picture of the triple act of re-imagination that I have in mind—re-imagining the regulatory environment, re-imagining traditional legal values, and re-imagining traditional legal rules.

Prologue

1

IN THE YEAR 2061

From law to technological management

I Introduction

In the year 2061—just 100 years after the publication of HLA Hart's *The Concept of Law*[1]—I imagine that few, if any, hard copies of that landmark book will be in circulation. The digitisation of texts has already transformed the way that many people read; and, as the older generation of hard copy book lovers dies, there is a real possibility that their reading (and text-related) preferences will pass away with them.[2] Still, even if the way in which *The Concept of Law* is read is different, should we at least assume that Hart's text will remain an essential part of any legal education? Perhaps we should; perhaps the book will still be required reading. However, my guess is that the jurists and legal educators of 2061 will view Hart's analysis as being of limited interest; the world will have moved on; and, just as Hart rejects the Austinian command model of law as a poor representation of twentieth-century legal systems, so history will repeat itself. In 2061, I suggest that Hart's rule model will seem badly out of touch with the use of modern technologies as regulatory instruments and, in particular, with the pervasive use of 'technological management' in place of what Hart terms the 'primary' rules (namely, duty-imposing rules that are directed at the conduct of citizens).[3]

---

1 Oxford: Clarendon Press, 1961; second edition 1994.
2 I am not sure, however, that I agree with Kevin Kelly's assertion that 'People of the Book favour solutions by laws, while People of the Screen favour technology as a solution to all problems' (Kevin Kelly, *The Inevitable* (New York: Penguin, Books, 2017) 88).
3 Compare Scott Veitch, 'The Sense of Obligation' (2017) 8 *Jurisprudence* 415, 430–432 (on the collapse of obligation into obedience).

Broadly speaking, by 'technological management' I mean the use of technologies—typically involving the design of products or places, or the automation of processes—with a view to managing certain kinds of risk by excluding (i) the possibility of certain actions which, in the absence of this strategy, might be subject only to rule regulation, or (ii) human agents who otherwise might be implicated (whether as rule-breakers or as the innocent victims of rule-breaking) in the regulated activities.[4] Anticipating pervasive reliance on technological infrastructures (and, by implication, reliance on technological management) Will Hutton says that we can expect 'to live in smart cities, achieve mobility in smart transport, be powered by smart energy, communicate with smart phones, organise our financial affairs with smart banks and socialise in ever smarter networks'.[5] It is, indeed, 'a dramatic moment in world history';[6] and I agree with Hutton that 'Nothing will be left untouched.'[7] Importantly, with nothing left untouched, we need to understand that there will be major implications for law and regulation.

Already, we can see how the context presupposed by Hart's analysis is being disrupted by new technologies. As a result, some of the most familiar and memorable passages of Hart's commentary are beginning to fray. Recall, for example, Hart's evocative contrast between the external and the internal point of view in relation to rules (whether these are rules of law or rules of a game). Although an observer, whose viewpoint is external, can detect some regularities and patterns in the conduct of those who are observed, such an (external) account misses out the distinctively (internal) rule-guided dimension of social life. Famously, Hart underlines the seriousness of this limitation of the external account in the following terms:

> If ... the observer really keeps austerely to this extreme external point of view and does not give any account of the manner in which members of the group who accept the rules view their own regular behaviour, his description of their life cannot be in terms of rules at all, and so not in the terms of the rule-dependent notions of obligation or duty. Instead, it will be in terms of observable regularities of conduct, predictions, probabilities, and signs ... His view will be like the view of one who, having observed the working of a traffic signal in a busy street for some time, limits himself to saying that when the light turns red there is a high probability that the traffic will stop ... In so doing he will miss out

---

4 See, further, Chapter Two, Part II. Compare Ugo Pagallo, *The Laws of Robots* (Dordrecht: Springer, 2013) 183–192, differentiating between environmental, product and communication design and distinguishing between the design of 'places, products and organisms' (185).

5 Will Hutton, *How Good We Can Be* (London: Little, Brown, 2015) 17.

6 *Ibid.*

7 *Ibid.*

a whole dimension of the social life of those whom he is watching, since for them the red light is not merely a sign that others will stop: they look upon it as a *signal for* them to stop, and so a reason for stopping in conformity to rules which make stopping when the light is red a standard of behaviour and an obligation.[8]

To be sure, even in 2061, the Hartian distinction between an external and an internal account will continue to hold good *where it relates to a rule-governed activity*. However, to the extent that rule-governed activities are overtaken by technological management, the distinction loses its relevance; for, where activities are so managed, the appropriate description will no longer be in terms of rules and rule-dependent notions.

Consider Hart's own example of the regulation of road traffic. In 1961, the idea that driverless cars might be developed was the stuff of futurology.[9] However, today, things look very different.[10] Indeed, it seems entirely plausible to think that, before too long, rather than being seen as 'a beckoning rite of passage', learning to drive 'will start to feel anachronistic'—for the next generation, driving a car or a truck might be comparable to writing in longhand.[11] At all events, by 2061, in the 'ubiquitous' or 'smart' cities[12] of that time, if the movement of vehicles is controlled by anything resembling traffic lights, the external account will be the only account; the practical reason and actions of the humans inside the cars will no longer be material. By 2061, it will be each vehicle's on-board technologies that will control the movement of the traffic—on the roads of 2061, technological management will have replaced road traffic laws.[13]

---

8 Hart (n 1), second edition, 89–90.
9 See Isaac Asimov, 'Visit to the World's Fair of 2014' *New York Times* (August 16, 1964) available at www.nytimes.com/books/97/03/23/lifetimes/asi-v-fair.html (last accessed 1 November 2018). According to Asimov:

   Much effort will be put into the designing of vehicles with 'Robot-brains' vehicles that can be set for particular destinations and that will then proceed there without interference by the slow reflexes of a human driver. I suspect one of the major attractions of the 2014 fair will be rides on small roboticized cars which will maneuver in crowds at the two-foot level, neatly and automatically avoiding each other.

10 See, e.g., Erik Brynjolfsson and Andrew McAfee, *The Second Machine Age* (New York: W.W. Norton and Co, 2014) Ch.2.
11 Jaron Lanier, *Who Owns the Future?* (London: Allen Lane, 2013) 349.
12 See, e.g., Jane Wakefield, 'Building cities of the future now' BBC News Technology, February 21, 2013: available at www.bbc.co.uk/news/technology-20957953 (last accessed November 1, 2018); and the introduction to the networked digital city in Adam Greenfield, *Radical Technologies* (London: Verso, 2017) 1–8.
13 For the relevant technologies, see Hod Lipson and Melba Kurman, *Driverless: Intelligent Cars and the Road Ahead* (Cambridge, Mass.: MIT Press, 2016). For the testing of fully driverless cars in the UK, see Graeme Paton, 'Driverless cars on UK roads this year after rules relaxed' *The Times*, March 17, 2018, p.9.

These remarks are no an ad hominem attack on Hart; they are of general application. What Hart says about the external and internal account in relation to the rules of the road will be seen as emblematic of a pervasive mistaken assumption made by twentieth-century jurists—by Hart and his supporters as much as by their critics. That mistake of twentieth-century jurists is to assume that rules and norms are the exclusive keys to social ordering. By 2061, rules and norms will surely still play some part in social ordering; and, some might still insist that all conceptions of law should start with the Fullerian premise that law is the enterprise of subjecting human conduct to the governance of rules.[14] But, by 2061, if the domain of jurisprudence is restricted to the normative (rule-based) dimension of the regulatory environment, I predict that this will render it much less relevant to our cognitive interests in the legitimacy and effectiveness of that environment.

Given the present trajectory of modern technologies, it seems to me that technological management (whether with driverless cars, the Internet of Things, blockchain, or bio-management) is set to join law, morals and religion as one of the principal instruments of social control. To a considerable extent, technological infrastructures that support our various transactions and interactions will structure social order. The domain of today's rules of law—especially, the 'primary' rules of the criminal law and the law of torts—is set to shrink. And this all has huge implications for a jurisprudence that is predicated on the use of rules and standards as regulatory tools or instruments. It has implications, for example, for the way that we understand the virtue of legality and the Rule of Law; it bears on the way that we understand (and value) regulatory coherence; and it calls for some re-focusing of those liberal critiques of law that assume that power is exercised primarily through coercive rules. To bring these issues onto the jurisprudential agenda, we must enlarge the field of interest; and I suggest that we should do this by developing a concept of the regulatory environment that accommodates both rules and technological management—that is to say, that facilitates inquiry into both the normative and the non-normative dimensions of the environment. With the field so drawn, we can begin to assess the changing complexion of the regulatory environment and its significance for traditional legal values as well as the communities who live through these transformative times.

II Golf carts and the direction of regulatory travel

At the Warwickshire Golf and Country Club, there are two championship 18-hole golf courses, together with many other facilities, all standing (as the club's website puts it) in '456 flowing acres of rolling Warwickshire

---

14 Lon L. Fuller, *The Morality of Law* (New Haven: Yale University Press, 1969).

countryside'.[15] The club also has a large fleet of golf carts. However, in 2011, this idyllic setting was disturbed when the club began to experience some problems with local 'joy-riders' who took the carts off the course. In response, the club used GPS technology so that 'a virtual geo-fence [was created] around the whole of the property, meaning that tagged carts [could not] be taken off the property.'[16] With this technological fix, anyone who tries to drive a cart beyond the geo-fence will find that it is simply immobilised. In the same way, the technology enables the club to restrict carts to paths in areas which have become wet or are under repair and to zone off greens, green approaches, water hazards, bunkers, and so on. With these measures of technological management, the usual regulatory pressures were relieved.

Let me highlight three points about the technological fix applied at the Warwickshire. First, to the extent that the activities of the joy-riders were already rendered 'illegal' by the criminal law, the added value of the use of technological management was to render those illegal acts 'impossible'. In place of the relatively ineffective rules of the criminal law, we have effective technological management. Secondly, while the signals of the criminal law sought to engage the prudential or moral reason of the joy-riders, these signals were radically disrupted by the measures of technological management. Following the technological fix, the signals speak only to what is possible and what is impossible; technological management guarantees that the carts are used responsibly but they are no longer used in a way for which users are responsible. Thirdly, regulators—whether responsible for the public rules of the criminal law or the private rules of the Warwickshire—might employ various kinds of technologies that are designed to discourage breach of the rules. For example, the use of CCTV surveillance and DNA profiling signals that the chance of detecting breaches of the rules and identifying those who so breach the rules is increased. However, this leaves open the possibility of non-compliance and falls short of full-scale technological management. With technological management, such as geo-fencing, nothing is left to chance; there is no option other than 'compliance'.

One of the central claims of this book is that the direction of regulatory travel is towards technological management. Moreover, there are two principal regulatory tracks that might converge on the adoption of technological management. One track is that of the criminal justice system. Those criminal laws that are intended to protect person and property are less than entirely effective. In an attempt to improve the effectiveness of these laws, various technological tools (of surveillance, identification, detection and correction) are employed. If it were possible to sharpen up these tools so that they became

---

15 www.thewarwickshire.com (last accessed 1 November 2018).

16 See www.hiseman.com/media/releases/dsg/dsg200412.htm (last accessed 1 November 2018). Similarly, see 'intelligent cart' systems used by supermarkets: see gatekeepersystems.com/us/cart-management.php (for, inter alia, cart retention) (last accessed 2 November 2018).

instruments of full-scale technological management (rendering it impossible to commit the offences), this would seem like a natural step for regulators to take. Hence, we should not be surprised that there is now discussion about the possible use of geo-fencing around target buildings and bridges in order to prevent vehicles being used for terrorist attacks.[17] The other track focuses on matters such as public welfare, health and safety, and conservation of energy. With the industrialisation of societies and the development of transport systems, new machines and technologies presented many dangers which regulators tried to manage by introducing health and safety rules[18] as well as by applying not always appropriate rules of tort law.[19] In the twenty-first century, we have the technological capability to replace humans with robots in some dangerous places, to create safer environments where humans continue to operate, and to introduce smart energy-saving devices. However, the technological management that we employ in this way can also be employed (pervasively so in on-line environments) to prevent acts that those with the relevant regulatory power regard as being contrary to their interests or to the interests of those for whom they have regulatory responsibility. It might well be the case that whatever concerns we happen to have about the use of technological management will vary from one regulatory context to another, and from public to private use; but, if we do nothing to articulate and engage with those concerns, there is reason to think that a regulatory pre-occupation with finding 'what works', in conjunction with a 'risk management' mind set, will conduce to more technological management.

III What price technological management?

Distinctively, technological management seeks to design out harmful options or to design in protections against harmful acts. In addition to the cars and

---

17 See, Graeme Paton, 'Digital force fields to stop terrorist vehicles' *The Times*, July 1, 2017, p.4.

18 Compare, e.g., Susan W. Brenner, *Law in an Era of 'Smart' Technology* (New York: Oxford University Press, 2007) (for the early US regulation of bicycles). At 36–37, Brenner says:

> Legislators at first simply banned bicycles from major thoroughfares, including sidewalks. These early enactments were at least ostensibly based on public safety considerations. As the North Carolina Supreme Court explained in 1887, regulations prohibiting the use of bicycles on public roads were a valid exercise of the police power of the state because the evidence before the court showed 'that the use of the bicycle on the road materially interfered with the exercise of the rights and safety of others in the lawful use of their carriages and horses in passing over the road'.

19 See, Kyle Graham, 'Of Frightened Horses and Autonomous Vehicles: Tort Law and its Assimilation of Innovations' (2012) 52 *Santa Clara Law Review* 1241, 1243–1252 (for the early automobile lawsuits and the mischief of frightened horses). For a review of the responses of a number of European legal systems to steam engines, boilers, and asbestos, see Miquel Martin-Casals (ed), *The Development of Liability in Relation to Technological Change* (Cambridge: Cambridge University Press, 2010).

carts already mentioned, a well-known example of this strategy in relation to products is so-called digital rights management, this being employed with a view to the protection, or possibly extension, of IP rights.[20] While IP proprietors might try to protect their interests by imposing contractual restrictions on use as well as by enlisting the assistance of governments or ISPs and so on, they might also try to achieve their purposes by designing their products in ways that 'code' against infringement.[21] Faced with this range of measures, the end-user of the product is constrained by various IP-protecting rules but, more importantly, by the technical limitations embedded in the product itself. Similarly, where technological management is incorporated in the design of places—for instance, in the architecture of transport systems such as the Metro—acts that were previously possible but prohibited (such as riding on the train without a ticket) are rendered impossible (or, at any rate, for all practical purposes, impossible). For agents who wish to ride the trains, it remains the case that the rules require a ticket to be purchased but the 'ought' of this rule is overtaken by the measures of technological management that ensure that, without a valid ticket, there will be no access to the trains and no ride.

Driven by the imperatives of crime prevention and risk management, technological management promises to be the strategy of choice for public regulators of the present century.[22] For private regulators, too, technological management has its attractions—and nowhere more so, perhaps, than in those on-line environments that will increasingly provide the platform and the setting for our everyday interactions and transactions (with access being controlled by key intermediaries and their technological apparatus). Still, if technological management proves effective in preventing crime and IP infringement, and the like; and if, at the same time, it contributes to human health and safety as well as protecting the environment, is there any cause for concern?

In fact, the rise of technological management in place of traditional legal rules might give rise to several sets of concerns. Let me briefly sketch just four kinds of concern: first, that the technology cannot be trusted, possibly leading to catastrophic consequences; secondly, that the technology will diminish our autonomy and liberty; thirdly, that the technology will have difficulty in

---

20 Compare, e.g., Case C-355/12, *Nintendo v PC Box*.
21 Seminally, see Lawrence Lessig, *Code and Other Laws of Cyberspace* (New York: Basic Books, 1999).
22 Compare Andrew Ashworth and Lucia Zedner, *Preventive Justice* (Oxford: Oxford University Press, 2014). Although Ashworth and Zedner raise some important 'Rule of Law' concerns about the use of preventive criminal measures, they are not thinking about technological management. Rather, their focus is on the extended use of coercive rules and orders. See, further, Deryck Beyleveld and Roger Brownsword, 'Punitive and Preventive Justice in an Era of Profiling, Smart Prediction, and Practical Preclusion: Three Key Questions' (2019) *International Journal of Law in Context* (forthcoming), and Ch.9 below.

reflecting ethical management and, indeed, might compromise the conditions for any kind of moral community; and, fourthly, that it is unclear how technological management will impact on the law and whether it will comport with its values.

(i) Can the technology be trusted?

For those who are used to seeing human operatives driving cars, lorries, and trains, it comes as something of a shock to learn that, in the case of planes, although there are humans in the cockpit, for much of the flight the aircraft actually flies itself. In the near future, it seems, cars, lorries, and trains too, will be driving themselves. Even though planes seem to operate safely enough on auto-pilot, some will be concerned that the more general automation of transport will prove to be a recipe for disaster; that what Hart's observer at the intersection of roads is likely to see is not the well-ordered technological management of traffic but chaos.

Such concerns can scarcely be dismissed as ill-founded. For example, in the early years of the computer-controlled trains on the Californian Bay Area Rapid Transit System (BART), there were numerous operational problems, including a two-car train that ran off the end of the platform and into a parking lot, 'ghost' trains that showed up on the system, and real trains that failed to show on the system because of dew on the tracks and too low a voltage being passed through the rails.[23] Still, teething problems can be expected in any major new system; and, given current levels of road traffic accidents, the concern that technologically managed transportation systems might not be totally reliable, hardly seems a sufficient reason to reject the whole idea. However, as with any technology, the risk does need to be assessed; it needs to be managed; and the package of risk management measures that is adopted needs to be socially acceptable. In some communities, regulators might follow the example of section 3 of the District of Columbia's Automated Vehicle Act 2012 where a human being is required to be in the driver's seat 'prepared to take control of the autonomous vehicle at any moment', or they might require that a duly licensed human 'operator' is at least present in the vehicle.[24] In all places, though, where there remains the possibility of technological malfunction (whether arising internally or as a result of external intervention) and consequent injury to persons or damage to their property, the agreed package of risk management measures is likely to provide for compensation.[25]

---

23 See www.cs.mcgill.ca/~rwest/wikispeedia/wpcd/wp/b/Bay_Area_Rapid_Transit.htm (last accessed 1 November, 2018).

24 On (US) State regulation of automated cars, see John Frank Weaver, *Robots Are People Too* (Santa Barbara, Ca: Praeger, 2014) 55–60.

25 See, further, Ch.10.

These remarks about the reliability of technological management and acceptable risk management measures are not limited to transportation. For example, where profiling and biometric identification technologies are employed to manage access to both public and private places, there might be concerns about the accuracy of both the risk assessment represented by the profiles and the biometric identifiers. Even if the percentage of false positives and false negatives is low, when these numbers are scaled up to apply to large populations there may be too many errors for the risks of misclassification and misidentification to be judged acceptable—or, at any rate, the risks may be judged unacceptable unless there are human operatives present who can intervene to override any obvious error.[26]

To take another example, there might be concerns about the safety and reliability of robots where they replace human carers. John Frank Weaver poses the following hypothetical:

> [S]uppose the Aeon babysitting robot at Fukuoka Lucle mall in Japan is responsibly watching a child, but the child still manages to run out of the child-care area and trip an elderly woman. Should the parent[s] be liable for that kid's intentional tort?[27]

As I will suggest in Part Three of the book, there are two rather different ways of viewing this kind of scenario. The first way assumes that before retailers, such as Aeon, are to be licensed to introduce robot babysitters, and parents permitted to make use of robocarers, there needs to be a collectively agreed scheme of compensation should something 'go wrong'. It follows that the answer to Weaver's question will depend on the agreed terms of the risk management package. The second way, characteristic of traditional tort law, is guided by principles of corrective justice: liability is assessed by reference to what communities judge to be fair, just and reasonable—and different communities might have different ideas about whether it would be fair, just and reasonable to hold the parents liable in the hypothetical circumstances.[28] Provided that it is clear which of these ways of attributing liability is applicable, there should be no great difficulty. However, we should not assume that regulatory environments are always so clear in their signalling.

---

26 Generally, see House of Commons Science and Technology Committee, *Current and future use of biometric data and technologies* (Sixth Report of Session 2014–15, HC 734).

27 Weaver (n 24), 89.

28 On different ideas about parental liability, see Ugo Pagallo (n 4) at 124–130 (contrasting American and Italian principles). On the contrast between the two general approaches, compare F. Patrick Hubbard, '"Sophisticated Robots": Balancing Liability, Regulation, and Innovation' (2014) 66 *Florida Law Review* 1803. In my terms, while the traditional tort-based corrective justice approach reflects a 'coherentist' mind-set, the risk-management approach reflects a 'regulatory-instrumentalist' mind-set. See, further, Ch.8.

If technological management malfunctions in ways that lead to personal injury, damage to property and significant inconvenience, this will damage trust. Trust will also be damaged if there is a suspicion that personal data is being covertly collected or applied for unauthorised purposes. None of this is good; but some might be concerned that things could be worse, much worse. For example, some might fear that, in our quest for greater safety and well-being, we will develop and embed ever more intelligent devices to the point that there is a risk of the extinction of humans—or, if not that, then a risk of humanity surviving 'in some highly suboptimal state or in which a large portion of our potential for desirable development is irreversibly squandered.'[29] If this concern is well founded—if the smarter and more reliable the technological management, the less we should trust it—then communities will need to be extremely careful about how far and how fast they go with intelligent devices.

(ii) Will technological management diminish our autonomy and liberty?

Whether or not technological management might impact negatively on an agent's autonomy or liberty depends in part on how we conceive of 'autonomy' and 'liberty' and the relationship between them. Nevertheless, let us assume that, other things being equal, agents value (i) having more rather than fewer options and (ii) making their 'own choices' between options. So assuming, agents might well be concerned that technological management will have a negative impact on the breadth of their options as well as on making their own choices (if, for example, agents become over-reliant on their personal digital assistants).[30]

Consider, for example, the use of technological management in hospitals where the regulatory purpose is to improve the conditions for patient safety.[31] Let us suppose that we could staff our hospitals in all sections, from the kitchens to the front reception, from the wards to the intensive care unit, from accident and emergency to the operating theatre, with robots. Moreover, suppose that all hospital robot operatives were entirely reliable and were programmed (in the spirit of Asimov's laws) to make patient safety their

---

29 See, Nick Bostrom, *Superintelligence* (Oxford University Press, 2014), 281 (n 1); and, Martin Ford, *The Rise of the Robots* (London: Oneworld, 2015) Ch.9.

30 See, e.g., Roger Brownsword, 'Disruptive Agents and Our Onlife World: Should We Be Concerned?' (2017) 4 *Critical Analysis of Law* (symposium on Mireille Hildebrandt, *Smart Technologies and the End(s) of Law*) 61; and Jamie Bartlett, *The People vs Tech* (London: Ebury Press, 2018) Ch.1.

31 For discussion, see Roger Brownsword, 'Regulating Patient Safety: Is it Time for a Technological Response?' (2014) 6 *Law, Innovation and Technology* 1; and Ford (n 29), Ch.6.

top priority.[32] Why should we be concerned about any loss of autonomy or liberty?

First, the adoption of nursebots or the like might impact on patients who prefer to be cared for by human nurses. Nursebots are not their choice; and they have no other option. To be sure, by the time that nursebots are commonplace, humans will probably be surrounded by robot functionaries and they will be perfectly comfortable in the company of robots. However, in the still early days of the development of robotic technologies, many humans will not feel comfortable. Even if the technologies are reliable, many humans may prefer to be treated in hospitals that are staffed by humans—just as the Japanese apparently prefer human to robot carers.[33] Where human carers do their job well, it is entirely understandable that many will prefer the human touch. However, in a perceptive commentary, Sherry Turkle, having remarked on her own positive experience with the orderlies who looked after her following a fall on icy steps in Harvard Square,[34] goes on to showcase the views of one of her interviewees, 'Richard', who was left severely disabled by an automobile accident.[35] Despite being badly treated by his human carers, Richard seems to prefer such carers against more caring robots. As Turkle reads Richard's views,

> For Richard, being with a person, even an unpleasant, sadistic person, makes him feel that he is still alive. It signifies that his way of being in the world has a certain dignity, even if his activities are radically curtailed. For him, dignity requires a feeling of authenticity, a sense of being connected to the human narrative. It helps sustain him. Although he would not want his life endangered, he prefers the sadist to the robot.[36]

While Richard's faith in humans may seem a bit surprising, his preferences are surely legitimate; and their accommodation does not necessarily present a serious regulatory problem. In principle, patients can be given appropriate

---

32 According to the first of Asimov's three laws, 'A robot may not injure a human being or, through inaction, allow a human being to come to harm.' See en.wikipedia.org/wiki/Three_Laws_of_Robotics (last accessed 1 November 2018). Already, we can see reflections of Asimov's laws in some aspects of robotic design practice—for example, by isolating dangerous robots from humans, by keeping humans in the loop, and by enabling machines to locate a power source in order to recharge its batteries: see F. Patrick Hubbard (n 28) 1808–1809.

33 See, Michael Fitzpatrick, 'No, robot: Japan's elderly fail to welcome their robot overlords' *BBC News*, February 4, 2011: available at www.bbc.co.uk/news/business-12347219 (last accessed 1 November 2018).

34 Sherry Turkle, *Alone Together* (New York: Basic Books, 2011) 121–122.

35 *Ibid.*, 281–282.

36 *Ibid.*

choices: some may elect to be treated in a traditional robot-free hospital (with the usual warts and waiting lists, and possibly with a surcharge being applied for this privilege), others in 24/7 facilities that involve various degrees of robotics (and, in all likelihood, rapid admission and treatment). Accordingly, so long as regulators are responsive to the legitimate different preferences of agents, autonomy and liberty are not challenged and might even be enhanced.

Secondly, the adoption of nursebots can impact negatively on the options that are available for those humans who are prospective nurses. Even if one still makes one's own career choices, the options available are more restricted. However, it is not just the liberty of nurses that might be so diminished. Already there are a number of hospitals that utilise pharmacy dispensing robots[37]—including a robot named 'Phred' (Pharmacy Robot-Efficient Dispensing) at the Queen Elizabeth Hospital Birmingham[38]—which are claimed to be faster than human operatives and totally reliable; and, similarly, RALRP (robotic-assisted laparoscopic radical prostatectomy) is being gradually adopted. Speaking of the former, Inderjit Singh, Associate Director of Commercial Pharmacy Services, explained that, in addition to dispensing, Phred carries out overnight stock-taking and tidying up. Summing up, he said:

> 'This sort of state-of-the-art technology is becoming more popular in pharmacy provision, both in hospitals and community pharmacies. It can dispense a variety of different medicines in seconds—at incredible speeds and without error. This really is a huge benefit to patients at UHB.'

If robots can make the provision of pharmacy services safer—in some cases, even detecting cases where doctors have mis-prescribed the drugs—then why, we might wonder, should we not generalise this good practice?

Indeed, why not? But, in the bigger picture, the concern is that we are moving from designing products so that they can be used more safely by humans (whether these are surgical instruments or motor cars) to making the product even more safe by altogether eliminating *human* control and use. So, whether it is driverless cars, lorries,[39] or Metro trains,[40] or robotic

---

37 See Christopher Steiner, *Automate This* (New York: Portfolio/Penguin, 2012) 154–156.

38 See www.uhb.nhs.uk/news/new-pharmacy-robot-named.htm. (last accessed 1 November 2018). For another example of an apparently significant life-saving use of robots, achieved precisely by 'taking humans out of the equation', see 'Norway hospital's "cure" for human error' BBC News, May 9, 2015: available at www.bbc.co.uk/news/health-32671111 (last accessed 1 November 2018).

39 The American Truckers Association estimates that, with the introduction of driverless lorries, some 8.7 million trucking-related jobs could face some form of displacement: see Daniel Thomas, 'Driverless convoy: Will truckers lose out to software?' *BBC News*, May 26, 2015, available at www.bbc.com/news/business-32837071 (last accessed 1 November 2018).

40 For a short discussion, see Wendell Wallach and Colin Allen, *Moral Machines* (Oxford: Oxford University Press, 2009) 14. In addition to the safety considerations, robot-controlled trains are more flexible in dealing with situations where timetables need to be changed.

surgeons, or Nursebots, humans are being displaced—the pattern of employment and the prospects for both skilled and unskilled workers are seriously
affected.[41] Where these technological developments, designed for safety, are
simply viewed as further options, this enhances human liberty; but, where
they are viewed as de rigueur, there is a major diminution in the conditions
for autonomous living. In this way, Nursebots are emblematic of a dual disruptive effect of technology.[42] It is not just a matter of not accommodating
patients who prefer human carers to Nursebots, *it is reducing the options for
those humans who would like to be carers.*[43]

(iii) Technological management and moral concerns

There are many ways in which technological management might elicit moral
concerns. For example, we may wonder whether the moral judgment of
humans can ever be replicated even in intelligent devices, such that taking
humans out of the loop (especially where life and death decisions are made)
is problematic. Similarly, in hard cases where one moral principle has to be
weighed against another, or where the judgment that is called for is to do the
lesser of two moral evils, we might pause before leaving the decision to the
technology—hence the agonised discussion of the manner in which autonomous vehicles should deal with the kind of dilemma presented by the trolley

---

41 For what is now an urgent debate about the disruptive impact of automation on the pros-
    pects for workers and the patterns of employment, see: Ford (n 29); Geoff Colvin, *Humans
    are Underrated* (London: Nicholas Brealey Publishing, 2015); Andrew Keen, *The Internet is
    Not the Answer* (London: Atlantic Books, 2015); Jaron Lanier, *Who Owns the Future?* (Lon-
    don: Allen Lane, 2013); and Kenneth Dau-Schmidt, 'Trade, Commerce, and Employment:
    the Evolution of the Form and Regulation of the Employment Relationship in Response to
    the New Information Technology' in Roger Brownsword, Eloise Scotford, and Karen Yeung
    (eds), *The Oxford Handbook of Law, Regulation, and Technology* (Oxford: Oxford University
    Press, 2017) 1052. For those who are interested in a personal prognosis, see *BBC News*,
    'Will a robot take your job?' (11 September, 2015): available at www.bbc.co.uk/news/
    technology-34066941 (last accessed 1 November 2018).
42 Compare the perceptive commentary in Jaron Lanier (n 11) which is full of insightful dis-
    cussion including, at 89–92, some reflections on the disruptive effects of relying on robot
    carers.
43 In some caring professions, the rate of reduction might be quite slow because the experience
    of robotics experts is that it is easier to programme computers to play chess than to fold
    towels or to pick up glasses of water. So, as Erik Brynjolfsson and Andrew McAfee (n 10)
    conclude, at 241:

    [P]eople have skills and abilities that are not yet automated. They may be automatable at
    some point but this hasn't started in any serious way thus far, which leads us to believe
    that it will take a while. We think we'll have human data scientists, conference organizers,
    divisional managers, nurses, and busboys for some time to come.

    Compare, however, Colvin (n 41) who cautions against underestimating the speed with
    which robotic physical skills will be perfected and, at the same time, underestimating the
    significance that we humans attach to interaction with other humans (even to the point of
    irrationality).

problem (where one option is to kill or injure one innocent human and the only other option is to kill or injure more than one innocent human)[44] or by the tunnel problem (where the choice is between killing a passenger in the vehicle and a child outside the vehicle).[45]

We might also be extremely concerned about the deployment of new technologies in the criminal justice system, not only to identify potential hot-spots for crime but to risk-assess the potential criminality or dangerousness of individuals. It is not just that technological management in the criminal justice system might eliminate those discretions enjoyed by human policemen or prosecutors that can be exercised to respond to morally hard cases. The worry is that technological profiling, prediction and prevention could lead to a revival of illiberal character-based (rather than capacity-based) responsibility, unethical bias, a lack of transparency, and an insouciance about false positives.[46] For moralists who subscribe to liberal values, due process rights and the like, there is plenty to be concerned about in the prospect of algorithmic or automated criminal justice.[47] However, let me highlight a more fundamental and pervasive concern, namely that the use of technological management might compromise the conditions for any aspirant moral community.

I take it that the fundamental aspiration of *any* moral community is that regulators and regulatees alike should try to do the right thing. However, this presupposes a process of moral reflection and then action that accords with one's moral judgment. In this way, agents exercise judgment in trying to do the right thing and they do what they do for the right reason in the sense that they act in accordance with their moral judgment. Of course, this does not imply that each agent will make the same moral judgment or apply the same reasons. A utilitarian community is very different from a Kantian community; but, in both cases, these are moral communities; and it is their shared aspiration to do the right thing that is the lowest common denominator.[48]

---

44 For the original, see Judith Jarvis Thomson, 'The Trolley Problem' (1985) 94 *Yale Law Journal* 1395.

45 For the tunnel problem in relation to autonomous vehicles, see Jason Millar, 'You should have a say in your robot car's code of ethics' *Wired* 09.02.2014 (available at: www.wired.com/2014/09/set-the-ethics-robot-car/) (last accessed 1 November 2018). See, further, Chapter Ten.

46 Generally, see Roger Brownsword, 'From Erewhon to Alpha Go: For the Sake of Human Dignity Should We Destroy the Machines?' (2017) 9 *Law, Innovation and Technology* 117, 138–146; and, on the question of false positives, see Andrea Roth, 'Trial by Machine' (2016) 104 *Georgetown Law Journal* 1245, esp. at 1252. See, further, Chapter Nine.

47 Seminally, see Bernard E. Harcourt, *Against Prediction* (Chicago: The University of Chicago Press, 2007).

48 See Roger Brownsword, 'Human Dignity, Human Rights, and Simply Trying to Do the Right Thing' in Christopher McCrudden (ed), *Understanding Human Dignity* (Proceedings of the British Academy 192) (Oxford: The British Academy and Oxford University Press, 2013) 345.

There is more than one way in which the context for moral action can be compromised by the use of technological regulatory strategies. In the case of measures that fall short of technological management—for example, where CCTV surveillance or DNA profiling are used in support of the rules of the criminal law—prudential signals are amplified and moral reason might be crowded out.[49] However, where full-scale technological management takes over, what the agent actually does may be the only act that is available. In such a context, even if an act accords with the agent's own sense of the right thing, it is not a paradigmatic moral performance: the agent is no longer freely doing the right thing, and no longer doing it for the right reason. As Ian Kerr has evocatively expressed it, moral virtue is simply not something that can be automated.[50]

If technological management that compels an agent to do x is morally problematic even where the agent judges that doing x is the right thing to do, it is at least equally problematic where the agent judges that doing x is morally wrong. Indeed, we may think that the use of technological management such that an agent is compelled to act against his or her conscience is an even more serious compromising of moral community. At least, in a normative order, there is the opportunity for an agent to decline to act in a way that offends their conscience.

It is an open question how far an aspirant moral community can tolerate technological management. In any such community, there must remain ample opportunity for humans to engage in moral reflection and then to do the right thing. So, the cumulative effect of introducing various kinds of surveillance technologies as well as adopting hard technological fixes needs to be a standing item on the regulatory agenda.[51] If we knew just how much space a moral community needs to safeguard against both the automation of virtue and the compulsion to act against one's conscience, and if we had some kind of barometer to measure this, we might be able to draw some regulatory red lines. It may well be that the technological measures that are adopted by the golf club neither make a significant difference to the general culture of the community nor materially reduce the opportunities that are presented for (authentic) moral action. On the other hand, if the community is at a 'tipping point', these regulatory interventions may be critical. Accordingly, taking a precautionary approach, we may reason that, as regulators discover, and then

---

49 For an interesting study, see U. Gneezy and A. Rustichini, 'A Fine is a Price' (2009) 29 *Journal of Legal Studies* 1; and, for relevant insights about the use of CCTV, see, Beatrice von Silva-Tarouca Larsen, *Setting the Watch: Privacy and the Ethics of CCTV Surveillance* (Oxford: Hart, 2011).

50 Ian Kerr, 'Digital Locks and the Automation of Virtue' in Michael Geist (ed), *From 'Radical Extremism' to 'Balanced Copyright': Canadian Copyright and the Digital Agenda* (Toronto: Irwin Law, 2010) 247.

51 Karen Yeung, 'Can We Employ Design-Based Regulation While Avoiding Brave New World?' (2011) 3 *Law, Innovation and Technology* 1.

increasingly adopt, measures of technological management, a generous margin for the development of moral virtue, for moral reflection, and for moral action needs to be maintained.[52]

(iv) Technological management and the 'Rule of Law'

If driverless cars mean that a huge swathe of road traffic laws are rendered redundant; if safer transportation systems backed by social insurance schemes mean that tort lawyers have less work; if GPS-enabled golf carts mean that fewer thefts are committed and fewer criminal defence lawyers are needed, then should we be concerned? Given that new info-technologies are already disrupting legal practice,[53] should we be concerned about the further impacts if technological management means that we need fewer lawyers and fewer laws? Perhaps we should. Nevertheless, I think that we can be fairly confident that, beyond legal quarters, the loss of some lawyers and the retirement of some laws is unlikely to be a matter of either general or great concern.

Nevertheless, the idea that governance is being handed over to technologists and technocrats may well be a prospect that does cause some concern. After all, technological management is 'instrumentalism' writ large; and this, some might worry, is likely to be a blank cheque for the interests of the powerful to be advanced in a way that would otherwise be constrained in old-fashioned legal orders.[54] The Rule of Law may not guarantee a perfectly just social order, but it does impose some constraints on those who govern. Perhaps, in a world of technological management there is still a role to be played by lawyers, in particular, a role in ensuring that there is a proper social licence for the adoption of regulatory technologies and then acting on complaints that the terms of the licence have not been observed. Indeed, in Part Two of the book, I will suggest that this is precisely the role that lawyers should be ready to play; and, at the same time, it will become apparent that, even if some of the Hartian primary rules are displaced, the authorising function of the secondary rules in relation to the use of technological management becomes critical.

So, what price technological management? Possibly, golf carts, Google cars, Nursebots, and the like can be adopted in the best democratic fashion and without any significant cost, either to a context that aligns with the preferences of agents or the options available to them, or to the community's

---

52 See Roger Brownsword, 'Lost in Translation: Legality, Regulatory Margins, and Technological Management" (2011) 26 *Berkeley Technology Law Journal* 1321.

53 See, e.g., Colvin (n 41) 17–19.

54 Compare e.g., Brian Z. Tamanaha, *Law as a Means to an End* (Cambridge: Cambridge University Press, 2006).

moral aspirations.[55] Nevertheless, it would be reckless to proceed with technological management as if no such risk could possibly exist. We need to pause so that we can ponder the potentially disruptive effects of technological management. It is at this point, and with this thought, that we should turn to the domain of jurisprudence.

IV Redrawing the domain of jurisprudence

For jurists, it is the 'law' that is the object of their inquiry; and, the standard assumption is that, however one fine-tunes one's conception of law, the aim of the legal enterprise is to subject human conduct to the governance of *rules*. It follows that, whether one argues for a legal positivist or a legal idealist conception of law, whether one's conception of law is narrowly restricted to the operations of the high institutions of nation states (as in the 'Westphalian' view of law) or ranges more broadly and pluralistically across the ordering of social life, it is agreed that law is about rules, about prescription, about normativity; in all conceptions, law is a normative enterprise, the rules prescribing what ought and ought not to be done.

From the fact that law is conceived of as a normative phenomenon, it does not follow that the ambit of jurisprudence should be limited to legal norms. Nevertheless, once law is conceived of in this way, and given that law is the object of jurisprudential inquiry, it is understandable that the domain of jurisprudence should be drawn in a way that stays pretty close to legal norms and normative legal orders. However, while this might seem to facilitate an efficient and clear division of labour between, on the one hand, jurists and, on the other, philosophers, sociologists, political theorists, economists, and so on, and while this specification of the domain of jurisprudence gives inquirers a clear and apparently coherent focus, it suffers from two major limiting tendencies.

The first limitation applies most obviously where a Westphalian conception of law is adopted. On this view, a limited set of norms (constitutional rules, legislation, precedents, and the like) is isolated from a larger context of (as the Westphalians would have it, non-legal) normative regulation. To be sure, this isolation does not preclude juristic inquiry beyond the boundaries of Westphalian law, but it hardly encourages it. By contrast, where a more pluralistic conception of law is adopted, the inhibition against broader

---

55 For example, at the South Glasgow University Hospital, a fleet of 26 robots does the work of 40 people in delivering food, linen, and medical supplies to wards and then taking away waste for disposal. This frees up the clinical staff; none of the existing staff were laid off; new staff were hired to operate and control the robots; and because humans have been removed from work that leads to many injuries, there are major health and safety gains. See Georgie Keate, 'Robot hospital porters get the heavy lifting done' *The Times*, April 25, 2015, 19.

juristic inquiry is eased—albeit at the risk of some loss of conceptual focus.[56] The second limitation applies to all rule-based conceptions of law, including the most inclusive kind of normative pluralism. On this view, normativity is treated as an essential feature of law. Again, the isolation of normative regulatory approaches does not preclude juristic inquiry into non-normative regulatory strategies; but, equally, it hardly encourages it.[57] We can say a little more about each of these limiting tendencies, each of which creates a barrier to seeing the bigger regulatory picture.

(i) The first limitation: the isolation of legal norms

Hart's conceptualisation of law invites jurists to focus their inquiries on high-level national rules of the kind that we find in legislation, codes, and the case law. While this licenses a broad range of juristic inquiries, the fact of the matter is that 'law' in this sense is by no means the only kind of normative order that we find in societies. Religious and (secular) moral codes are normative, as are the relatively formal codes of conduct that guide the practice of professional people (including lawyers themselves) and the less formal codes that are observed in a myriad of social settings. From Eugen Ehrlich's 'living law'[58] found in the customs and practices of provincial Bukowina (then part of the Austro-Hungarian empire) to Robert Ellickson's study of the informal norms of 'neighbourliness' and 'live and let live' recognised by the close-knit group of ranchers and farmers of Shasta County, California,[59] through to Stewart Macaulay's seminal paper on the way that business people relate to the law in planning their transactions and settling their disputes,[60] there is a literature that charts the norms that actually guide human conduct. Yet, in this ocean of normativity, Hartian-inspired jurisprudence invites inquiry directed at just one island of norms (the island that it characterises as 'law').

---

56 See the caveats in Simon Roberts, 'After Government? On Representing Law Without the State' (2005) 68 *Modern Law Review* 1.

57 Compare Roger Brownsword, 'Comparatively Speaking: "Law in its Regulatory Environment"' in Maurice Adams and Dirk Heirbaut (eds), *The Method and Culture of Comparative Law* (Festschrift for Mark van Hoecke) (Oxford: Hart, 2014) 189.

58 Eugen Ehrlich, *Fundamental Principles of the Sociology of Law* (New Brunswick: Transaction Publishers, 2001 [1913]). For a useful introductory overview, see Neil O. Littlefield, 'Eugen Ehrlich's Fundamental Principles of the Sociology of Law' (1967) 19 *Maine Law Review* 1.

59 Robert C. Ellickson, *Order Without Law* (Cambridge, Mass.: Harvard University Press, 1991). Although the rural group is close-knit, there are significant sub-groups—for example, there is a contrast between the 'traditionalists' who let their cattle roam, and the 'modernists' who 'keep their livestock behind fences at all times in order to increase their control over their herds' (at 24).

60 Stewart Macaulay, 'Non Contractual Relations in Business: A Preliminary Study' (1963) 28 *American Sociological Review* 55.

Where the domain of inquiry is restricted in this way, jurisprudence has nothing to say about the way that the many other islands of normativity might function to maintain a particular kind of social order nor about the way in which they might interact with or disrupt the operation of (Westphalian) legal norms.[61]

If it were the case that non-legal norms were of marginal significance, this might not be too serious a restriction. However, if (as there is ample reason to believe) these other norms are at least as important as legal norms in the daily lives of people, then this is a serious limitation on our ability to understand not only what makes things tick in the social world but, more particularly, how legal norms fit into the full array of normative signals. To understand our legal and social world, we need a wide-lens approach.

Exemplifying such an approach, Orly Lobel, in her excellent discussion of the optimal regulatory conditions for innovation, takes into account not only several relevant strands of law (especially intellectual property law, competition law, and contract law), but also a range of social norms that operate alongside (and interact with) the law.[62] In many spheres of competition and innovation, Lobel points out, it is the foreground social norms which impose informal restraints that are more important than the formal restraints (if any) found in the background laws. Thus:

> Competitive environments from fashion to magic, comedy to cuisine have managed to protect the process of innovation without resorting to the most restrictive controls. The distinction between control and freedom does not have to be a binary one, that is, one of negative spaces free of controls contrasted with formal legal systems saturated with the big sticks of protections and restrictions … In every industry, we find

---

61 Again, note the caveat in Roberts (n 56). As Roberts puts it, at 12:

> We can probably all now go along with some general tenets of the legal pluralists. First, their insistence on the heterogeneity of the normative domain seems entirely uncontroversial. Practically any social field can be fairly represented as consisting of plural, interpenetrating normative orders/systems/discourses. Nor would many today wish to endorse fully the enormous claims to systemic qualities that state law has made for itself and that both lawyers and social scientists have in the past too often uncritically accepted. Beyond that, consensus is more difficult … Will all the normative orders that the legal pluralists wish to embrace necessarily be comfortable with their rescue as 'legal' orders?

> See, too, the impressive overview in Neil Walker, *Intimations of Global Law* (Cambridge: Cambridge University Press, 2015).

62 Orly Lobel, *Talent Wants to Be Free* (New Haven: Yale University Press, 2013). One of the critical variables here is whether regulators take a 'Californian' view of restraint of trade clauses or a 'Massachusetts' view, the former signalling a reluctance to keep employees out of the market, the latter being more supportive of the employer's interest in restraining ex-employees.

practices of self-imposed confidentiality alongside practices of openness and sharing; we find places where norms substitute for law. Reputation replaces litigation. Carrots alternate with sticks.[63]

It follows that we are likely to go badly wrong—whether we are trying (as regulators) to channel conduct or simply seeking to understand compliance or non-compliance by regulatees—if we focus on a narrow class of legal norms.

In this context, we can recall the frequent references that are made in public life to the importance of getting the 'regulatory environment' right—right for banking and financial services, right for innovation, right for health care and patient safety, right for the European single market, right for small businesses, right for privacy and press freedom, right for online intermediaries and platforms, and so on.[64] Typically, these references are a common starting point for a public 'post-mortem' following a crisis, a scandal, a collapse, or a catastrophe of some kind. While this may be a good place to start, the remarks made above indicate that we will go badly wrong if we try to reduce the regulatory environment to just *one* area of law, or indeed to several areas of (Hartian) *law*. The regulatory environment is more complex than that and it is layered. Accordingly, if we think that the regulatory environment is 'broken', our attempts at repair and renewal are unlikely to come to much if they are limited to replacing one set of legal norms with another. Or, to put this another way, it is one thing to grasp that the law is relatively ineffective in channelling conduct but, unless we open our inquiry to the full range of norms in play, we will not understand why law suffers from this shortcoming.[65] As Ellickson concludes, 'lawmakers who are unappreciative of the social conditions that foster informal cooperation are likely to create a world in which there is both more law and less order.'[66]

---

63 *Ibid.*, 239. For an illuminating analysis of the relationship between copying and innovation (with, in some contexts, imitation encouraging and supporting innovation to the advantage of both those who copy and those who are copied), see Kal Raustiala and Christopher Sprigman, *The Knockoff Economy* (Oxford: Oxford University Press, 2012).

64 For the last mentioned, see, e.g., the European Commission, *Synopsis Report on the Public Consultation on the Regulatory Environment for Platforms Online Intermediaries and the Collaborative Economy* (May 25, 2016): available at ec.europa.eu/digital-single-market/en/news/full-report-results-public-consultation-regulatory-environment-platforms-online-intermediaries, accessed 2 November 2018.

65 It is also important, of course, to be aware of the full range of strategies available to regulators in trying to tweak the 'choice architecture' within which agents act: see, e.g., Cass R. Sunstein, *Simpler: The Future of Government* (New York: Simon and Schuster, 2013) 38–39 for an indicative list of possible 'nudges'.

66 Ellickson (n 59) 286.

(ii) The second limitation: the exclusivity of the normative

Even if the domain of jurisprudence is expanded to include the full set of normative instruments, it is still limited so long as it treats norms, *and only norms*, as within its field of inquiry. So limited, jurists are disabled from assessing the significance of non-normative instruments such as technological management. Back in 1961, this was not a significant limitation. However, once regulators adopt strategies of technological management (such as golf carts that are immobilised or motor vehicle ignition systems that lock unless a seatbelt is worn), the ocean of normativity contains potentially significant new currents.[67] To restrict the field of inquiry to the exclusively normative is to miss a sea-change in social ordering. To give ourselves the chance of understanding and assessing a radical transformation in the way that the state channels and confines human conduct, we need to work with a notion of the regulatory environment that includes both normative and non-normative instruments.

There is no denying that, by including non-normative instruments in the domain of jurisprudence, we are locating legal inquiry in a much larger and very different ball-park. To advocate a shift in focus from 'law' to 'regulation' might meet with resistance; but, at least, mainstream regulatory theorists conceive of regulation as starting with the setting of standards and, thus, as normative. If the 'regulatory environment' adopted this conception of 'regulation', it would still be limited to normative signals; and many jurists might be comfortable with this. However, the comfort of jurists is not our concern. Our cognitive interest is in a perceived shift to technological management and, with that, the development of a pervasive risk management mentality. This is quite different from the traditional legal approach and legal mentality but the function of channelling human conduct (one of the principal 'law jobs' as Karl Llewellyn puts it[68]) is at least one thread of connection. To understand what is happening with regard to the channelling of conduct within their own narrowly circumscribed domain, jurists have to broaden their horizons, uncomfortable though this might be. When, at the Warwickshire, technological management is used to respond to a failure of normative governance, the lesson is not simply one to be taken by golf clubs; the lesson is a general one: namely, that law is not the only way of managing risk and, in some cases, a technological fix will be far more effective.[69]

---

67 Compare Lee Tien, 'Architectural Regulation and the Evolution of Social Norms' (2004) 9 *International Journal of Communications Law and Policy* 1.

68 See Karl N. Llewellyn, 'The Normative, the Legal, and the Law-Jobs: The Problem of Juristic Method' (1940) 49 *Yale Law Journal* 1355.

69 A similar lesson might be drawn in relation to dispute-resolution, one of the other principal law-jobs: see, e.g., David Allen Larson, 'Artificial Intelligence: Robots, Avatars, and the

If the prognosis of this book is accepted, in future, normative regulatory environments will co-exist and co-evolve with technologically managed environments—but not always in a tidy way. For jurists to turn away from the use of technological instruments for regulatory purposes is to diminish the significance of their inquiries and to ignore important questions about the way that power is exercised and social order maintained.[70]

Yet, should we accept my prognosis? Technological management will assume increased significance only if it 'works'; but will it prove to be any more effective than normative regulatory instruments? Before we proceed, this is a question that merits attention.

V Technological management and regulatory effectiveness

The premise, and the prognosis, of this book is that regulators (both public and private; and, quite possibly, with a power struggle between the two) will be attracted to make use of technological management because it promises to be an effective form of regulation. Whether this promise will be fulfilled, we do not yet know. However, we do know that many traditional normative regulatory interventions are relatively ineffective; and we do have a stock of particular case studies concerning the impact of particular laws that offer rich accounts of ineffective interventions or of unintended negative effects.[71] Moreover, we also know that the cross-boundary effects of the online provision of goods and services have compounded the challenges faced by regulators. If we synthesise this body of knowledge, what do we understand about the conditions for regulatory effectiveness?

To begin with, we should note a radical difference between the expectations that we might have with regard to the effectiveness of technological management, where the promise is of perfect control, as against traditional regulatory regimes, where the expectation is of better but far from perfect control. In relation to the latter, we can agree with Francis Fukuyama[72] that:

> [N]o regulatory regime is ever fully leak-proof ... But this misses the point of social regulation: no law is ever fully enforced. Every country makes murder a crime and attaches severe penalties to homicide, and yet murders nonetheless occur. The fact that they do has never been a reason for giving up on the law or on attempts to enforce it.[73]

Demise of the Human Mediator' (2010) 25 *Ohio State Journal on Dispute Resolution*. Available at SSRN:http://ssrn.com/abstract=1461712 (last accessed 1 November 2018).
70 Compare Veitch (n 3).
71 For a very readable recent overview, see Tom Gash, *Criminal: The Truth about Why People Do Bad Things* (London: Allen Lane, 2016).
72 Francis Fukuyama, *Our Posthuman Future* (London: Profile Books, 2002).
73 *Ibid.*, 189.

It follows that, while the sponsors of technological management might set the bar for regulatory effectiveness at the level of complete achievement of the regulators' objectives, we know that this is a totally unrealistic target for traditional normative regulation. Moreover, we know that, in the case of the latter, it is not only murderers, thieves, and vagabonds that stand between regulators and complete effectiveness. The keys to effective regulation are far more complex.

Placing our inquiries within the context of the regulatory environment, we can simplify the complexity by focusing on the three sets of factors that will determine the relative effectiveness or ineffectiveness of an intervention. These factors concern the acts of the regulators themselves, the responses of their regulatees, and the acts of third-parties.

First, where regulators are corrupt (whether in the way that they set the standards, or in their monitoring of compliance, or in their responses to non-compliance),[74] where they are captured,[75] or where they are operating with inadequate resources,[76] the effectiveness of the intervention will be compromised. To the extent that technological management takes human regulators right out of the equation, there should be fewer problems with corruption and capture, but if the resourcing for the technology is inadequate it might prove to be unreliable, leading to pressure for its withdrawal.[77]

Turning to the second set of factors, commentators generally agree that regulators tend to do better when they act with the backing of regulatees (with a consensus rather than without it).[78] The general lesson of the well-known Chicago study, for example, is that compliance or non-compliance

---

74 See, e.g., Ben Bowling, *Policing the Caribbean* (Oxford: Oxford University Press, 2010). According to Bowling, the illicit drugs economy in the Caribbean thrives on 'widespread corruption, ranging from the junior Customs officer paid to 'look the other way' when baggage handlers are packing aircraft luggage holds with parcels of cocaine at the international airport, to the senior officials who take a cut of the cash generated on their watch' (at 5).

75 For an excellent study of the capture of the FDA by the pharmaceutical companies (concerning the Fen-Phen diet drug), see Alicia Mundy, *Dispensing with the Truth* (New York: St Martin's Press, 2001).

76 For example, according to Ian Walden, when the Internet Watch Foundation identified some 7,200 UK persons who were involved with 'Landslide' (a major child pornography site hosted in Texas), this was a level of offending with which the criminal justice system simply could not cope (inaugural lecture, 'Porn, Pipes and the State: Regulating Internet Content', February 3, 2010, at Queen Mary University London). More recently, and to similar effect, see Andrew Ellson, 'Scandal over police failure to pursue millions of online frauds', *The Times*, September 24, 2015, p.1.

77 Compare Michael J. Casey and Paul Vigna, *The Truth Machine* (London: Harper Collins, 2018) 176–177 (on blockchain as a technological response to the corruptibility of registrars and record-keepers).

78 The idea that regulators do best when they 'work with the grain' is emphasised in Iredell Jenkins, *Social Order and the Limits of Law* (Princeton, NJ: Princeton University Press, 1980); see, too, Phillipe Sands, *Lawless World* (London: Penguin, 2005) 56, for the eminently generalisable piece of regulatory wisdom that 'there exists in diplomatic circles a

hinges not only on self-interested instrumental calculation but also (and significantly) on the normative judgments that regulatees make about the morality of the regulatory standard, about the legitimacy of the authority claimed by regulators, and about the fairness of regulatory processes.[79] It follows that, if regulatees do not perceive the purpose that underlies a particular regulatory intervention as being in either their prudential or their moral interests (let alone in both their prudential and moral interests), the motivation for compliance is weakened. The use of marijuana as a recreational drug is the textbook example. Thus:

> The fact remains ... that marijuana use continues to be illegal in most parts of the world, even as people continue to break these laws with apparent impunity. And there is no resolution in sight. The persistence of marijuana use remains a prime example of how our legal system is based on an implicit social contract, and how the laws on the books can cease to matter when a large percentage of people decide they want to do something that may not be acceptable under the law.[80]

Similarly, experience (especially in the United States) with regulatory prohibitions on alcohol suggests not only that legal interventions that overstep the mark will be ineffective but pregnant with corrupting and secondary criminalising effects.

To put this another way, regulatee resistance can be traced to more than one kind of perspective. Business people (from producers and retailers through to banking and financial service providers) may respond to regulation as rational economic actors, viewing legal sanctions as a tax on certain kinds of conduct;[81] professional people (such as lawyers, accountants, and doctors) tend to favour and follow their own codes of conduct; the police are stubbornly guided by their own 'cop culture';[82] consumers can resist by declining to buy; and, occasionally, resistance to the law is required as a matter of conscience—witness, for example, the peace tax protesters, physicians

---

strongly held view that if a treaty cannot be adopted by consensus its long-term prospects are crippled.'

79 See Tom R. Tyler, *Why People Obey the Law* (Princeton: Princeton University Press, 2006).

80 Stuart Biegel, *Beyond Our Control?* (Cambridge Mass.: MIT Press, 2003) 105.

81 To a considerable extent, rational economic man operates on both sides of the regulatory fence—for example, in both the licit and the illicit drugs market. Compare Nichola Dorn, Tom Bucke, and Chris Goulden, 'Traffick, Transit and Transaction: A Conceptual Framework for Action Against Drug Supply' (2003) 42 *Howard Journal of Criminal Justice* 348, 363, according to whom, it seems likely that 'only interventions causing traffickers to perceive a significant risk of capture leading to imprisonment have a worthwhile deterrent effect, lower-impact interventions providing for traffickers no more than the expected "costs of doing business".'

82 See, e.g., Tom Cockcroft, *Police Culture: Themes and Concepts* (Abingdon: Routledge, 2013); and Robert Reiner, *The Politics of the Police* 4th edn (Oxford: Oxford University Press, 2010) Ch.4.

who ignore what they see as unconscionable legal restrictions, members of religious groups who defy a legally supported dress code, and the like.

In all these cases, the critical point is that regulation does not act on an inert body of regulatees: regulatees will respond to regulation—sometimes by complying with it, sometimes by ignoring it, sometimes by resisting or repositioning themselves, sometimes by relocating, and so on. Sometimes those who oppose the regulation will seek to overturn it by lawful means, sometimes by unlawful means; sometimes the response will be strategic and organised, at other times it will be chaotic and spontaneous.[83] But, regulatees have minds and interests of their own; they will respond in their own way; and the nature of the response will be an important determinant of the effectiveness of the regulation.[84]

Now, technological management promises to eliminate the option of non-compliance that is preferred, for a variety of reasons, by many regulatees. It is hard to believe, however, that the attitudes of regulatees will change to come smoothly into alignment with the constraints imposed by new regimes of technological management. For example, while there might be many persons who are happy to travel in driverless cars, we cannot be so confident about attempts to use technological management in driven cars. Famously, in one well-documented example in the United States, the so-called interlock system (immobilising vehicles if the driver's seatbelt was not engaged) was withdrawn after pressure from the motoring lobby.[85] Although the (US) Department of Transportation estimated that this particular example of technological management would save 7,000 lives per annum and prevent 340,000 injuries, 'the rhetoric of prudent paternalism was no match for visions of technology and "big brotherism" gone mad.'[86] As Jerry Mashaw and David Harfst take stock of the legislative debates of the time:

> Safety was important, but it did not always trump liberty. [In the safety lobby's appeal to vaccines and guards on machines] the freedom fighters saw precisely the dangerous, progressive logic of regulation that

---

83 Compare Tim Wu, 'When Code Isn't Law' (2003) 89 *Virginia LR* 679; and, according to Jamie Bartlett (n 30), 'tech is just the latest vehicle for very rich people to use well-tested techniques of buying political influence, monopolistic behaviour and regulation avoidance, to help them become even richer' (at 156).

84 For an illuminating account of the illicit GM seed trade in India, by-passing both the local bio-safety regulations and Mahyco-Monsanto's premium prices, see Ronald J. Herring and Milind Kandlikar, 'Illicit Seeds: Intellectual Property and the Underground Proliferation of Agricultural Biotechnologies' in Sebastian Haunss and Kenneth C. Shadlen (eds), *Politics of Intellectual Property* (Cheltenham: Edward Elgar, 2009) 56. At 74, the authors remark: 'Stealth seeds reflect the same kind of agency as urban appropriation of pharmaceuticals and software, films and music—the same anarchic capitalism at the grass roots—with similar risks and rewards.'

85 Jerry L. Mashaw and David L. Harfst, *The Struggle for Auto Safety* (Cambridge, Mass.: Harvard University Press 1990).

86 *Ibid.*, 135.

they abhorred. The private passenger car was not a disease or a workplace, nor was it a common carrier. For Congress in 1974, it was a private space.[87]

More generally, regulatee resistance to technological management might be expressed in various attempts to damage, disrupt, or circumvent the technology; and, while laws might be introduced to criminalise such attempts, not only would this rather defeat the regulatory purpose in employing technological management, such laws would be unlikely to be any more effective than their predecessors.

The third set of factors brings into the reckoning various kinds of external distortion or interference with the regulatory signals. Some kinds of third-party interference are well known—for example, regulatory arbitrage (which is a feature of company law and tax law) is nothing new. However, even where regulatory arbitrage is not being actively pursued, the effectiveness of local regulatory interventions can be reduced as regulatees take up more attractive options that are available elsewhere.[88]

Although externalities of this kind continue to play their part in determining the fate of a regulatory intervention, it is the emergence of the Internet that has most dramatically highlighted the possibility of interference from third parties. As long ago as the closing years of the last century, David Johnson and David Post predicted that national regulators would have little success in controlling extra-territorial on-line activities, even though those activities have a local impact.[89] While national regulators are not entirely powerless,[90] the development of the Internet has dramatically changed the regulatory environment, creating new vulnerabilities to cybercrime and cyberthreats, as well as new on-line suppliers, and community cultures. The net effect is that local regulators are left wondering how they can control access to drugs, or alcohol, or gambling or direct-to-consumer genetic testing services, when Internet pharmacies, or on-line drinks suppliers or casinos or the like, all of which are hosted on servers that are located beyond the national borders, direct their goods and services at local regulatees.[91]

While technological management will surely draw heavily on new information and communication technologies, the more that smart environments rely

---

87 *Ibid.*, 140.

88 See, e.g., Arthur J. Cockfield, 'Towards a Law and Technology Theory' (2004) 30 *Manitoba Law Journal* 383, 391–395.

89 David R. Johnson and David Post, 'Law and Borders—The Rise of Law in Cyberspace' (1996) 48 *Stanford Law Review* 1367.

90 Compare Jack Goldsmith and Tim Wu, *Who Controls the Internet?* (Oxford: Oxford University Press, 2006).

91 A case in point is the growing online market for non-invasive prenatal testing: see Roger Brownsword and Jeff Wale, 'Testing Times Ahead: Non-Invasive Prenatal Testing and the Kind of Community that We Want to Be' (2018) 81 *Modern Law Review* 646.

on such infrastructures, the more important it is that these systems are resilient. As we have said, we do not know whether regulatee resistance might take the form of attempting to disable or disrupt these systems; and, in a context of volatile international relations, there must be a real concern about the extent to which it will be possible to protect these systems against cyberattacks.[92]

Summing up, because technological management controls many of the variables that distort and obstruct compliance with rules, the thought is encouraged that social control may be more effective if new technologies are utilised as regulatory instruments. However, until we have some experience of more general technological management, we do not know whether its promissory notes for more effective regulation will be honoured. It may be that, in the year 2061, technological management will be widely regarded as a flawed and failed regulatory experiment; and, as in Samuel Butler's *Erewhon*,[93] a once sophisticated technological society will have reverted to a much simpler form of life. My guess, though, is that this is not our future.

VI The new domain and the questions to be addressed

If we adopt the idea of 'the regulatory environment' as setting the field of jurisprudential inquiry, and if we embrace the notion that, in this environment, there will be both normative and non-normative dimensions, we facilitate an understanding of how legal norms relate to technological management. However, this is just the beginning: with this new domain for juristic inquiry, new questions abound as we reflect on the significance of technological management in the regulatory environment.

In an attempt to engage with some of these new questions, the book is divided into three principal parts, each involving a major exercise in re-imagination—re-imagining the idea of the regulatory environment, re-imagining the application and meaning of key legal ideals such as the Rule of Law, the coherence of the law, and the protection of liberty, and re-imagining

---

92 Following the devastating DDoS attacks on Estonia in 2007, the question of the vulnerability of critical information infrastructures in Europe rapidly moved up the political agenda: see, House of Lords European Union Committee, *Protecting Europe Against Large-Scale Cyber-Attacks* (Fifth Report, Session 2009–2010); and *The UK Cyber Security Strategy* (Protecting and promoting the UK in a digital world) (London: Cabinet Office, November 2011). Generally, see Susan W. Brenner, *Cyberthreats: the Emerging Fault Lines of the Nation State* (Oxford: Oxford University Press, 2009); David Patrikarakos, *War in 140 Characters* (New York: Basic Books, 2017); Lawrence Freedman, *The Future of War: A History* (London: Allen Lane, 2017) 223; Sascha-Dominik Dov Bachmann and Anthony Paphiti, 'Russia's Hybrid war and its Implications for Defence and Security in the United Kingdom' (2016) 44 *Scientia Militaria, South African Journal of Military Studies* 28; and Sascha-Dominik Dov Bachmann and Håkan Gunneriusson, 'Russia's Hybrid Warfare in the East: The Integral Nature of the Information Sphere' (2015) 16 *Georgetown Journal of International Affairs* 198.

93 Samuel Butler, *Erewhon* (London: Penguin, 1970; first published 1872).

the relevance of such basic doctrinal modules as the criminal law, the law of torts, and the law of contract.

In the first part, the focus is on three organising ideas: the regulatory environment itself, the 'complexion' of the regulatory signals that constitute that environment, and the full range of regulatory responsibilities relative to which we assess the fitness of the regulatory environment. With regard to the first of these ideas, which is the key framing notion, there is still much work to be done. For example, there are questions about the scope of the regulatory environment, about the significance of the source of the regulatory intervention, about the relevance of the intent that accompanies technological management, and about the conflict between competing regulatory interventions.[94] With regard to the second—the still unfamiliar idea of the 'complexion' of the regulatory environment[95]—we need to be clear about why this matters. Why does it matter, for example, that agents comply with the law, or other norms, for prudential rather than moral reasons? Or, why does it matter whether regulatees comply freely with the regulatory purposes or because, with technological management, they have no other choice? The story of the carts at the golf club begins to answer these questions, but there is more to be said. Finally, in relation the third idea, a three-tiered scheme of regulatory responsibilities is elaborated. The first and paramount responsibility is to ensure that the preconditions (the commons) for any form of human social existence, for any kind of human community, are maintained and protected; the second is to articulate and respect the distinctive fundamental values of the community; and, the third is to maintain an acceptable balance between the legitimate (but potentially competing and conflicting) interests of members of the community (for example, an acceptable balance between the interest in innovation and technological progress and the interests in human health and safety, protection of the environment, preservation of employment options, and so on).

In the second part of the book, the spotlight falls on a number of 'ideals' that we assume to be, if not intrinsic to the legal enterprise, at least associated with best practice. At the head of this list is the Rule of Law; but we will also examine the ideal of coherence (which is highly valued by private lawyers) and the Millian harm principle (which is regularly relied on to test the legitimacy of particular uses of the criminal law).[96] In each case, the adoption of technological management seems to be disruptive of these ideals and the

---

94 The seeds of these questions can be found in Roger Brownsword, *Rights, Regulation and the Technological Revolution* (Oxford: Oxford University Press, 2008); and Roger Brownsword and Morag Goodwin, *Law and the Technologies of the Twenty-First Century* (Cambridge: Cambridge University Press, 2012) Ch.2.

95 Compare Roger Brownsword (n 52).

96 See John Stuart Mill, 'On Liberty', in J.S. Mill, *Utilitarianism* (edited by Mary Warnock) (Glasgow: Collins/Fontana, 1962) (first published 1859); and, for a sketch of the necessary

question is whether we can re-engineer them for a regulatory environment that employs both normative and non-normative signals. To illustrate one of the issues, consider the well-known Fullerian principles of legality and, concomitantly, the notion of the Rule of Law as the publication (promulgation) of rules and then the congruent administration of those rules. Can these principles be stretched across to non-normative strategies or do they again presuppose rules (and a normative dimension)? On the face of it, the Fullerian principles do presuppose rules (that the rules should be published, that the rules should be prospective, that the rules should be clear and relatively constant, that the rules should not be contradictory, and so on) and, indeed, they seem to be particularly focused on the rules of the criminal law (or other duty-imposing rules). If this is correct, perhaps the ideal of legality remains relevant but its focus shifts to (i) the processual public law values of transparency, accountability, inclusive participation, and the like, together with (ii) the controls exerted by background fundamental values (such as compatibility with respect for human rights and human dignity). In this way, while the zone regulated *directly* by legal norms (that is, by Hartian primary rules) might shrink, the significance of the Hartian secondary rules of law and the ideal of legality (as a check on technological management) remains. More generally, the question for this second part of the book is this: if we make some generous assumptions about the effectiveness of technological management, what should we treat as the criteria for its legitimate use and how do traditional ideals of legality and coherence bear on this matter?

In the third part of the book, on the assumption that there will be a major technological impingement on the way in which we interact and transact in the future, attention switches to the fate of those bodies of law—criminal law, tort law, and contract law, in particular—that have hitherto regulated such activities. To what extent will laws of this kind be rendered redundant and replaced by technological management (and how problematic is this); to what extent will existing laws remain relevant but in need of renewal; to what extent will the regulatory focus be displaced from one legal doctrine to another; and to what extent will technological management breathe new life into old puzzles (such as the criminality of attempting not just the happenstance impossible but the systematically technologically managed impossible)?[97] Smart cars, as we have said, might make much of road traffic laws redundant; and, where the technology fails, if tort law is still relevant for compensatory purposes, there might be changes as to who is held liable

---

re-working of Millian liberalism, see Roger Brownsword, 'Criminal Law, Regulatory Frameworks and Public Health' in A.M. Viens, J. Coggon and A. Kessel (eds) *Criminal Law, Regulatory Frameworks and Public Health* (Cambridge: Cambridge University Press, 2013) 19.

97 On the last-mentioned, see George P. Fletcher, *Basic Concepts of Criminal Law* (Oxford: Oxford University Press, 1998) Ch.10.

(producers rather than owners/operators) as well as to the basis of liability (strict rather than fault-based).[98] Similarly, if technological management in hospitals and workplaces ensures that patients and employees are safe, how much of both the criminal law and the law of torts is likely to be side-lined and how might the law otherwise be disrupted—for example, how would vicarious liability and assumption of risk operate in such a context?[99]

Where we are relying on technological management for health and safety purposes, it will be largely 'regulatory' criminal law that is rendered redundant. Given that this is a body of law that features strict and absolute liability offences, and about which we do not feel entirely comfortable, perhaps the retirement of these laws is no bad thing.[100] To be sure, to the extent that technological management replaces those rule-based regulatory offences that do allow for a defence of due diligence (or those enforcement practices that recognise due diligence as an excuse), regulatees will lose the occasion and the opportunity to plead that they were good citizens, that they did actually exercise reasonable care. However, it is unclear that this loss significantly compromises the conditions for moral community. Counter-intuitive though this might seem, is it where technological management is employed to eliminate the possibility of committing the classical crimes of intent ('true' or 'real' crimes, so to speak) that our moral concerns should be most acutely raised?

Turning to contracts and the law of the marketplace, if the technological infrastructure for transactions manages much of the process, what does this mean for the leading principles of a jurisprudence that is heavy with case law from earlier centuries? Rather than being an occasion for the celebration

---

98 Compare Hubbard (n 28) at 1858–1859; and Gary E. Marchant and Rachel A. Lindor, 'The Coming Collision Between Autonomous Vehicles and the Liability System' (2012) 52 *Santa Clara Law Review* 1321, 1326–1330. But NB Graham (n 19) 1270:

> When suits against the manufacturers of autonomous automobiles first appear, they likely will sound in a failure to warn of some danger associated with vehicle use, as opposed to a design defect. For a plaintiff to reach a jury on a design-defect claim, she may have to engage in a searching review of the computer code that directs the movement of these vehicles.

99 On the latter, see Marchant and Lindor (n 98), 1336–1337.
100 See, e.g., *R. v Sault Ste. Marie* [1978] 2 SCR 1299, esp. at 131–1312. In *Sault Ste. Marie*, the Canadian Supreme Court differentiated between crimes where (i) the prosecution must prove both actus reus and mens rea, (ii) the prosecution need prove only actus reus but the defendant has the opportunity to show that due diligence was taken ('strict liability'), and (iii) the prosecution need prove only actus reus and it is irrelevant whether the defendant has exercised due diligence ('absolute liability'). So far as 'public welfare' offences are concerned (such as pollution in the case itself), the court treats the second category as the default classification of the offence (thus, giving defendants the chance to escape conviction by proving on a balance of probabilities that they took reasonable care). See, further, Chapters 8 and 9; and for a critical appreciation of the reasoning in *Sault Ste. Marie*, see Allan C. Hutchinson, '*Sault Ste. Marie, Mens Rea* and the Halfway House: Public Welfare Offences Get a Home of their Own' (1979) 17 *Osgoode Hall Law Journal* 415.

of new forms of contract, does this signify 'the end of contracts'?[101] Does contract law have a future when—anticipating the first Amazon Go store in Seattle[102]—thanks to 'RFID tags and smart payment cards, together with a reader able to link the information together and some biometric security, [customers can] merrily wheel [their] trolley[s] out of the supermarket without either queuing or being arrested'[103]? Does contract have a future when potential breaches are forestalled by technological interventions? Given that in Europe (if not in all parts of the world), in the sphere of routine consumption, regulatory law has long since displaced traditional contract law together with the concerns for doctrinal coherence that are characteristic of much common law thinking about contracts,[104] would it be a cause for concern if that regulatory law were itself to be displaced by technological management?

The backcloth to the discussion in Part Three is the claim that technology disrupts the law in two principal ways: first, it disrupts the content of the rules; and, secondly, it disrupts the use of rules as the strategy for regulating human conduct. Moreover, these disruptions affect the ways in which we think about the law. Before the first disruption, lawyers recur to a simple moral thread—enjoining agents to respect the physical integrity of others, to act reasonably, and to honour promises and agreements ('transactionalism')—that runs through criminal, tort, and contract law. Valuing coherence in the law, the focus is on maintaining the integrity of the rules. However, as new (and often dangerous) technologies reshape workplaces, redefine transport and communication, revolutionise medicine and reproduction, and so on, communities demand regulatory interventions that cover the most worrying risks. This provokes a quite different, regulatory-instrumentalist, mindset. In the twenty-first century, there is a second disruption as a number

---

101 Shoshana Zuboff, 'Big Other: Surveillance Capitalism and the Prospects of an Information Civilization' (2015) 30 *Journal of Information Technology* 75, 86 (and, for elaboration of such important matters as the compromising of conditions for making promises and for trusting others in transactions, see *ibid.*, 81–83).

102 On the opening of which, see: www.theguardian.com/business/2018/jan/21/amazons-first-automated-store-opens-to-public-on-monday (last accessed 1 November 2018).

103 Kieron O'Hara and Nigel Shadbolt, *The Spy in the Coffee Machine* (London: Oneworld Publications, 2008) 193.

104 See, e.g., Roger Brownsword, 'Regulating Transactions: Good Faith and Fair Dealing' in Geraint Howells and Reiner Schulze (eds), *Modernising and Harmonising Consumer Contract Law* (Munich: Sellier, 2009) 87; 'The Theoretical Foundations of European Private Law: A Time to Stand and Stare' in Roger Brownsword, Hans Micklitz, Leone Niglia, and Steven Weatherill (eds), *The Foundations of European Private Law* (Oxford: Hart, 2011) 159; 'The Law of Contract: Doctrinal Impulses, External Pressures, Future Directions' (2014) 31 *Journal of Contract Law* 73; 'The E-Commerce Directive, Consumer Transactions, and the Digital Single Market: Questions of Regulatory Fitness, Regulatory Disconnection and Rule Redirection' in Stefan Grundmann (ed), *European Contract Law in the Digital Age* (Antwerp: Intersentia, 2017) 165; and 'After Brexit: Regulatory Instrumentalism, Coherentism and the English Law of Contract' (2017) 34 *Journal of Contract Law* 139.

of emerging technologies offer themselves to regulators as instruments to assess and manage the risks. For those who already think in a regulatory-instrumentalist way, it is a short step to a technocratic mind-set that looks for a technological fix rather than a new rule. The existence of these threads and twists invites a potentially confusing range of regulatory responses to new technologies—for example, trying to sustain (in a coherentist way) the traditional moral threads, taking a regulatory-instrumentalist approach to the management of risk, and resorting to technological management. The challenge—one that is vividly present in relation to the regulation of privacy and fair processing of data in a context of pervasive data-gathering and data-giving—is to understand which of these responses is appropriate, when and why. Hence, should we be concerned if our informational interests are to be reduced to a balance of acceptable risks and desired benefits, or if, employing technological management, regulators seek to 'design in' appropriate protections?

Finally, in a short Epilogue, I offer some reflections on the possible bio-management of human conduct by the year 2161. Thus far, although advances in biotechnologies have received at least as much attention as developments in information and communication technologies, their penetration into daily life has been much more modest and their utility as regulatory instruments much less obvious.[105] Genetics and genomics are extremely complex. To be sure, there is a major investment in big biobanking projects that seek to unravel these complexities in a way that improves health care; but, even if major strides are made in biobank and in genetic research, by 2061, the chances are that *behavioural* genetics will be extremely primitive.[106] However, given another one hundred years of research and development, by 2161, it may be a different story. If so, and if bio-management operates through internal signalling mechanisms which we (humans) know to be operative but of which we are individually not conscious, there is a new internal dimension to the regulatory environment. In 2161, that environment may still feature signals that are 'external' to regulatees and to which regulatees respond but it will also feature signals that are 'internal' to regulatees. In other words, if we move towards such a scenario, we will need to frame our inquiries by reference to a regulatory environment that has not only normative and non-normative dimensions but also external and internal dimensions.

Where does all this lead? What is the message of the book? In practice, technologies are often introduced to save money but, in principle, technological management can be employed for many different purposes: to improve human health and safety, to protect the natural environment, to produce

---

105 This is not to suggest that there have not been extremely significant developments in biotechnology: see, e.g., Robert Carlson, 'Biotechnology's possibilities' in Daniel Franklin (ed), *Megatech* (London: Profile Books, 2018) 41.

106 For an excellent summary of the (then) state of the art, see Nuffield Council on Bioethics, *Genetics and Human Behaviour: the Ethical Context* (London, September 2002).

goods and services more cheaply and reliably, to respect values such as privacy, to control criminal acts and those with 'dangerous' dispositions, and so on. In itself, technological management does not point to either dystopian or utopian futures. There is no reason to become a technophobe, categorically rejecting the development and use of new technologies. Nevertheless, what becomes of the law and the ideal of legality in the regulatory environment of 2061—let alone in 2161—is a matter of concern not just to legal practitioners but to the community as a whole. Accordingly, before technological management has already become an unquestionable and irreversible part of the infrastructure and a pervasive part of the regulatory environment, it is important to think through the implications of this prospective transformation in social ordering.[107]

Who is to do this? No doubt, some of the work will be done by politicians and their publics. However, as the historian, Yuval Harari, has remarked, radically new technologies are seemingly stealing a march on politics. While these technologies have the potential to 'destroy the very foundations of liberal society, individual choice and the human economy as we know it', they are

> hardly a blip on our political radar. Public debates are still dominated by 20th-century arguments between left and right, socialism and capitalism, democracy and authoritarianism. These arguments are certainly relevant and they still have a huge impact on our lives, but the 21st–century technological revolutions are creating entirely new questions, which cannot be easily pigeonholed into this or that 20th-century drawer.[108]

Even if Harari overstates the risk that today's technologies are moving so rapidly that they are escaping the control of politicians and electors, it seems to me that he is on the right track.[109] Technologies and technological management should not be allowed to run out of public control and, unless

---

107 Compare Daniel Franklin, 'Introduction: meet megatech' in Franklin (ed) (n 105) at 8:

> [T]here is nothing inevitable about what lies ahead. The impact of technology is only partly a matter of the innovations of scientists, geeks and entrepreneurs. The outcome by 2050 will be shaped by the decisions of governments, the strategies of companies and the choice of individuals.

108 Yuval Harari, 'Sorry, everyone; technology has decided you don't matter' *The Sunday Times News Review* (September 13, 2015) p.7; and, compare, Evgeny Morozov, 'The state has lost control: tech firms now run western politics' *The Guardian*, March 27, 2016, available at www.theguardian.com/commentisfree/2016/mar/27/tech-firms-run-western-politics-evgeny-morozov (last accessed 1 November 2018).

109 Compare, too, Jamie Bartlett (n 30). Anticipating the failure of democracy, Bartlett suggests that the 'looming dystopia to fear is a shell democracy run by smart machines and a new elite of "progressive" but authoritarian technocrats. And the worst part is that lots of people will prefer this, since it will probably offer them more prosperity and security than what we have now' (at 7). Similarly, see Franklin Foer, *World Without Mind* (London:

jurists are to stick their heads in the sand, they have a vanguard role to play in ensuring that there is no such loss of control. The central message of the book is that, for today's jurists, some of the issues can be glimpsed and put on the agenda; but it will fall to tomorrow's jurists to rise to the challenge by helping their communities to grapple with the many questions raised by the accelerating transition from law (especially from the primary rules of law) to technological management.

Jonathan Cape, 2017) 127, claiming that the problem is that '[tech companies] that are indifferent to democracy have acquired an outsized role in it'.

PART ONE
Re-imagining the regulatory environment

2

THE REGULATORY ENVIRONMENT

An extended field of inquiry

I Introduction

If technologically managed environments are to be a significant part of our regulatory future, then the argument so far is that the domain of jurisprudence needs to be extended in a way that facilitates inquiry into both law as a normative regulatory strategy and the use of non-normative technological instruments. Law, that is to say, needs to be set in a broader signalling and steering context: first, in a context that takes full account of the variety of norms that impact on, and influence, human behaviour; and, secondly, in a context that recognises the channelling and constraining effect of technological management. In order to do this, the suggestion is that we should broaden the field for juristic inquiry by operating with a notion of the regulatory environment that accommodates both normative and non-normative approaches.

What would such a regulatory environment look like? Famously, Clifford Shearing and Phillip Stenning highlighted the way in which, at Disney World, the vehicles that carry visitors between locations act as barriers (restricting access).[1] However, theme parks are no longer a special case. We find similar regulatory environments in many everyday settings, where along with familiar laws, rules, and regulations, there are the signs of technological management—for example, we find mixed environments of this kind in homes and offices where air-conditioning and lighting operate automatically,

---

1 Clifford D. Shearing and Phillip C. Stenning, 'From the Panopticon to Disney World: the Development of Discipline' in Anthony N. Doob and Edward L. Greenspan (eds), *Perspectives in Criminal Law: Essays in Honour of John LL.J. Edwards* (Toronto: Canada Law Book, 1985) 335.

in hotels where the accommodation levels can be reached only by using an elevator (and where the elevators cannot be used and the rooms cannot be accessed without the use of security key cards), and perhaps par excellence at airports. On arrival at a modern terminal building, while there are many airport rules to be observed—for example, regulations concerning parking vehicles, smoking in the building, or leaving bags unattended, and so on—there is also a distinctive architecture that creates a physical track leading from check-in to boarding. Along this track, there is nowadays an 'immigration and security zone', dense with identifying and surveillance technologies, through which passengers have little choice other than to pass. In this conjunction of architecture and surveillance technologies, we have the non-normative dimensions of the airport's regulatory environment—the fact of the matter is that, if we wish to board our plane, we have no practical option other than to follow the technologically managed track.

If we treat the regulatory environment as essentially a signalling and steering environment, then each such environment operates with a distinctive set of regulatory signals that are designed to channel the conduct of regulatees within, so to speak, a regulated sphere of possibility. Of course, one of the benefits of technologies is that they can expand our possibilities; without aircraft, we could not fly. Characteristically, though, the kind of technological management that we are contemplating is one that restricts or reduces existing human possibilities (albeit, in some cases, by way of a trade-off for new possibilities). In other words, while normative regulation is directed at actions that are possible—and that remain possible—technological management engages with spheres of possibility but in ways that restructure those regulatory spaces and redefine what is and is not possible.

This brief introduction to an extended conception of the regulatory environment needs more detail. First, we need to make a few schematic remarks about technological management as a regulatory option. Then, to fill in some of the detail, we can start by mapping the salient features of regulatory environments that are purely normative before extending it to include environments that are technologically managed.

II Technological management as a regulatory option

In the previous chapter, the general idea of technological management was introduced and some examples were given. We said that technological management might employ a variety of measures, including the design of products (such as golf carts or computer hardware and software) and processes (such as the automated production and driving of vehicles, or the provision of consumer goods and services), or places (such as the Metro, or theme parks and airports) and people. Typically, such measures are employed with a view to managing certain kinds of risk by excluding (i) the possibility of certain actions which, in the absence of this strategy, might be subject only to rule regulation, or (ii) human agents who otherwise might be implicated (whether as rule-breakers or as the innocent victims of rule-breaking) in the regulated activities. Moreover, technological management might be employed by both public regulators or by private self-regulating agents (such as corporations protecting their IP rights or golf clubs protecting their property).

Schematically, where the use of technological management is available as a regulatory option, the process can be presented in the following terms:

- Let us suppose that a regulator, R, has a view about whether regulatees should be required to, permitted to, or prohibited from doing x (the underlying normative view),
- R's view could be expressed in the form of a rule that requires, permits, or prohibits the doing of x (the underlying rule),
- but, R uses (or directs others to use) technological management rather than a rule,
- and R's intention in doing so is to translate the underlying normative view into a practical design that ensures that regulatees do or do not do x (according to the underlying rule),
- the ensuing outcome being that regulatees find themselves in environments where the immediate signals relate to what can and cannot be done, to possibilities and impossibilities, rather than to the underlying normative pattern of what ought or ought not to be done.

This description, however, is highly schematic and what such a process actually amounts to in practice—in particular, how transparent the process is, how much debate there is about the underlying normative view and then about the use of technological measures—will vary from one context to another, from public to private regulators, between one public regulator and another, and between one private regulator and another.

It also should be emphasised that the ambition of technological management is to replace the rules by controlling the practical options that are open to regulatees. In other words, technological management goes beyond technological assistance in support of the rules. Of course, regulators might first turn to technological instruments that operate in support of the rules. For example, in an attempt to discourage shoplifting, regulators might require or encourage retailers to install surveillance and identification technologies, or technologies that sound an alarm should a person carry goods that have not been paid for through the exit gates. However, this is not yet full-scale technological management. Once technological management is in operation shoppers will find that it is simply not possible to take away goods without having paid for them. For example, they will find that there are technologies operating at both the entrance and the exit points of retail outlets to ensure that those who have poor profiles for ability to pay or honesty do not enter and that those who have not paid are not able to exit (instead of sensors triggering alarms, they control the opening of the otherwise closed exit doors).

III Law in context: normative regulatory environments

Conceiving of a regulatory environment as a kind of signalling environment, much in the way that we think about the setting for road traffic or for the railways, then regulators may signal (normatively) that certain conduct is prohibited, or permitted, or required. These signals may be legal or ethical, formal or informal; and, in the world that Hart and others contemplated, these signals were normative in character, speaking to regulatees in terms of what ought or ought not to be done.

In modern regulatory settings, the legal signals might be supported by various kinds of advisory or assistive technologies. For example, on the roads, technologies might be employed to advise drivers that they are either non-compliant or at risk of non-compliance—as when roadside signs might flash a warning that the vehicle is exceeding the speed limit, or an overhead motorway gantry might advise that it is time to take a break; or a vehicle equipped with the appropriate sensors might caution the driver against proceeding when affected by drink or drugs; and various kinds of roadside or on-board surveillance technologies might signal that non-compliance will be detected. In all these cases, although technological instruments are used, the technological signals are normative (various acts ought or ought not to be done) and, of course, they complement or supplement normative legal signals.

There is much more to be said, however, about this relatively straightforward picture of a normative signalling environment (whether with or without technological support). In particular, there are questions about the identity of the regulators, about what counts as regulation, about the extent of the regulatory repertoire, about the identity of the regulatees, about conflicting signals, and about the boundaries of any particular regulatory environment. However, before we respond to each of these questions, we need to note a broad distinction between two types of normative regulatory environment.

(i) Two types of regulatory environment

On the roads, motorists are notorious for developing their own unofficial rules, the drivers of each country being caricatured for respecting the official traffic signals more than the lives of fellow humans or vice versa, as well as employing their own local signalling systems based on hand gestures or the use of their car headlights. The co-existence of these unofficial codes with the official traffic laws can be a source of confusion as well as, potentially, an accident waiting to happen. In some cases, motorists go one step further, taking the law into their own hands. For example, Jonathan Zittrain reminds his readers about the experiment in the Dutch city of Drachten when traffic signs, parking meters, and parking spaces were removed.[2] The sign-free ('ver-keersbordvrij') environment, far from inducing dangerous and disorderly driving, generated responsible and communicative conduct on the part of motorists, cyclists, and pedestrians. As Zittrain reflects, the Drachten experience suggests that when 'people can come to take the welfare of one another seriously and possess the tools to readily assist and limit each other, even the most precise and well-enforced rule from a traditional public source may be less effective than that uncompelled goodwill'.[3] In Drachten, as in Shasta County,[4] there can be order without law.

These remarks invite the drawing of a crude distinction between 'top-down regulator-initiated' and 'bottom-up' (Drachten or Shasta County-type) regulation. Whereas, in top-down regulatory environments, there is likely to be a significant formal legal presence, along with a clear distinction between regulators who initiate the norms and regulatees to whom the norms are addressed, in bottom-up self-regulatory environments, on both scores, this is less likely to be the case—there will be a much less clear-cut distinction between regulators and regulatees as well as a certain amount of informality. For convenience, we can mark this distinction by referring to 'Type 1' (top-down with significant hard law elements) and 'Type 2' (bottom-up with fewer hard law elements) regulatory environments.

It should be emphasised, however, that this is not only a crude distinction, it is also far from exhaustive. For example, top-down government regulators might enlist the aid of non-governmental intermediaries (such as Internet service providers or platform providers) or they might adopt a co-regulatory approach setting general targets or objectives for regulatees but leaving them to determine how best to comply.[5] With new technologies occupying and disrupting regulatory spaces, regulators need to re-imagine how best to regulate. As Albert Lin says, in his analysis of new distributed innovative technologies (such as DIYbio, 3D printing, and the platforms of the share economy) these new forms of dynamic activity 'confound conventional regulation'.[6] In response, Lin argues, it turns out that '[g]overnance of distributed innovation ... must be both distributed and innovative'.[7] There is no one size fits all; and the regulatory environment that is most acceptable and effective is likely to have elements of both Type 1 and Type 2 approaches together with elements that fit neither of these types.

---

2 Jonathan Zittrain, *The Future of the Internet* (London: Penguin Books, 2008).

3 *Ibid.*, 129.

4 See Robert C. Ellickson, *Order Without Law* (Cambridge, Mass.: Harvard University Press, 1991).

5 Compare Julia Black, 'De-centring Regulation: Understanding the Role of Regulation and Self-Regulation in a "Post-Regulatory" World' (2001) 54 *Current Legal Problems* 103.

6 Albert C. Lin, 'Herding Cats: Governing Distributed Innovation' (2018) 96 *North Carolina Law Review* 945, 965.

7 *Ibid.*, 1011.

(ii) Who counts as a 'regulator'?

In principle, who counts as a 'regulator'? How broad is this concept? Is the class of regulators restricted to public functionaries—or might we also find private 'regulators' contributing to the normative signals of the regulatory environment?

To some extent, the distinction between public and private initiators of regulation correlates with the distinction between the two types of regulatory environment. In 'Type 1' environments, the sources of top-down regulation will often be governments or other public agencies; by contrast, in 'Type 2' environments—for example, in the Drachten case or at the Warwickshire Golf and Country Club—the sources of bottom-up regulation will often be private bodies. The mapping, however, is not neat and tidy: private bodies might resort to command and control strategies and public bodies might encourage bottom-up governance as well as entering into co-regulatory alliances with stakeholders in the private sector. It follows that the category of regulators is not self-selecting.

This being so, we can say that, for some purposes, it might be convenient to limit 'regulators' to public functionaries, so that 'regulation' is an activity that the State or its delegates carry out. However, where (as in this book) the purpose is to examine technological management, and when so many of the relevant initiatives emanate from the private sector (recall, for example, Monsanto's so-called 'terminator gene' and the technologies of digital rights management), it makes no sense to confine the class of regulators to those who act in a public capacity. Moreover, if one of the core objectives of jurisprudence is to track, explain, and evaluate the exercise of power, we should allow for the concept of a 'regulator' to be interpreted broadly as covering parties in both public and private positions.

(iii) What counts as 'regulation'?

A common starting point is that the concept of 'regulation' signifies,

> the sustained and focused attempt to alter the behaviour of others according to standards or goals with the intention of producing a broadly identified outcome or outcomes, which may involve mechanisms of standard-setting, information-gathering and behaviour-modification.[8]

---

8 Julia Black, 'What is Regulatory Innovation?' in Julia Black, Martin Lodge, and Mark Thatcher (eds), *Regulatory Innovation* (Cheltenham: Edward Elgar, 2005) 1, 11. For two background visions of regulation (one as an infringement of private autonomy justified only by considerations of economic efficiency, the other as a much broader collaborative enterprise), see Tony Prosser, *The Regulatory Enterprise* (Oxford: Oxford University Press, 2010) Ch.1.

Nevertheless, regulation is an unwieldy and a contested concept.[9] In an earlier work, Morag Goodwin and I took a broad approach, treating regulation as 'encompassing any instrument (legal or non-legal in its character, governmental or non-governmental in its source, direct or indirect in its operation, and so on) that is designed to channel group behaviour'[10]; and we emphasised that it should not be assumed that 'regulation' is co-extensive with 'law' any more than it should be assumed that 'regulators' are restricted to those who are authorised to issue legal directives.

There is a problem, however, with interpreting regulation in a broad way. As Karen Yeung has pointed out, such an approach

> seems to encompass a plethora of varied activities and relationships which would not typically be regarded as regulatory, either by scholars of regulatory governance or in ordinary usage, such as attempts by parents to encourage their children to eat more fruit and vegetables, or the marketing and sales techniques of commercial firms to encourage consumers to purchase their products in greater quantities. While both these examples involve the use of power by one party to control or influence the behaviour of others, and thus provoke concerns about the use and abuse of power and have been the focus of considerable academic examination, they are not issues which regulatory scholars would regard as typically falling within the scope of their inquiry or expertise.[11]

Yeung's response is to propose a narrower approach, restricting 'our understanding of regulation to intentional attempts to alter the behaviour of others in order to address a collective issue or problem'.[12] If we adopt this proposal, we will continue to treat government-initiated regulation as paradigmatic and this will serve as a benchmark for our willingness to extend the range of regulation.

So, at what point should we treat standard-setting and channelling as beyond the zone of regulation? For present purposes, we do not need to resolve this conundrum. What matters is not that our usage of the term 'regulation' fits with the usage of a particular group of scholars or common convention, nor that we can be precise about the point at which we leave the

---

9 See, e.g., Julia Black (n 5).

10 Roger Brownsword and Morag Goodwin, *Law and the Technologies of the Twenty-First Century* (Cambridge: Cambridge University Press, 2012) Ch.2.

11 Karen Yeung, 'Are Design-Based Regulatory Instruments Legitimate?' King's College London Dickson Poon School of Law Research Paper Series: Paper No 2015–27, p.12, available at papers.ssrn.com/sol3/papers.cfm?abstract_id=2570280 (last accessed 1 November 2018).

12 *Ibid.*

realm of regulation and enter some other normative zone; rather the priority is that jurists should find a way of giving themselves the room to be able to engage with *non-normative* channelling strategies that compete with, or complement, or simply supersede Hartian legal norms.

(iv) The regulatory repertoire

For regulators, whether operating in Type 1 or Type 2 regulatory environments, the choice is not simply between enacting a law (or an analogue of law) or doing nothing, or between enacting one kind of law rather than another. The regulatory repertoire extends well beyond legal norms. Famously, Lawrence Lessig identifies four regulatory modalities (or modes of regulation) that are available: namely, the law, social norms, the market, and architecture (or, code).[13] Although technological management (in the form of the coding of software or hardware) already appears in this typology of regulatory modalities, and although the general point of this chapter is precisely to suggest that we should treat technological management as a 'regulatory' phenomenon, my present point is much less contentious. It is simply to note that, within Lessig's list, we see a variety of *normative* strategies that regulators might employ. The law is one such strategy; but so, too, is the use of social norms and, in many cases, the use of market signals. Even smart cars might be equipped with technologies that operate in a normative way (for example, by advising against driving when under the influence of drink or drugs, or against not wearing a seat belt, and so on).

Smart regulators should seek out the optimal mix of the various instruments that they have in this extended repertoire.[14] Moreover, this 'smartness' includes being aware of the full range of normative signals in play, including the norms that already are foregrounded in the regulatee community. Chiming in with this thought, in his inaugural lecture,[15] Colin Scott has suggested that

> wherever governments are considering a policy problem—be it unsafe food, passive smoking or poor quality university research— what they are considering is an existing regime which cannot be swept away and replaced by a regulatory agency. A more fruitful approach would be to seek to understand where the capacities lie within the existing regimes,

---

13 Lawrence Lessig, *Code and Other Laws of Cyberspace* (New York: Basic Books, 1999), Ch.7; and Lessig, 'The Law of the Horse: What Cyberlaw Might Teach' (1999) 113 *Harvard Law Review* 501, 507–514.

14 Generally, see Neil Gunningham and Peter Grabosky, *Smart Regulation* (Oxford: Clarendon Press, 1998); and Ian Ayres and John Braithwaite, *Responsive Regulation* (Oxford: Oxford University Press, 1992).

15 Colin Scott, 'Regulating Everything' UCD Geary Institute Discussion Paper Series (Dublin, inaugural lecture, February 26, 2008).

and perhaps to strengthen those which appear to pull in the right direction and seek to inhibit those that pull the wrong way. In this way the regulatory reform agenda has the potential to address issues of regulatory fragmentation in a manner that recognizes both the limits of governmental capacity and the potential of reconceptualizing regulation in other ways, for example that invoke non-state actors and alternative mechanisms to hierarchy.[16]

Scott argues that the core idea of this proposed 'meta-regulatory' approach is that 'all social and economic spheres in which governments or others might have an interest in controlling already have within them mechanisms of steering—whether through hierarchy, competition, community, design or some combination thereof'; and he sees the principal challenges for regulators as being, first, 'to observe and identify, to some approximation, the variety of mechanisms of regulation at play' and, secondly, 'to work out ways to key into those mechanisms to steer them, to the extent they are not already securing desired outcomes.'[17] Even without the possible utilisation of technological management, then, it is apparent that the regulatory repertoire is extensive and that making a successful regulatory intervention represents a sophisticated challenge (although it is a challenge, as Scott emphasises, that does not necessarily call for a totally fresh start).

(v) The regulators' perspective: tensions within the regulators' own sphere

In principle, there can be three sets of tensions in a normative field. *First*, there are the often conspicuous 'vertical' tensions between the 'official' norms signalled by regulators and the 'unofficial' norms followed by regulatees—tensions of the kind highlighted by Ehrlich, Ellickson, Macaulay, and others.[18] *Secondly*, however, there is also the possibility of less visible 'horizontal' tensions between different sections of the regulatory body. As Brian Simpson explains in his retelling of the great Victorian case of *R v Dudley and Stephens*,[19] the shipwrecked sailors' killing of young Richard Parker, the cabin boy, highlighted more than a conflict between domestic criminal law and the customs of the sea, there was also a cultural clash between government departments.[20] While the Home Office strongly favoured a prosecution in order to demonstrate public condemnation of the customs of the seas, the Board of Trade took a different view. However, within the regulators' own

---

16 *Ibid.*, 25.
17 *Ibid.*, 27.
18 See Chapter One.
19 14 QBD 273 (1884).
20 A.W. Brian Simpson, *Cannibalism and the Common Law* (Chicago: University of Chicago Press, 1984).

domain, it is not only differences of this kind that can create tensions, there might also be tensions between the 'official' formal norms and the norms that guide the working practices of inspectorates and regulatory agencies. Sometimes, these tensions arise from a lack of resources but there may also be resistance in the agencies. In the criminal justice system, for example, there is a notorious tension between the norms that lay jurors bring into the court-room and the formal rules of the law[21]—indeed, this was precisely the situation in *Dudley and Stephens* where the rescued seamen were welcomed home, not as 'murderers', but as heroes. *Thirdly*, there might be further 'horizontal' tensions between different sections of the regulatee community. While some sections of the community might support the 'official' norms, others might oppose them; and, of course, it might be that both sections reject the official norms and yet find themselves at odds in supporting rival norms.

To draw out tensions of the second kind, consider the modern regulation of assisted suicide. As the well-known case of *Purdy v DPP*[22] highlights, pros-ecution policy might be an important gloss on the law. Sometimes, citizens know well enough where they stand—motorists know, for example, how the traffic police are likely to view a slight breach of the speed limit. However, in *Purdy*, the problem was that citizens did not know where they stood, it being argued that the criminal law on assisted suicide lacked clarity because pros-ecutions were subject to an agency discretion that lacked the necessary trans-parency. At one level, the legal position was reasonably clear, section 2(1) of the Suicide Act 1961 providing that 'A person who aids, abets, counsels or procures the suicide of another, or an attempt by another to commit suicide, shall be liable on conviction on indictment to imprisonment for a term not exceeding fourteen years.' To be sure, some nice points of interpretation might be taken on the language and scope of the provision; but, the main point in *Purdy* concerned the operationalisation of section 2(4) of the Act, according to which no proceedings are to be instituted for an offence under section 2(1) except with the consent of the DPP. However, the argument was that, even though the DPP had published a Code for Crown Prosecutors giving general guidance on the exercise of prosecutorial discretion, it was not clear that this enabled citizens to identify the considerations that would be treated as relevant in deciding whether or not to prosecute a party who assisted another to commit suicide—and particularly so where assistance was given by one family member to another and where the assistance involved transporting the latter to Zurich where they would avail themselves of the

---

21 For a revealing account, see Ely Devons, 'Serving as a Juryman in Britain' (1965) 28 *Modern Law Review* 561; and, for a celebrated example, see Clive Ponting, '*R v Ponting*' (1987) 14 *Journal of Law and Society* 366.

22 [2009] UKHL 45.

services offered by Dignitas. Giving the leading opinion for the Law Lords, Lord Hope concluded:

> 54. The Code will normally provide sufficient guidance to Crown Prosecutors and to the public as to how decisions should or are likely to be taken whether or not, in a given case, it will be in the public interest to prosecute. This is a valuable safeguard for the vulnerable, as it enables the prosecutor to take into account the whole background of the case. In most cases its application will ensure predictability and consistency of decision-taking, and people will know where they stand. But that cannot be said of cases where the offence in contemplation is aiding or abetting the suicide of a person who is terminally ill or severely and incurably disabled, who wishes to be helped to travel to a country where assisted suicide is lawful and who, having the capacity to take such a decision, does so freely and with a full understanding of the consequences. There is already an obvious gulf between what section 2(1) says and the way that the subsection is being applied in practice in compassionate cases of that kind.
>
> 55. The cases that have been referred to the Director are few, but they will undoubtedly grow in number. Decisions in this area of the law are, of course, highly sensitive to the facts of each case. They are also likely to be controversial. But I would not regard these as reasons for excusing the Director from the obligation to clarify what his position is as to the factors that he regards as relevant for and against prosecution in this very special and carefully defined class of case. How he goes about this task must be a matter for him, as also must be the ultimate decision as to whether or not to prosecute. But, as the definition which I have given may show, it ought to be possible to confine the class that requires special treatment to a very narrow band of cases with the result that the Code will continue to apply to all those cases that fall outside it.
>
> 56. I would therefore allow the appeal and require the Director to promulgate an offence-specific policy identifying the facts and circumstances which he will take into account in deciding, in a case such as that which Ms Purdy's case exemplifies, whether or not to consent to a prosecution under section 2(1) of the 1961 Act.

In the event, this led to the publication of a draft set of guidelines, on which there was public consultation, and then a final set of prosecution guidelines for this particular offence.[23]

---

23 See, *Policy for Prosecutors in Respect of Cases of Encouraging or Assisting Suicide* (DPP, February 2010). For reflections on *Purdy* in the context of 'hidden' and 'intermediate' lawmakers,

Margaret Brazier and Suzanne Ost underline the significance of these guidelines when they remark that:

> In the context of death…[it is] crucial to any account of the criminal law, medicine and bioethics to look at the whole of the criminal process and not just the 'law in the books'. First, in cases relating both to 'medical manslaughter' and to assisted suicide, the principal actor has become the CPS, with the CPS's practice in relation to charging doctors with manslaughter and the DPP's policy on assisting suicide playing the major role in defining the doctor's vulnerability to criminal responsibility.[24]

However, we could also turn this round somewhat to underline one of the key points of this chapter. This is that, if we are to understand the regulatory environment in relation to medical manslaughter and assisted suicide, we need to look at the whole of the criminal process (including the norms that guide prosecution policy) but also the norms that guide medical practitioners in end-of-life situations and the norms articulated by bioethics committees. Moreover, my point is that if we hope to understand the dynamics of this field of human action, it is imperative that we engage with the regulatory environment in this comprehensive way.

(vi) The regulatees' perspective

Not surprisingly, the perspectives and attitudes of regulatees might be different to those of the regulators. This can account for individual acts of defiance, or non-compliance, or technical compliance—witness, those many complaints about the tax that is allegedly (under)paid by global corporations and the expenses (over)claimed by politicians where the rights and wrongs of particular actions hinge not only on the distinction between 'avoidance' and 'evasion' but also on whether compliance with the rules suffices.[25] In some cases, the story is about an isolated individual act; but, in many cases, the acts that come to light reflect much deeper and wider cultural differences expressing rival normative orders.

Such systematic differences are particularly prevalent in Type 1 regulatory environments: here, while regulators resort to a range of normative interventions, regulatees act on and respond to these norms in a context that itself

---

see Jonathan Montgomery, Caroline Jones, and Hazel Biggs, 'Hidden Law-Making in the Province of Medical Jurisprudence' (2014) 77 *Modern Law Review* 343, esp 367–370.

24 Margaret Brazier and Suzanne Ost, *Bioethics, Medicine and the Criminal Law Volume III: Medicine and Bioethics in the Theatre of the Criminal Process* (Cambridge: Cambridge University Press, 2013) 69.

25 See, e.g., Patrick Wintour, 'HSBC boss: I'm ashamed but not culpable' *The Guardian*, February 26, 2015, p.1.

can be rich with non-legal normative elements—as when the context reflects a particular professional, business, or working culture.[26] As Lobel, Ellickson, and others have pointed out, in many cases, these elements will be the ones that are in the forefront of regulatees' minds or will be the ones with which regulatees habitually comply. When these norms are in tension with the legal norms, the latter will have limited effect and, indeed, they might be counter-productive. Examples of such ineffectiveness are legion, ranging from laws that try to control the supply and consumption of alcohol or recreational drugs through to legal attempts to control serial infringements of copyright by millions of file-sharers.[27] Even when the official norms are offered as options or defaults, it is notorious that the target regulatees might systematically ignore these offerings and act on their own norms.[28]

By contrast, in Type 2 regulatory environments—at any rate, where there is a commonality between regulators and regulatees—we might expect the tensions to be less serious. Even here, however, there will be majority and minority views as well as 'insiders' and 'outsiders'.[29] Particularly for those who are 'outsiders', Type 2 regulatory environments might present themselves as imposed, their norms being seen as unacceptable. In these circumstances, a counter-culture might develop in opposition to the insiders' normative regime; and, this process of differentiation might repeat itself as the regulatory environment becomes increasingly fractured and complex. Without doubt, there is an interesting analysis to be undertaken, charting the circumstances in which regulatee norms (and, similarly, intermediate normative orders) complement or supplement formal top-down norms, and in which circumstances the lack of coincidence between norms or the incompatibility of norms creates practical problems.[30] However, this is not a task to be undertaken here.[31]

---

26 See, e.g., John C. Coffee Jr, '"No Soul to Damn, No Body to Kick": An Unscandalized Inquiry into the Problem of Corporate Punishment' (1981) 79 *Michigan Law Review* 386.

27 See Roger Brownsword and Morag Goodwin, *Law and the Technologies of the Twenty-First Century* (Cambridge: Cambridge University Press, 2012) Ch.13.

28 Famously, see Stewart Macaulay, 'Non-Contractual Relations in Business' (1963) 28 *American Sociological Review* 55.

29 Compare Julia Black, 'Regulation as Facilitation: Negotiating the Genetic Revolution' (1998) 61 *Modern Law Review* 621.

30 On which, see, e.g., Zelia A. Gallo, 'Punishment, authority and political economy: Italian challenges to western punitiveness' (2015) 17 *Punishment and Society* 598. At 611, Gallo contrasts 'Legal Italy' and 'real Italy' in the following evocative way:

> 'Legal Italy' is the nation as imagined by Italy's laws; 'real Italy' is Italy as it operates in fact, including via the myriad intermediate orders that the nation contains ... The relationship between real and legal 'Italies' should be understood as a tension or a dualism, rather than a relationship of mutual exclusivity.

31 For a range of interesting remarks about the theoretical questions presented by informal systems of social control, see Ellickson (n 4) esp. Ch.7.

(vii) The boundaries of a particular regulatory environment

One of the many (and more obvious) questions that we might ask about our embryonic concept of the regulatory environment is: how do we draw the limits or boundaries of any particular regulatory environment? Should we draw them from the perspective of would-be regulators or from the perspective of regulatees?[32] Or, is there some external vantage point from which we can draw them?

To start with the regulators' view, their standpoint will be that they want (i) a class of regulatees (it could be a broad class or quite a narrow one) (ii) to act (or to desist from acting) in a certain way (iii) in some defined zone (which again could be broad or narrow). For example, while regulators want their regulatees, whoever they are and wherever they are, to respect the physical well-being of one another, they will have more focused schemes for safety on the roads, for health and welfare at work, and for the safety of patients in hospitals, and so on. Accordingly, in many cases, the regulatory environment will be multi-level: in the normative dimension, there will be background standards of general application but, in many cases, the foreground standards will be specific to particular persons or particular places, or both. On this account, the limits of the regulatory environment are set by the sum of the signals given by the regulators.

If, however, we construct this same regulatory environment from the standpoint of regulatees, which of the regulators' signals are to count? Do regulatees need to be aware of the signals that the regulators have given? Do they need to regard the regulators as 'authorised' to give signals? Do regulatees need to act on the signals? Different criteria of knowledge, legitimacy, and effectiveness will generate differently constituted regulatory environments, each with its own limits.[33] Moreover, regulatees might have their own rules and regulations, such as those represented by the unofficial rules of the road. This can present some drivers with a dilemma: do they follow the official rules and their signals or do they follow the unofficial code and its signals? What we have here is not a competition between two top-down regulators but an exercise in bottom-up self-regulation coming into tension with the background top-down standards and signals. This is a phenomenon that is far from uncommon: we see it, time and again, whenever a regulatee

---

32 Compare Chris Reed, *Making Laws for Cyberspace* (Oxford: Oxford University Press, 2012). One of the principal themes of Reed's critique of current (failed) attempts to legislate for cyberspace is that lawmakers need to put themselves in the position of regulatees in order to assess which interventions will gain the respect of the latter. Although Reed focuses on law rather than the larger regulatory environment, his discussion invites constructing the regulatory environment from the standpoint of regulatees.

33 Exactly in the way that Reed appreciates for cyberspace (n 32).

community sets up its own rival code or culture (whether this is drivers, file-sharers, bankers, doctors, or sports people).[34] In these cases, the limits of the regulatory environment are unclear and, depending on the balance of power and influence, there is likely to be a degree of top-down regulatory ineffectiveness.

How, then, might a detached observer capture the essence of a 'regulatory environment' (if, indeed, this is a meaningful question)? Although we can say that we are conceiving of a signalling environment that is designed to channel group conduct, it needs to be recognised that there is a great deal of variety in particular instantiations. Whilst some regulatory environments are reasonably stable and well formed, others are unstable, overlapping, conflictual, and so on. In the latter cases, the way in which we draw the boundaries of any particular regulatory environment will depend on whether we adopt the perspective of the regulators or that of the regulatees. Which perspective should prevail?

Perhaps the best way to address the current question is to relate it to our particular cognitive interest. For example, if our cognitive interest is in the relative effectiveness of the exercise of top-down regulatory power, we will start with a notion of the regulatory environment that reflects the regulators' view; but, to account for a lack of effectiveness, we will almost certainly construct an account of the regulatory environment that reflects the practice of regulatees. However, if our interest is in the *legitimate* exercise of regulatory power, we might consider the issue from the standpoints of both regulators and regulatees before trying to develop a detached view that then offers an ideal-typical construct of the regulatory environment. This takes us deep into issues of conceptual methodology. The approach that I am proposing for my present purposes is 'instrumental' and 'pragmatic'; it makes no claim to specifying the concept of the regulatory environment in some rationally compelling form. This is not to say that the methodological issues are unimportant.[35] However, in the present context, they takes us away from the central task of sketching a working concept of the regulatory environment that will recognise the full set of normative signals as well as the non-normative signals given by technological management.

---

34 On the last-mentioned, see Roger Brownsword, 'A Simple Regulatory Principle for Performance-Enhancing Technologies: Too Good to be True?' in Jan Tolleneer, Pieter Bonte, and Sigird Sterckx (eds) *Athletic Enhancement, Human Nature and Ethics: Threats and Opportunities of Doping Technologies* (Dordrecht: Springer, 2012) 291. And, for a rare insight into the world of professional cycling, see Tyler Hamilton and Daniel Coyle, *The Secret Race* (London: Bantam Books, 2012).

35 Compare pp 60–61 below and Roger Brownsword, 'Field, Frame and Focus: Methodological Issues in the New Legal World' in Rob van Gestel, Hans Micklitz, and Ed Rubin (eds), *Rethinking Legal Scholarship: A Transatlantic Interchange* (New York: Cambridge University Press, 2016) 112.

IV Law in context: non-normative regulatory environments

In its hard form, a technologically managed environment will simply design out what otherwise would be a practical option (including by excluding human agents from the 'loop' or the process). However, we should not assume that all technological instruments function in a non-normative way; it bears repetition that some such instruments remain normative. This point is very clearly drawn out in Mireille Hildebrandt's distinction between 'regulative' (normative) and 'constitutive' (non-normative) technological features.[36] By way of an illustrative example, Hildebrandt invites readers to imagine a home that is enabled with a smart energy meter:

> One could imagine a smart home that automatically reduces the consumption of energy after a certain threshold has been reached, switching off lights in empty rooms and/or blocking the use of the washing machine for the rest of the day. This intervention [which is a case of a 'constitutive' technological intervention] may have been designed by the national or municipal legislator or by government agencies involved in environmental protection and implemented by the company that supplies the electricity. Alternatively [this being a case of a 'regulative' technological intervention], the user may be empowered to program her smart house in such a way. Another possibility [again, a case of a 'regulative' technological intervention] would be to have a smart home that is infested with real-time displays that inform the occupants about the amount of energy they are consuming while cooking, reading, having a shower, heating the house, keeping the fridge in function or mowing the lawn. This will allow the inhabitants to become aware of their energy consumption in a very practical way, giving them a chance to change their habits while having real-time access to the increasing eco-efficiency of their behaviour.[37]

Similarly, Pat O'Malley charts the different degrees of technological control that might be applied to regulate the speed of motor vehicles:

> In the 'soft' versions of such technologies, a warning device advises drivers they are exceeding the speed limit or are approaching changed traffic regulatory conditions, but there are progressively more aggressive versions. If the driver ignores warnings, data—which include calculations of the excess speed at any moment, and the distance over

---

36 Mireille Hildebrandt, 'Legal and Technological Normativity: More (and Less) than Twin Sisters' (2008) 12.3 *TECHNE* 169.

37 *Ibid.*, 174.

which such speeding occurred (which may be considered an additional risk factor and *thus* an aggravation of the offence)—can be transmitted directly to a central registry. Finally, in a move that makes the leap from perfect detection to perfect prevention, the vehicle can be disabled or speed limits can be imposed by remote modulation of the braking system or accelerator.[38]

Accordingly, whether we are considering smart cars, smart homes, or smart regulatory styles, we need to be sensitive to the way in which the regulatory environment engages with regulatees, whether it directs normative signals at regulatees enjoining them to act in particular ways, or whether the technology of regulation simply imposes (non-normatively) a pattern of conduct upon regulatees irrespective of whether they would otherwise choose to act in the way that the technology now dictates.

(i) Technological management as part of the regulatory environment

When we turn to technological management, we are thinking about technologies that are (in Hildebrandt's terminology) 'constitutive', that prevent, or disable, or compel certain actions.[39] On modern railway trains, for example, it is not possible for passengers to open the carriage doors until the guard has released the central locking system.[40] Similarly, smart cars might be disabled on sensing alcohol or drugs in the driver; and driverless cars, with human agents out of the operating loop, will be designed so that they always comply with what are safe speed limits.[41] Where environments are technologically controlled in this way, the regulatory signal is no longer normative. In other words, in place of (normative) prescription, we have only (non-normative) possibility and impossibility.[42] As O'Malley puts it, in the context of criminal justice, the adoption of technological management signals the move from 'perfect detection to perfect prevention'.[43]

---

38 Pat O'Malley, 'The Politics of Mass Preventive Justice' in Andrew Ashworth, Lucia Zedner, and Patrick Tomlin (eds), *Prevention and the Limits of the Criminal Law* (Oxford: Oxford University Press 2013) 273, 280.
39 See, e.g., Mireille Hildebrandt (n 36); and Roger Brownsword, *Rights, Regulation and the Technological Revolution* (Oxford: Oxford University Press, 2008).
40 See Jonathan Wolff, 'Five Types of Risky Situation' (2010) 2 *Law, Innovation and Technology* 151.
41 See Eric Schmidt and Jared Cohen, *The New Digital Age* (New York: Alfred A. Knopf, 2013) 25: 'Google's fleet of driverless cars, built by a team of Google and Stanford University engineers, has logged hundreds of thousands of miles without incident ...'
42 Roger Brownsword, 'Whither the Law and the Law Books: From Prescription to Possibility' (2012) 39 *Journal of Law and Society* 296.
43 O'Malley (n 38), 280.

Nevertheless, the technologically managed environment is still recognisably a regulatory environment. There are still regulators (employing technological management) and regulatees (at whom measures of technological management are directed). In any regulatory environment, we can say that, in principle, so many actions (x actions) are available to regulatees. Where a normative strategy is employed, regulators will seek to channel the behaviour of their regulatees by steering them towards one or more (required or desired) actions or away from one or more (prohibited or undesired) actions. In the former case, the steer is positive, in the latter it is negative. Where a non-normative strategy is employed, the steering follows a similar pattern: in order to steer regulatees towards a desired action, other options might be closed off or the regulatee is somehow 'locked in' to the required action; and, in order to steer regulatees away from an undesired action, that action is simply closed off (instead of x actions being available, it is now x-1 actions that are available). Quite possibly, in a non-normative managed environment, the use of positive and negative steering techniques is more complex than in a normative regulatory environment; and, because technological management leaves regulatees with no choice in relation to the regulated matter itself, there is an important sense in which regulatees will always be less free in managed environments. Emphatically, though, this is not to suggest that technologically managed environments should be equated with total control; the range of actions (x actions) actually available in a particular managed environment might leave regulatees with significant choice—albeit a necessarily narrower range of options, other things being equal, than in a normative regulatory environment where regulatees retain the practical option of non-compliance with the rules. All that said, in both normative and non-normative environments—most obviously where the purpose is that of crime control—the starting point is that regulators seek to direct, shape, or structure some aspects of their regulatees' conduct and they do this by employing various positive and negative channelling instruments.

These similarities notwithstanding, some instances of technological management might seem to fall beyond the concept of regulation as an intervention that is (1) intended to (2) channel the behaviour of regulatees (3) either positively or negatively (4) relative to some regulatory purpose. While there is no doubt that technological management is designed to serve some regulatory purpose (whether it is crime control, protecting human health and safety, protection of the environment, simply cutting costs, or protecting values such as privacy, and so on), and that it might seek to do this by positive or negative means, is it so clear that technological management is always aimed at channelling (or guiding or directing) behaviour? Further, can we assume that, when regulators introduce technological management, they always intend to influence human behaviour—perhaps they are simply thinking about improving safety so that particular products and places represent fewer risks to the physical well-being of human agents?[44]

(ii) The channelling question

To pick up on the first of these questions, Karen Yeung[45] makes the excellent point that some 'regulatory interventions'

> operate by seeking to mitigate the harm associated with particular activities rather than with seeking to provoke a change in user behaviour, such as the installation of air-bags in motor vehicles to reduce the impact of a collision on vehicle occupants, the installation of shatter-proof glass in buildings that might be vulnerable to intentional or accidental damage or the fluoridation of community water supplies to reduce the incidence and severity of dental caries.

To accommodate such interventions, Yeung proposes that the definition of 'regulation' should be constructed around the desired social outcomes which the regulator intentionally seeks to bring about. Regulation would then be defined as 'sustained and focused attempts intended to produce a broadly defined outcome or outcomes directed at a sphere of social activity according to defined standards or purposes that affect others'.[46] If we accept this accommodation, we will accept that, in some cases, regulators may achieve their objectives without in any way seeking to modify the behaviour of their regulatees.

Now, those who are reluctant to make any concession to the idea that 'regulation' necessarily involves an attempt to steer conduct might argue that, in the examples given by Yeung, there is actually an attempt to steer the conduct of motor car manufacturers, builders, and water companies—these being, in Yeung's scenarios, the relevant regulatees. However, even if the facts support this analysis, why should we be reluctant to accept that 'regulatory' interventions and the regulatory environment can include measures of the

---

44 Compare Ugo Pagallo, *The Laws of Robots* (Dordrecht: Springer, 2013) 183–192. In a perceptive discussion, Pagallo, having pointed out that the shaping of the world might be influenced by environmental, product, and communication design, and having distinguished between the design of places, products and organisms, then notes that the aim of design might be (i) to encourage the change of social behaviour, (ii) to decrease the impact of harm-generating behaviour, or (iii) to prevent harm-generating behaviour. In relation to harm reduction (the second aim), Pagallo says that, instead of thinking about speed bumps or the like (which are examples of design being used for the first purpose), we should think about air bags.

45 Karen Yeung (n 11) p.5.

46 *Ibid.*, p.5.

kind represented by technological management where there is sometimes no attempt to steer or to re-channel behaviour? The fact of the matter is that, in cases of the kind suggested by Yeung, the purpose of regulators is to improve health and safety and their strategy for doing this (let us suppose) is by setting standards that are intended to channel the conduct of motor car manufacturers, builders, and water companies. Why should we exclude from our regulatory scrutiny the consequential impact that this has on the practical options available to the end users? Their conduct may not be the direct object of the channelling (they may not be guided or directed as such) but the contexts in which they act are the targets for regulators. Moreover, if (contrary to our assumption) the motor car manufacturers, builders, and water companies, rather than complying with norms set by a public regulator, took it upon themselves to improve health and safety in this way, why should their use of technological management be treated as somehow less 'regulatory'?

Applying this analysis, we can treat any particular regulatory environment as having (in principle) two dimensions. First, there is the normative dimension: here, regulators seek to guide the conduct of regulatees; standards are set; prohibitions, permissions, and requirements are signalled; regulators employ prudential or moral or combined prudential-moral registers to engage the practical reason of regulatees; and various kinds of technology might be used to reinforce the normative standards and the signals—in our times, such reinforcement typically involves surveillance, tracking, and identifying technologies that increase the likelihood of detection in the event of non-compliance. Secondly, there is the non-normative dimension: here, regulators target a certain pattern of behaviour, sometimes intentional conduct, at other times unintentional acts; various kinds of technology (product design, architecture, and the like) are used to channel regulatees in the desired way or to place regulatees in contexts that are more desired; and the only register employed is that of practicability/possibility. As this dimension becomes more sophisticated, the technologies that are employed are embedded in our physical surroundings (in the way, for example, that sensors detect a human presence in a room and switch on the lights), or possibly in our bodies, and even perhaps by modifications to our genetic coding. Although, in the technologically managed environments of the future, a regulatee might explore what can and cannot be done, in many respects this will no longer seem like a signalling environment: for regulatees who are habituated to this kind of environment, it will just seem to be the way things are. In such environments, the regulatory 'ought' will be embedded in the regulatory 'is'.

(iii) Regulatory intent

This leaves the question of regulatory intent. When regulators employ technological management (rather than rules), their intention is to serve some purpose. The fact that they may not intend to re-channel conduct, as we have said, is significant; but it does not speak against the 'regulatory' character of the intervention. However, what if, contrary to the regulatory intention, an intervention that is intended merely to restructure the context in which regulatees act does have some unintended re-channelling effects?

As expressed, this is not a question about the case for extending the notion of the regulatory environment to include measures of technological management. Rather, it is a question about the accountability of regulators who employ such measures. Perhaps the best-known example that raises such a question is the long-running debate about whether the design of Robert Moses' bridges on the New York parkways was intended to have the (racially discriminatory) effect of making it more difficult for the poor, mainly black, population to reach the beaches on Long Island.[47] From the point of view of prospective beach-users, it made little difference whether the bridges had been designed with this intent—in practice, the bridges had the regulative effect of making the beaches more difficult to access. Nevertheless, if we are to hold regulators (designers) to account, is it not the case that their intentions remain important?

The paradigm, as we have said, is one in which regulators have certain purposes (whether for crime control or health and safety and so on), and they put in place a rule framework or a design[48] that is intended to have a particular effect. In such a case, it is perfectly fair to ask regulators to justify both their purposes and the instruments (the rules or the designs) that they have adopted. However, even the best-laid regulatory plans can go awry and, indeed, a common problem with regulatory interventions is that they generate unintended effects.[49] Clearly, when regulators are held to account, they must answer for both the intended and the unintended effects of the regulatory environments that they have put in place.

---

47 See Noëmi Manders-Huits and Jeroen van den Hoven, 'The Need for a Value-Sensitive Design of Communication Infrastructures' in Paul Sollie and Marcus Düwell (eds), *Evaluating New Technologies* (Dordrecht: Springer, 2009) 51, 54.

48 Smart regulators will be astute to learn from the 'design against crime' approach, using the processes and products of design to reduce crime and promote community safety;: see Paul Ekblom, 'Designing Products Against Crime' at 203; available at www.gripp aclip.com/wp-content/uploads/Designing-Products-Against-Crime.pdf (last accessed 2 November 2018).

49 Seminally, see Sally Falk Moore, 'Law and Social Change: The Semi-Autonomous Social Field as an Appropriate Subject of Study' (1973) 7 *Law and Society Review* 719, 723:

[I]nnovative legislation or other attempts to direct change often fail to achieve their intended purposes; and even when they succeed wholly or partially, they frequently carry with them unplanned and unexpected consequences. This is partly because new laws are thrust upon going social arrangements in which there are complexes of binding obligations already in existence. Legislation is often passed with the intention of altering the going social arrangements in specified ways. The social arrangements are often effectively stronger than the new laws.

Having said this, the case of the New York parkway bridges might seem rather different. In defence of the bridge designers, it might be argued that there was no regulatory plan as such, simply an attempt to strengthen the bridges. To be sure, in practice, the newly constructed bridges might have had a regulative impact, but this was an unintended effect of the design. Once upon a time, such a defence might have been adequate; but, nowadays, regulators will not get off the hook quite so easily. For, as it becomes increasingly clear that design can matter (potentially, having both negative and positive effects), so it is no longer acceptable for regulators to plead a lack of intent, or attention, with regard to such technical details. And, where regulators are using design against crime, then as Paul Ekblom rightly says, it is important that they

> consider whether their design violates privacy or unacceptably constrains freedom in some way—for example, a mobile phone which reports on someone's movements, whether tracking him or her for his or her own good or for other people's, without his or her awareness or free consent.[50]

While inattention may lead to regulatory environments that are detrimental to, say, the health or the privacy of regulatees, smart regulatory action can have the opposite impact (for example, by requiring or encouraging architects and technologists to default to health-promoting or privacy-enhancing designs).[51] In short, although the paradigmatic regulatory environment is the product of intentional design, regulators need to answer for both the intended and the unintended channelling effects of their actions as well as for their omissions.[52]

(iv) Field, frame and focus

If, in the preceding remarks, I seem to be taking a fairly 'cavalier' approach to what counts as regulation, how we understand channelling, and so on, forcing the phenomena in which I have declared a cognitive interest to fit into the frame of the regulatory environment, let me try briefly to be as explicit and self-reflective as I can about the nature of the exercise.

First, I am saying that the field in which lawyers and jurists currently pursue their inquiries is too limited. The field needs to be extended so that

---

50 Ekblom (n 48), 216.

51 See, e.g., Manders-Huits and van den Hoven (n 47); and Peter-Paul Verbeek, 'The Moral Relevance of Technological Artifacts', in Paul Sollie and Marcus Düwell (eds), *Evaluating New Technologies* (Dordrecht: Springer, 2009) 63.

52 Compare Bibi van den Berg, 'Robots as Tools for Techno-Regulation' (2011) 3 *Law, Innovation and Technology* 319.

technological management falls within the scope of inquiry. All lawyers and jurists are interested in laws; and some lawyers and jurists are interested in regulation and governance. However, the field tends to be restricted to normative phenomena. Where the norms are complemented, opposed or replaced by technological management, we should be interested in what is going on; and the field of inquiry needs to be adjusted. After all, what sense does it make to say that normative attempts to influence the conduct of car drivers, such as rules requiring the wearing of seat belts, or rules requiring manufacturers to fit seat belts, or informal industry codes that require seat belts to be fitted, and the like, fall within the field of inquiry but technological management applied to seat belts (immobilising the vehicle unless the seat belt is engaged) falls outside that field? To make sense of this, we do not need to insist that 'code is law'; but we do need to have technological management on our regulatory radar.

Secondly, as a mechanism to frame inquiries within the field, I have suggested that we employ the concept of the regulatory environment. To cover the field, this framing has to allow for both normative and non-normative dimensions in the regulatory environment. Because the concept of regulation has been developed in normative settings, we sense some strain when we stretch the concept to include non-normative strategies. However, the alternative to doing this is to risk diverting inquiry back to normative strategies rather than taking seriously the non-normative elements that are of acute interest.

Thirdly, by presenting the complexion of the regulatory environment as a focus for our inquiries (as we do in the next chapter), we do put the spotlight on the relevant differences between normative and non-normative regulatory strategies. Indeed, the object of the exercise is not at all to suppress these differences, by covering them in a broadly defined field and framing of our inquiries. To the contrary, the object of the exercise is precisely to interrogate these differences and to assess their significance for both regulators and regulatees.

V Three generations of regulatory environment

Gathering up the strands of the discussion in this chapter, we might construct a typology highlighting three ideal-typical generations of regulatory environment. In a first-generation regulatory environment, regulators would rely exclusively on normative signals. This is not necessarily a technology-free environment; but the technology would signal in a normative register—for example, in the way that surveillance technologies signal that the likelihood of detection is increased and, hence, regulatees ought (if only for prudential reasons) to comply. In a second-generation regulatory environment, regulators would rely on the design of products and places (architecture) as well as the automation of processes that result in humans being taken out of the loop. Where regulators rely on such a design strategy, the signal is no longer normative; instead, the design features signal what is practicable or possible. Moreover, at this stage, the line between infrastructural hard-wiring (that has regulatory effects) and actions to be regulated starts to shift dramatically. Finally, in a third-generation regulatory environment, regulators would go beyond traditional normative signals and design of products and places by incorporating the regulatory design within regulatees themselves (for example, by controlling their genetic coding or modulating their neuro-signals). Where design is embedded in regulatees in such a way that it channels their behaviour, it is likely to be much less apparent to regulatees that they are being regulated—if the design is reliable, regulatees will simply behave (like products) in accordance with their specification.

It should be emphasised that these are ideal-types, possibly capturing the general direction of regulatory travel, but not at any time perfectly instantiated. So, in the year 2061, we may find that the regulatory environment is still predominantly first-generation. However, if technological management increases in the way that this book anticipates, by that time there will be significant second-generation elements in the regulatory environment. Indeed, in some contexts—for example, the regulation of road traffic[53]—the co-existence of first and second-generation dimensions will require some imaginative articulation. While in the year 2061, third-generation regulation may still be experimental, it is possible that by the year 2161, it too will make a more general contribution. At that point in the evolution of the regulatory environment, all strands of normative ordering and technological management will be on the radar. In these three generations, I suggest, albeit as ideal-types, we see our regulatory past, present and future.

---

53 Compare, Ronald Leenes and Federica Lucivero, 'Laws on Robots, Laws by Robots, Laws in Robots: Regulating Robots' Behaviour by Design' (2014) 6 *Law, Innovation and Technology* 193.

3

THE 'COMPLEXION' OF THE REGULATORY ENVIRONMENT

I Introduction

In place of the Warwickshire Golf and Country Club that we met in Chapter One, imagine a fictitious golf club, 'Westways'. The story at Westways begins when some of the older members propose that a couple of carts should be acquired for use by members who otherwise have problems in getting from tee to green. There are sufficient funds to make the purchase but the green-keeper expresses a concern that the carts might cause damage to Westways' carefully manicured greens. The proposers share the green-keeper's concerns and everyone is anxious to avoid causing such damage. Happily, this is easily solved. The proposers, who include most of the potential users of the carts, act in a way that is respectful of the interests of all club members; they try to do the right thing; and this includes using the carts in a responsible fashion, keeping them well clear of the greens.

For a time, the carts are used without any problem. However, as the membership of Westways changes—and, particularly, as the older members leave—there are some 'incidents' of irresponsible cart use. The green-keeper of the day suggests that the club needs to take a firmer stance. In due course, the club adopts a rule that prohibits taking carts onto the greens and that penalises members who break the rule. Unfortunately, this intervention does not help; indeed, if anything, the new rule aggravates the situation. While the rule is not intended to license irresponsible use of the carts (on payment of a fine), this is how some members perceive it; and the effect is to weaken the original 'moral' pressure to respect the greens. Moreover, members know that, in some of the more remote parts of the course, there is little chance of rule-breakers being detected.

Taking a further step to discourage breaches of the rule, it is decided to install a few CCTV cameras around the course at Westways. However, not only is the coverage patchy (so that in some parts of the course it is still relatively easy to break the rule without being seen), the man who is employed to watch the monitors at the surveillance control centre is easily distracted and members soon learn that he can be persuaded to turn a blind eye in return for the price of a couple of beers. Once again, the club fails to find an effective way of channelling the conduct of members so that the carts are used in a responsible fashion.

It is at this juncture that the club turns to a technological fix. The carts are modified so that, if a member tries to take the cart too close to one of the greens (or to take the cart off the course) they are warned and, if the warnings are ignored, the cart is immobilised.[1] At last, thanks to technological management, the club succeeds in realising the benefits of the carts while also protecting its greens.

As we trace the particular history at our fictitious club, Westways, we see that the story starts with an informal 'moral' understanding. In effect, just as in the early days of eBay, regulation rests on the so-called Golden Rule:[2] that is to say, the rule is that members should use the carts (or the auction site) in the way that they would wish others to use them. It then tries to reinforce the moral signal with a rule (akin to a law) that sends a prudential signal (namely, that it is in the interests of members to comply with the rule lest they incur the penalty). However, the combination of a prudential signal with a moral signal is not altogether a happy one because the former interferes with the latter.[3] When CCTV cameras are installed, the prudential signals are amplified to the point that they are probably the dominant (but still not fully effective) signals. Finally, with technological management, the signals change into a completely different mode: once the carts are redesigned, it is no longer for members to decide on either moral or prudential grounds to use the carts responsibly; at the end of the story, the carts cannot be driven onto the greens and the signals are entirely to do with what is possible and impossible.

What the story at Westways illustrates is the significant changes that take place in the 'complexion' of the regulatory environment; with each regulatory initiative, the 'signalling register' changes from moral, to prudential, to what is possible. With each move, the moral register is pushed further into the background.

---

1 Compare, too, the next generation carts introduced to the Celtic Manor golf course in Wales: see www.ispygolf.com/courses/news/article/Next_Generation_Golf_Carts_Arrive.html (last accessed 2 November 2018).

2 As recounted by Jack Goldsmith and Tim Wu, *Who Controls the Internet?* (Oxford: Oxford University Press, 2006).

3 Compare the findings in U. Gneezy and A. Rustichini, 'A Fine is a Price' (2000) 29 *Journal of Legal Studies* 1.

In this chapter, we can begin to equip ourselves to think more clearly about the changing complexion of the regulatory environment—and particularly the changes introduced by technological management. Starting with the idea of the 'regulatory registers' (that is, the ways in which regulators seek to engage the practical reason of their regulatees) and their relevance, we find that technological management relies on neither of the traditional normative registers (moral and prudential). With technological management, there is a dramatic change in the complexion of the regulatory environment and the question is whether the use of such a regulatory strategy can be justified. This question is posed in relation to three kinds of regulatory purpose, namely: crime control; to improve health and safety; and to protect the essential infrastructure for a community of human agents. In each case, the regulatory purpose might be appropriate and defensible; and if standard normative strategies (employing a mix of appeals to prudential and moral reasons) were employed, there might be little to question. However, where technological management is employed, the question is whether the changes to the complexion of the regulatory environment can also be justified.

II The regulatory registers

By a 'regulatory register', I mean the kind of signal that regulators employ in communicating with regulatees. There are three such registers, each of which represents a particular way in which regulators attempt to engage the practical reason (in the broad and inclusive sense of an agent's reasons for action)[4] of regulatees. Thus:

(i) in the first register (the moral register), the coding signals that some act, x, categorically ought or ought not to be done relative to standards of right action—regulators thus signal to regulatees that x is, or is not, the right thing to do;

(ii) in the second register (the prudential register), the coding signals that some act, x, ought or ought not to be done relative to the prudential interests of regulatees—regulators thus signal to regulatees that x is, or is not, in their (regulatees') self-interest; and,

(iii) in the third register (the register of practicability or possibility), the environment is designed in such a way that it is not reasonably practicable (or even possible) to do some act, x—in which case, regulatees reason, not that x ought not to be done, but that x cannot be done.

In an exclusively moral environment, the primary normative signal (in the sense of the reason for the norm) is always moral; but the secondary signal,

---

4 In this broad sense, 'practical reason' encompasses both moral and non-moral reasons for action.

depending upon the nature of the sanction, might be more prudential. In traditional criminal law environments, the signals are more complex. Whilst the primary normative signal to regulatees can be either moral (the particular act should not be done because this would be immoral) or paternalistically prudential (the act should not be done because it is contrary to the interests of the regulatee), the secondary signal represented by the deterrent threat of punishment is prudential.[5] As the regulatory environment relies more on technological instruments, the strength and significance of the moral signal fades; here, the signals to regulatees tend to accentuate that the doing of a particular act is contrary to the interests of regulatees, or that it is not reasonably practicable, or even that it is simply not possible.[6] Where the signal is that a particular act is no longer a possible option, regulatee compliance is, so to speak, fully determined; in all other cases, and especially in the normative range, the conduct of regulatees is under-determined.

III The relevance of the registers

How does a framing of this kind assist our inquiries? Crucially, if we share the aspiration for legitimacy, then the framing of our inquiries by reference to the regulatory registers draws our attention to any technology-induced drift from the first to the second register (where moral reasons give way to prudence) and then from the second to the third register (where the signal is no longer normative). Sophisticated technologies of control surely will appeal to future classes of regulators in both the public and the private sectors; but communities with moral aspirations need to stay alert to the corrosion of the regulatory conditions that give meaning to their way of life.[7] The key point is that, in moral communities, and in all walks of life, the aspiration is to do the right thing; respecting persons is doing the right thing; and it matters

---

5 Compare Alan Norrie, 'Citizenship, Authoritarianism and the Changing Shape of the Criminal Law' in Bernadette McSherry, Alan Norrie, and Simon Bronitt (eds), *Regulating Deviance* (Oxford: Hart, 2009) 13. *Ibid.*, at 15, Norrie highlights three broad developments in recent British criminal law and justice, namely: (i) an increasing emphasis on notions of moral right and wrong and, concomitantly, on individual responsibility ('responsibilisation'); (ii) an increasing emphasis on dangerousness and, concomitantly, on the need for exceptional forms of punishment or control ('dangerousness'); and (iii) an increasing reliance on preventative orders and new forms of control ('regulation'). While the first of these developments is in line with the aspirations of moral community, it is the second and the third that such a community needs to monitor with care. In this light, see, in particular, Lucia Zedner, 'Fixing the Future? The Pre-emptive Turn in Criminal Justice' in McSherry, Norrie, and Bronitt (eds), op. cit, 35. See, further, Chapter Nine.

6 Compare Bert-Jaap Koops, 'Technology and the Crime Society: Rethinking Legal Protection' (2009) 1 *Law, Innovation and Technology* 93.

7 Compare Mireille Hildebrandt, 'A Vision of Ambient Law', in Roger Brownsword and Karen Yeung (eds), *Regulating Technologies* (Oxford: Hart, 2008) 175, and 'Legal and Technological Normativity: More (and Less) than Twin Sisters' (2008) 12.3 *TECHNE* 169.

to moral communities that such respect is shown freely and for the right reason—not because we know that we are being observed or because we have no practical option other than the one that the regulators have left for us.

While surveillance technologies, such as the use of CCTV at Westways, leave it to regulatees to decide how to act, technologies can be used in ways that harden the regulatory environment, by designing out or disallowing the choice that regulatees previously had. As we said in the previous chapter, such hard technologies speak only to what can and cannot be done, not to what ought or ought not to be done; these are non-normative regulatory interventions.[8] When regulatory environments are hardened in this way, it is not quite accurate to say, as Zygmunt Baumann says of post-panopticon ordering, that it involves 'rendering the permissible obligatory';[9] rather such hardening signifies that the impermissible (but, hitherto, possible) becomes impossible—'ought not' becomes 'cannot'.

To return to one of the remarks made in the introductory chapter of the book, it may well be that our concerns about technological management change from one context to another. As we intimated, it may be that moral concerns about the use of technological management are at their most acute where the target is *intentional* wrongdoing or *deliberately* harmful acts— even though, of course, some might argue that this is exactly where a community most urgently needs to adopt (rather than to eschew) technological management.[10] That said, it surely cannot be right to condemn all applications of technological management as illegitimate. For example, should we object to modern transport systems on the ground that they incorporate safety features that are intended to design out the possibility of human error or carelessness (as well as intentionally malign acts)?[11] Or, should we object to the proposal that we might turn to the use of regulating technologies to replace a failed normative strategy for securing the safety of patients who are taking medicines or being treated in hospitals?[12]

In the next three sections of the chapter, we will begin to develop a more context-sensitive appreciation of the use of technological management by

---

8 See Hildebrandt (n 7) ('Legal and Technological Normativity: More (and Less) than Twin Sisters').

9 Zygmunt Bauman and David Lyon, *Liquid Surveillance* (Cambridge: Polity Press, 2013) 80.

10 Compare Roger Brownsword, 'Neither East Nor West, Is Mid-West Best?' (2006) 3 *Scripted* 3–21 (available at papers.ssrn.com/sol3/papers.cfm?abstract_id=1127125 (last accessed 2 November 2018).

11 For a vivid example (slam-doors on trains), see Jonathan Wolff, 'Five Types of Risky Situation' (2010) 2 *Law, Innovation and Technology* 151.

12 For an example of an apparently significant life-saving use of robots, achieved precisely by 'taking humans out of the equation', see 'Norway hospital's "cure" for human error' *BBC News*, May 9, 2015: available at www.bbc.co.uk/news/health-32671111 (last accessed 2 November 2018).

differentiating between three classes of regulatory purpose, namely: classic crime control (reflected in the core offences of the criminal code); attempts to improve the conditions for human health and safety as well as to protect the integrity of the environment; and measures to protect the essential infrastructure for a community of human agents. As we do so, we might also begin to consider the relevance of the following: first, whether the target for technological management is an act that is intentional or reckless, or one that is negligent or unintentional—as Holmes famously quipped, even a dog knows the difference between being intentionally kicked and being tripped over accidentally; secondly, whether when technological management is adopted by an individual or by a group, it impacts negatively on others; and, thirdly, whether (as at Westways), before technological management is adopted, other regulatory strategies have been tried (in other words, whether technological management should be treated as a strategy of last resort).

IV Technological management and crime control

Technological management might be used, in place of the classic criminal laws, to prevent the infliction of intentional harm on persons, or their property, or to extract money from them; but it might also be used to replace regulatory crimes because it is now possible to use a technological fix to make workplaces and the like safer for humans (indeed, perfectly safe where automation of processes means that humans are taken out of the equation). In this section, we can focus on the use of technological management for the former purposes.

First, we can remind ourselves of the price that moralists might pay if technological management ensures that human agents do (what regulators and regulatees alike judge to be) the right thing. Secondly, we can consider the ways in which conscientious objectors and civil disobedients might find that their options are compromised by technological management (unlike rules). While, in the former case, the price is paid by individual agents, in the latter we are talking about a price that is paid both by the individuals and also by the community collectively.

(i) Compromising the possibility of doing the right thing

Where hard technologies are introduced in order to protect an agent against the risk of the intentionally (or wilfully) harmful acts of others, the pattern of 'respectful' non-harming conduct is not explained by the prudential choices of regulatees, even less by their moral judgments. In a moral community, it is when code and design leave regulatees with no option other than compliance that the legitimacy of the means employed by regulators needs urgent consideration. The problem here is that, even if we concede that the technology channels regulatees towards right action, the technologically secured pattern of right action is not at all the same as freely opting to do the right thing.[13] One agent might be protected from the potentially harmful acts of others, but moral virtue, as Ian Kerr protests, cannot be automated.[14]

Expressing this concern in relation to the use of 'digital locks', Kerr says:

> [A] generalized and unimpeded use of digital locks, further protected by the force of law, threatens not merely [various] legal rights and freedoms but also threatens to significantly impair our moral development. In particular, I express deep concern that digital locks could be used in a systematic attempt to 'automate human virtue'—programming people to 'do the right thing' by constraining and in some cases altogether eliminating moral behaviour through technology rather than ethics or law. Originally introduced to improve the human condition, digital locks and other automation technologies could, ironically, be used to control our virtual and physical environments in unprecedented ways, to eliminate the possibility for moral deliberation about certain kinds of action otherwise possible in these spaces by disabling the world in a way that ultimately disables the people who populate it. Not by eliminating their choices but by automating them—by removing people from the realm of moral action altogether, thereby impairing their future moral development.[15]

Applying this analysis to the case of honest action, Kerr detects the irony that

> a ubiquitous digital lock strategy meant to 'keep honest people honest' is a self-defeating goal since it impairs the development of *phronesis*, stunts moral maturity and thereby disables the cultivation of a deep-seated disposition for honesty. Woven into the fabric of everyday life, digital locks would ensure particular outcomes for property owners but would do so at the expense of the moral project of honesty.[16]

The point is that the shift from law (or ethics) to technological instruments changes the 'complexion' of the regulatory environment in a way that threatens to compromise the possibility of authentic moral action.[17]

---

13 Compare Evgeny Morozov, *To Save Everything, Click Here* (London: Allen Lane, 2013), 190–193 ('Why You Should Ride the Metro in Berlin').

14 Ian Kerr, 'Digital Locks and the Automation of Virtue' in Michael Geist (ed), *From 'Radical Extremism' to 'Balanced Copyright': Canadian Copyright and the Digital Agenda* (Toronto: Irwin Law, 2010) 247.

15 *Ibid.*, 254–255.

16 *Ibid.*, 292.

17 See, further, Roger Brownsword, 'Lost in Translation: Legality, Regulatory Margins, and Technological Management' (2011) 26 *Berkeley Technology Law Journal* 1321. Compare Charles Taylor, *The Ethics of Authenticity* (Cambridge, Mass: Harvard University Press,

The precise pathology of precluding an agent from doing the right thing for the right reason is, I confess, a matter for further debate. Arguably, the problem with a complete technological fix is that it fails to leave open the possibility of 'doing wrong' (thereby disabling the agent from confirming to him or herself, as well as to others, their moral identity); or it is the implicit denial that the agent is any longer the author of the act in question; or, possibly the same point stated in other words, it is the denial of the agent's responsibility for the act; or perhaps that it frustrates the drive for recognition and interferes with the complex relationship between respect and esteem.[18] At all events, it will be for each community with moral aspirations to determine how far it matters that (as Kantians might put it) agents should act 'out of duty' rather than merely 'in accordance with duty' and how important it is that agents are able to 'express' their moral virtue, and so on. Having clarified such matters, the community will then be in a position to consider whether, and in which particular way, technological management might compromise the possibility of moral agency.

Where the State is responsible for laying out the general framework for community life, it follows that moralists might have reservations about the use of technological management for general crime control purposes. If the State could eliminate the need for crimes against the person by, let us imagine, clothing its citizens in lightweight nanosuits or nanoskins of immense defensive strength, this would be hugely attractive; but it would not only reduce harm, it would reduce the scope for agents to do the right thing by choosing restraint over otherwise harmful acts of violence. However, it would not be only the State's use of technological management that would need to be scrutinised. Private acts of self-defence—such

---

1991) for a thoughtful critique of various strands of 'modernity' in relation to a supposed loss of 'authenticity'. Taylor argues persuasively that the real issue is not whether we are for or against authenticity, but '*about* it, defining its proper meaning' (at 73); and, for Taylor, authenticity is to be understood as a moral ideal that accords importance to agents fulfilling themselves by being true to themselves—but without this 'excluding unconditional relationships and moral demands beyond the self' (at 72–73).

18 I am grateful for this point to Marcus Düwell and others who participated in a symposium on 'Human Dignity: A Major Concept in Ethics?' held at the University of Tübingen on October 24, 2014. I am also grateful to Patrick Capps for drawing my attention to Frederick Neuhouser, *Rousseau's Theodicy of Self-Love: Evil, Rationality, and the Drive for Self-Recognition* (Oxford: Oxford University Press, 2008). The precise bearing of the latter on my concerns about technological management merits further reflection. However, we should certainly note: (i) Neuhouser's important observation that 'since respect alone can never wholly satisfy the human need for recognition, a complete solution to the problem of *amour-propre* will require a set of social—not merely political—institutions that, beyond insuring equal respect for all citizens, creates a space within which individuals can pursue their many and diverse needs for esteem without producing a world of enslavement, conflict, vice, misery, and alienation' (at 116); and (ii) his insightful discussion (in Ch.7) of the ways in which the drive for recognition from others might motivate the adoption of an impartial view that is characteristic of moral reason.

as the acts of self-defence taken by the Warwickshire—would also need to be scrutinised. This is not say that the acts of the joy-riders should be condoned or to deny that the Warwickshire had legitimate grounds for taking measures of self-defence; the point is that technological management, even in a good moral cause, comes at a moral price. Similarly, private uses of technological management that might be accused of being disproportionate or as overreaching need to be scrutinised—for example, the use of the 'Mosquito' (a device emitting a piercing high-pitched sound that is audible only to teenagers) in order to discourage young people from congregating outside shops or in other public places where they might interfere with private interests, or the use of digital rights management (DRM) to prevent what might actually be a legally protected 'fair use' of a digital product. In these cases, the use of technological management might be less likely to slip by unnoticed; however, once again, my point is not that questions might be raised about the particular purposes for which technological management is used but, rather, that the use of technological management to exclude the possibility of what some might view as intentional wrongs is problematic because of its impact on the complexion of the regulatory environment.

(ii) Compromising the possibility of conscientious objection and civil disobedience

Suppose that, having purchased some digital goods, one finds that the producers' measures of technological management constrain the use of the goods in ways that one thinks are unconscionable. Unlike an unjust legal rule which can always be met with acts of conscientious objection, technological management allows for no such act of protest. This puts the spotlight on another concern about the corrosion of the conditions for moral community, namely the risk that technological management might compromise the possibility of engaging in responsible moral citizenship.

Evgeny Morozov[19] identifies the particular problem with technological management when he considers how this regulatory strategy might restrict the opportunities for responsible acts of conscientious objection and civil disobedience. Recalling the famous case of Rosa Parks, who refused to move from the 'white-only' section of the bus, Morozov points out that this important act of civil disobedience was possible only because 'the bus and the sociotechnological system in which it operated were terribly inefficient. The bus driver asked Parks to move only because he couldn't anticipate how many people would need to be seated in the white-only section at the front; as the bus got full, the driver had to adjust the sections in real time, and Parks

---

19 Morozov (n 13).

happened to be sitting in an area that suddenly became "white-only".'[20] However, if the bus and the bus-stops had been technologically enabled, this situation simply would not have arisen—Parks would either have been denied entry to the bus or she would have been sitting in the allocated section for black people. Morozov continues:

> Will this new transportation system be convenient? Sure. Will it give us Rosa Parks? Probably not, because she would never have gotten to the front of the bus to begin with. The odds are that a perfectly efficient seat-distribution system—abetted by ubiquitous technology, sensors, and facial recognition—would have robbed us of one of the proudest moments in American history. Laws that are enforced by appealing to our moral or prudential registers leave just enough space for friction; friction breeds tension, tension creates conflict, and conflict produces change. In contrast, when laws are enforced through the technological register, there's little space for friction and tension—and quite likely for change.[21]

In short, technological management disrupts the assumption made by liberal legal theorists who count on acts of direct civil disobedience being available as an expression of responsible moral citizenship.

The argument that technological management might compromise the public life of an aspirant moral community is significant in its own right. However, some might see here a fresh angle of attack against those (liberal-minded) legal positivists who claim that their conceptualisation of law (as separate and distinct from morality) has certain *practical* advantages over its rivals.[22] The gist of the legal positivists' claim is that responsible citizens, thinking like legal positivists, will recognise that duly enacted legislation is legally valid even though they might have moral reservations about it. In this way, moral reservations do not muddy the waters in relation to what is or is not 'law'. However, so argue legal positivists, this is not the end of the matter. If the moral reservations in question are sufficiently serious, citizens retain the option of conscientious non-compliance or protest. In this way, the claim runs, we have certainty about what is and is not law and, at the same time, we have the benefits of critical citizenship.

Whether or not this line of legal positivist thinking is damaged by the emergence of technological management is an open question. Some legal positivists might think that they can nip the objection in the bud, either by

---

20 *Ibid.*, 204.

21 *Ibid.*, 205.

22 For discussion, see Roger Brownsword, '*Law as a Moral Judgment*, the Domain of Juris-prudence, and Technological Management' in Patrick Capps and Shaun D. Pattinson (eds), *Ethical Rationalism and the Law* (Oxford: Hart, 2016) 109.

saying that technological management falls outwith the scope of their conceptual debate (which is about the nature of legal *rules or norms*) or by arguing
that, so long as legal rules continue to form part of the regulatory environment, there are practical advantages in viewing such rules in a positivist way.
In other words, how we view law is one thing; how we view technological
management is a quite different matter. However, the more interesting question is whether critical citizenship is compromised by the employment of
technological management.

Returning to Morozov's example, suppose that legislation is introduced
that specifically authorises or mandates the use of a suite of smart technologies on and around buses in order to maintain a system of racial segregation
on public transport. Those who believe that the legislative policy is immoral
might have opportunities to protest before the legislation is enacted; they
might be able, post-legislation, to demonstrate at sites where the technology
is being installed; they might be able to engage in direct acts of civil disobedience by interfering with the technology; and they might have opportunities
for indirect acts of civil disobedience (breaking some other law in order to
protest about the policy of racial segregation on public transport). Regulators
might then respond in various ways—for example, by creating new criminal offences that are targeted at those who try to design round technological management.[23] Putting this more generally, technological management
might not altogether eliminate the possibility of principled moral protest.
The particular technology might not always be counter-technology proof
and there might remain opportunities for civil disobedients to express their
opposition to the background regulatory purposes indirectly by breaking
anti-circumvention laws, or by engaging in strategies of 'data obfuscation',[24]
or by initiating well-publicised 'hacks', or 'denial-of-service' attacks or their
analogues.

Nevertheless, if the general effect of technological management is to
squeeze the opportunities for traditional direct acts of civil disobedience, ways
need to be found to compensate for any resulting diminution in responsible
moral citizenship. By the time that technological management is in place, for
many this will be too late; for most citizens, non-compliance is no longer an
option. This suggests that the compensating adjustment needs to be ex ante:
that is to say, it suggests that responsible moral citizens need to be able to air
their objections before technological management has been authorised for
a particular purpose; and, what is more, the opportunity needs to be there

---

23 Compare, Directive 2001/29/EC on the harmonisation of certain aspects of copyright and
related rights in the information society, OJ L 167, 22.06.2001, 0010–0019 (concerning
attempts to circumvent technological measures, such as DRM).
24 See Finn Brunton and Helen Nissenbaum, 'Political and Ethical Perspectives on Data
Obfuscation' in Mireille Hildebrandt and Katja de Vries (eds), *Privacy, Due Process and the
Computational Turn* (Abingdon: Routledge, 2013) 171.

to challenge both an immoral regulatory purpose and the use of (morality-corroding) technological management.

Taking stock, it seems to me that, whether we are talking about individual moral lives or moral community, some precaution is in order. To repeat my remarks in the opening chapter, if we knew just how much space a moral community needs to safeguard against the automation of virtue, we might be able to draw some regulatory red lines. However, without knowing these things, a precautionary approach (according and protecting a generous margin of operational space for moral reflection, for moral reason, and for moral objection) looks prudent. Without knowing these things, the cumulative effect of adopting technological management—at any rate, in relation to intentional wrongdoing—needs to be a standing item on the regulatory agenda.[25]

V The legitimate application of technological management for health, safety, and environmental purposes

Where technological management is employed within, so to speak, the health and safety risk management track, there might be very real concerns of a prudential kind. For example, if the technology is irreversible, or if the costs of disabling the technology are very high, or if there are plausible catastrophe concerns, precaution indicates that regulators should go slowly with this strategy. Perhaps a prudential rule of thumb would be that technological management should be adopted only where it is designed for cheap and feasible discontinuity. The present question, however, is not about prudence; the question is whether there are any reasons for thinking that measures of technological management are an illegitimate way of pursuing policies that are designed to improve human health and safety or protect the environment. If we assume that the measures taken are transparent and that, if necessary, regulators can be held to account for taking the relevant measures, the legitimacy issue centres on the reduction of the options that are available to regulatees.

To clarify our thinking about this issue, we might start by noting that, in principle, technological management might be introduced by A in order to protect or to advance:

(i)   A's own interests;
(ii) the interests of some specific other, B; or
(iii) the general interest of some group of agents.

---

25 Karen Yeung, 'Can We Employ Design-Based Regulation While Avoiding Brave New World?' (2011) 3 *Law, Innovation and Technology* 1.

We can consider whether the reduction of options gives rise to any legitimacy concerns in any of these cases.

First, there is the case of A adopting technological management with a view to protecting or promoting A's own interests. While there might be some delegations of decision-making to personal digital assistants, or the like, that occasion concern (such as the delegation of moral decisions),[26] others seem innocuous. For example, suppose that A, wishing to reduce his home energy bills, adopts a system of technological management of his energy use. This seems entirely unproblematic. However, what if A's adoption of technological management impacts on others—for example, on A's neighbour B? Suppose that the particular form of energy capture, conversion, or conservation that A employs is noisy or unsightly. In such circumstances, B's complaint is not that A is using technological management per se but that the particular kind of technological management adopted by A is unreasonable relative to B's interest in peaceful enjoyment of his property (or some such interest). This is nothing new. In the days before clean energy, B would have made similar complaints about industrial emissions, smoke, dust, soot, and so on. Given the conflicting interests of A and B, it will be necessary to determine which set of interests should prevail; but the use of technological management itself is not in issue.

In the second case, A employs technological management in the interests of B. For example, if technological management is used to create a safe zone within which people with dementia or young children can wander, this is arguably a legitimate enforcement of paternalism. However, if B is a competent agent, A's paternalism is problematic. Quite simply, even if A correctly judges that exercising some option is not in B's best interest, or that the risks of exercising the option outweigh its benefit, how is A to justify this kind of interference with B's freedom?[27] If B consents to A's interference, that is another matter. However, in the absence of B's consent, and if A cannot justify such paternalism, then A certainly will not be able to justify his intervention. Although we should not underestimate the possibility that the adoption of technological management may not be as transparent as the adoption of a rule (so it might not be so obvious that B's liberty is being restricted),[28] and although the former is likely to be far more effective than a rule, the fundamental issue here is about paternalism, not about technological management as such.

---

26 See, e.g., Mireille Hildebrandt, *Smart Technologies and the End(s) of Law* (Cheltenham: Edward Elgar, 2015); and Roger Brownsword, 'Disruptive Agents and Our Onlife World: Should We Be Concerned?' (2017) 4 *Critical Analysis of Law* 61.

27 Compare the discussion in Roger Brownsword, 'Public Health Interventions: Liberal Limits and Stewardship Responsibilities' (2013) *Public Health Ethics* (special issue) doi: 10.1093/phe/pht030.

28 See, further, Chapter Seven.

In the third case, the use of technological management (in a generalised paternalistic way) may impinge on the legitimate preferences of more than one agent. Suppose, for example, that, in response to fatalities occasioned by truck drivers who, working long hours, fall asleep at the wheel, the employers require their drivers to take modafinil. This provokes a storm of controversy because the community is uncomfortable about the use of drugs or other technologies in order to 'enhance' human capacities. Faced with this conflict, the government steps in. Initially, it offers subsidies to employers who replace their lorries with new-generation driverless vehicles; and, in due course, it requires those few employers who have not abandoned their drivers and lorries to use only driverless vehicles. Of course, this means that many driving and trucking-related jobs are lost.[29] In the absence of consent by all those affected by the measure, technological disruption of this kind and on this scale is a cause for concern.[30] Against the increment in human health and safety, we have to set the loss of livelihood of the truckers. Possibly, in some contexts, it should be possible for regulators to accommodate the legitimate preferences of their regulatees—for example, for some time at least, it should be possible to accommodate the preferences of those who wish to drive their cars rather than be transported in driverless vehicles. However, if such preferences come at a cost or present a heightened risk to human health and safety, we might wonder how long governments and majorities will tolerate them.

Yet again, though, the issues are not about the impact of technological management on the complexion of the regulatory environment so much as about the impact on the particular interests of individuals and groups of human agents. The questions that arise are familiar ones about how to resolve competing or conflicting interests. Nevertheless, the more that technological management is used to secure and to improve the conditions for human health and safety, the less reliant we will be on background laws—particularly so-called regulatory criminal laws and some torts law—that have sought to encourage health and safety and to provide for compensation where accidents happen at work. The loss of these laws, and their possible replacement with some kind of compensatory scheme where (exceptionally) the technology fails, does have some impact on the complexion of the regulatory environment; but, if their loss does contribute to the corrosion of the conditions for moral community, it is far less obvious how this works than when the target

---

29 Daniel Thomas, 'Driverless convoy: Will truckers lose out to software?' *BBC News*, May 26, 2015, available at www.bbc.com/news/business-32837071 (last accessed 2 November 2018).

30 See, e.g., Kenneth Dau-Schmidt, 'Trade, Commerce, and Employment: the Evolution of the Form and Regulation of the Employment Relationship in Response to the New Information Technology', in Roger Brownsword, Eloise Scotford, and Karen Yeung (eds), *The Oxford Handbook of Law, Regulation and Technology* (Oxford: Oxford University Press) 1052.

of technological management is intentional or reckless infliction of harm. In other words, it is tempting to think that the adoption of technological management in order to improve human health and safety, even when disruptive of settled interests, is actually progressive.

VI Technological management and infrastructural protection

The concerns that have been expressed with regard to the complexion of the regulatory environment assume that the regulatory interventions in question are directed at human interactions in viable communities rather than at the generic infrastructure presupposed by such interactions. Although the drawing of the line between such interactions and the infrastructure itself are subject to debate, there is a view that regulators may legitimately use a technological fix where their intention is to protect the generic infrastructure.[31] Stated bluntly, the underlying thought is that, even if technological management will interfere with moral community, unless the basic infrastructure is in place, there is no prospect of any kind of human community.[32]

So long as our understanding of the generic infrastructure is restricted to the *essential* conditions for human existence (particularly to such vital matters as the availability of food and water, and the preservation of the supporting physical environment), there will remain a great deal of regulatory space in which to debate the legitimacy of the use of technological management. However, if each community adjusts its understanding of the infrastructural threshold in a way that reflects the stage of its technological development, this might involve characterising many safety features that are designed into, say, everyday transport and health-care facilities, as 'infrastructural'. A corollary of this is that, where infrastructural conditions are heavily engineered, there is a temptation simply to define away concerns about the legitimacy of technological management and its impact on the complexion of the regulatory

---

31 See, e.g., David A. Wirth, 'Engineering the Climate: Geoengineering as a Challenge to International Governance' (2013) 40 *Boston College Environmental Affairs Law Review* 413, 437: 'Lawmakers, both domestic and international would·be remiss not to seriously consider geoengineering proposals to mitigate the harm of global warming.' This view is also implicit in Gareth Davies, 'Law and Policy Issues of Unilateral Geoengineering: Moving to a Managed World' available at: papers.ssrn.com/sol3/papers.cfm?abstract_id=1334625 (last accessed 2 November 20218). See, too, Roger Brownsword, 'Crimes Against Humanity, Simple Crime, and Human Dignity' in Britta van Beers, Luigi Corrias, and Wouter Werner (eds), *Humanity across International Law and Biolaw* (Cambridge: Cambridge University Press, 2013) 87, 106–109.

32 See, further, Roger Brownsword, 'Responsible Regulation: Prudence, Precaution and Stewardship' (2011) 62 *Northern Ireland Legal Quarterly* 573, and 'Criminal Law, Regulatory Frameworks and Public Health' in A.M. Viens, John Coggon, and Anthony S. Kessel (eds) *Criminal Law, Philosophy and Public Health Practice* (Cambridge: Cambridge University Press, 2013) 19.

environment. In other words, if technological management may legitimately be used in order to protect infrastructural conditions, and if what is treated as an 'infrastructural condition' is interpreted expansively in line with the state of technological development in the community, it might become too easy for regulators to justify the use of technological management.

In an attempt to develop these primitive ideas, we can ask a number of questions, in particular: What might be the candidate ingredients or conditions for the generic infrastructure in this most fundamental sense? Where do we draw the line between the generic infrastructure and more specific infrastructures? And, where do we draw the line between those acts and activities that are harmful to the generic infrastructural conditions and those acts and activities that are harmful but not to the infrastructure as such? Having sketched some responses to these questions, we can outline the features of a special regulatory jurisdiction that, so it might be argued, should go with the defence of the generic infrastructure.

(i) The generic infrastructural conditions

What are the conditions that make up the generic infrastructure? According to Brett Frischmann, one of the characteristics of infrastructural resources—and these resources, Frischmann argues, tend to be undervalued—is that they are 'generic' rather than 'special purpose'.[33] So, for example, an electricity grid 'delivers power to the public, supporting an incredibly wide range of uses, users, markets, and technologies. It is not specially designed or optimized for any particular use, user, market, technology, or appliance; it provides non-discriminatory service for a toaster and a computer, for Staples and a pizzeria, and so on.'[34] Continuing in this vein, Freischmann suggests:

> Genericness implies a range of capabilities, options, opportunities, choices, freedoms. Subject to standardized compatibility requirements, users decide what to plug in, run, use, work with, play with. Users decide which roads to travel, where to go, what to do, who to visit. Users choose their activities; they can choose to experiment, to innovate, to roam freely. Users decide whether and what to build. Users decide how to use their time and other complementary resources. Infrastructure (providers) enable, support, and shape such opportunities.[35]

In these evocative remarks, Frischmann assumes that infrastructure 'users' are already in the position of a potentially functioning agent. His 'users' are

---

33 Brett M. Frischmann, *Infrastructure* (Oxford: Oxford University Press, 2012).
34 *Ibid.*, 65.
35 *Ibid.*, 65.

already in a position to decide how to use the infrastructural resources; they are already in a position to be able to choose their activities—whether to be innovators or free spirits, and so on. For my purposes, 'genericness' is to be understood as signifying those infrastructural resources and capabilities that are already possessed by Frischmann's users. What resources might these be?

The obvious place to start is with life itself. For, quite simply, the one thing that humans must have before they are capable of acting, transacting or interacting in the purposive (goal-directed) way that we associate with human agency is life itself.[36] However, to be alive, although necessary, is not sufficient for agency; there must also be conditions that are conducive to the enjoyment of a minimal level of health and well-being. For humans whose basic health and well-being are under threat, there is little prospect of actualising their agency.

That said, we need to differentiate between those conditions that are essential for *humans* to exist and those conditions that a community of (human) *agents* needs to secure if its members are to have any prospect of flourishing. When we ask whether the conditions on some planet other than Earth might be sufficient to support human life, when we ask whether there are signs of water or whether there is a supply of oxygen and so on, we are asking whether the infrastructure of the planet would be responsive to the biological needs that we humans happen to have. Here, on Earth, it is imperative that the planetary boundaries are respected.[37] In addition to their basic biological needs, humans are distinctively vulnerable—they can be wiped out by asteroids, volcanic eruptions, and plague; they can be struck down by a lightning strike, by tornados, and tsunamis; and, they can be savaged by lions and tigers, trampled underfoot by a herd of charging elephants, and so on. Even before we get to interactions with other humans, our world—albeit a world that is capable of supporting human life—is a dangerous place. If we can manage the risks to our lives, it is in our prudential interest to do so.

Into this mix of basic human need and vulnerability, we now add the fact that we humans—at any rate, we humans who populate the places in which the legitimacy and effectiveness of laws and regulation are debated—are also prospective *agents*. By this, I mean that we are capable of forming a view of our own needs, desires, and preferences, we are capable of forming our own purposes, we are capable of reasoning instrumentally (so that we understand which means will service the particular ends represented by our needs,

---

36 For the difference between goal-achieving, goal-seeking, and goal-directed behaviour, see David McFarland, *Guilty Robots, Happy Dogs—The Question of Alien Minds* (Oxford: Oxford University Press, 2008) 10–12.

37 See, J. Rockström, et al, 'Planetary Boundaries: Exploring the Safe Operating Space for Humanity' (2009) 14 *Ecology and Society* 32, available at www.ecologyandsociety.org/vol14/iss2/art32/ (last accessed 2 November 2018); and Kate Raworth, *Doughnut Economics* (London: Random House Business Books, 2017).

desires, preferences, and purposes), and we are capable of following rules (otherwise the normative regulatory enterprise would be even less successful than it is in guiding the conduct of regulatees). For the purposes of our prospective agency, it is not sufficient that we have the means of existing; we also need protection against human acts or practices that can compromise the prospects of our flourishing as agents.[38] Against the risks of human aggression, intrusion, repression and suppression, we also need some protection; if we can manage such risks, it is again in our prudential interest to do so. In other words, we each have prudential reasons to sign up to regulatory interventions that are designed to secure the common infrastructural conditions for agency—otherwise, the regulatory environment would leave us insecure and this would encourage practices of defensive agency thereby inhibiting interactions and transactions with others.

Now, these common infrastructural conditions for agency are 'generic' in the sense that, irrespective of the particular prudentially guided preferences or projects of agents, they are necessary for each and every agent to have a chance of translating its prospective purpose-fulfilment into an actuality. Even for a group of agents whose purpose is to establish a suicide club, these conditions need to be in place—it would not do, would it, for the life of a member of this group to be taken by an outsider? Moreover, for an agent to act in a way that compromises these generic conditions is not just contrary to the agent's deepest prudential interests, it fails to respect the generic interests of other agents (and, in this way, becomes contrary to both prudential and moral reason).[39] Suppose, for example, that three agents, A, B, and C are plugged into a single life support machine. If any disruption of power to the machine will be fatal to all three agents, they each have a straightforward prudential reason for not cutting the power supply. However, if each agent has its own line of support to the machine, and if cutting that line will be fatal to the agent in question, then it is possible for one agent to terminate the life of another without negatively impacting on its own life. In this context, what if A attempts to cut the life-line to B? A might claim that this is perfectly consistent with his prudential interests (assuming that A wants to end B's life). However, the generic conditions of agency are represented by the rules that secure the life of each agent against unwilled damage or interference. For A,

---

38 The point is nicely made in The Royal Society and British Academy, *Connecting Debates on the Governance of Data and its Uses* (London, December 2016) 5:

> Future concerns will likely relate to the freedom and capacity to create conditions in which we can flourish as individuals; governance will determine the social, political, legal and moral infrastructure that gives each person a sphere of protection through which they can explore who they are, with whom they want to relate and how they want to understand themselves, free from intrusion or limitation of choice.

39 Compare, Roger Brownsword, 'From Erewhon to Alpha Go: For the Sake of Human Dignity Should We Destroy the Machines?' (2017) 9 *Law, Innovation and Technology* 117.

B, and C to have any chance of realising their agency, they need a rule that prohibits any cutting of the life-lines. When A threatens to break the rule by cutting B's line, both B and C have more than contingent prudential reasons to oppose A's act; they rightly claim that A violates their generic interests by threatening to do so—and, of course, the same would hold, *mutatis mutandis*, if it were B who was threatening to cut the life-line to A or to C.[40]

Stated rather abstractly, we can say that, in the absence of the generic infrastructural conditions that relate to the biological needs and vulnerabilities of the species, humans simply will not be able to exist. Sadly, chronic shortages of food and water, of basic medicines, in conjunction with environmental pollution can be found in too many parts of our world and, following a natural disaster, we will often see some of these conditions in an acute form. Similarly, where there are serious deficiencies in the conditions that agents need if they are to realise their agency, human life might also be brutish and unproductive—in short, a failed state of agency unfulfilled. Again, sadly, chronic cases of this kind can be found in our world.

Arguably, this quest to identify and to protect the essential conditions of a human agent's well-being is analogous to Martha Nussbaum's specification of the threshold conditions for human dignity.[41] Or, again, we might try to analogise these conditions to the idea of global public goods, where such goods are commonly understood to relate to such matters as global climate stability, international financial stability, the control of communicable diseases, global peace and security, the infrastructure for international trade and finance, global communication and transportation systems, and global norms such as basic human rights.[42]

To comment very briefly on the former, as is well known, Nussbaum lists ten capabilities, a threshold level of each of which must be secured as a necessary condition of any decent political order. The ten capabilities are: life; bodily health; bodily integrity; senses, imagination, and thought; emotions; practical reason; affiliation; [concern for] other species; play; and control over one's environment. While the capabilities at the top of the list must be within the essential infrastructural set (because they pertain to the possibility of any kind of agency), they are less clearly so as we move down the list. In the event

---

40 My debts to Alan Gewirth, *Reason and Morality* (Chicago: University of Chicago Press, 1978) will be apparent. See, too, Roger Brownsword, 'Friends, Romans, and Countrymen: Is There a Universal Right to Identity?' (2009) 1 *Law Innovation and Technology* 223.

41 Martha C. Nussbaum, *Creating Capabilities* (Cambridge, Mass.: The Belknap Press of Harvard University Press, 2011).

42 From the extensive literature, see, e.g., Inge Kaul, Isabella Grunberg, and Marc A. Stern (eds), *Global Public Goods: International Cooperation in the 21st Century* (New York: Oxford University Press, 1999); and, Inge Kaul, 'Global Public Goods—A Concept for Framing the Post-2015 Agenda', German Development Institute, Bonn, Discussion Paper 2/2013; available at www.ingekaul.net/wp-content/uploads/2014/01/Internetfassung_DiscPaper_2_2013_Kaul1.pdf (last accessed 2 November 2018).

that it were agreed that all the capabilities listed were within the essential infrastructure, it would remain to determine the applicable threshold level of each capability. As these matters are debated, the one restriction to be observed is that the conditions must not be tied to *particular* projects that agents might have. What we are after is the infrastructure for purposive activity in a generic sense, not for particular purposes or projects.

(ii) Generic and specific infrastructures

How do we draw the line between the generic infrastructure and more particular, specific, infrastructures? In the light of what we have already said, I suggest that it is not too difficult to distinguish between generic and specific infrastructures. Stated shortly, the critical question is whether, if the infrastructure were to be compromised, this would affect the possibility of any kind of human activity or whether it would have more limited impact, affecting only a particular range of activities. To take Frischmann's example of the roads, these are important and they are enabling of many different uses and projects; they enhance agency but they are not essential to it. Human agency does not presuppose railway tracks, roads, or any other kind of transport infrastructure. In Frischmann's terminology, and for the purposes of his economic analysis these resources are 'generic'; but for my regulatory purposes, and particularly with a view to the application of technological management, these resources are not part of the *generic* infrastructure. Indeed, without the generic infrastructure, the construction of more specific infrastructures (such as the infrastructure for railways or road transport) would not be possible. It is not simply that some infrastructural conditions are more specific than others; it is that the human construction of specific infrastructures itself presupposes the generic infrastructure.

Even in our 'information societies', we might say much the same about the infrastructural technological elements.[43] Granted, cybercrime is particularly serious when it strikes at these infrastructural elements; and, for those communities that increasingly transact and interact online, this is an extremely serious matter.[44] Nevertheless, this is not part of the *generic* infrastructure—at any rate, not just yet.[45] We could also say much the same about the banking system, both offline and online. Without the banks, the conditions for modern commerce would be compromised. However, the crisis in the financial sector, although extremely serious for a certain class of human activities, does not come close to threatening the generic infrastructure itself. Although

---

43 For discussion by Frischmann, see (n 33) Ch 13.
44 See David R. Johnson and David Post, 'Law and Borders—The Rise of Law in Cyberspace' (1996) 48 *Stanford Law Review* 1367.
45 But, see Roger Brownsword, 'The Shaping of Our On-Line Worlds: Getting the Regulatory Environment Right' (2012) 20 *International Journal of Law and Information Technology* 249.

many thought that the collapse of key financial institutions would be 'the end of the world', they were some way off the mark. Moreover, those recent developments that are associated with blockchain technologies, cryptocurrencies, and distributed autonomous organisations are designed precisely to enable human agents to operate without the intermediation of traditional banks and financial institutions.[46]

(iii) Distinguishing between those acts and activities that are harmful to the generic infrastructure and those which are not

In the abstract, we can imagine a set of generic infrastructural conditions that support agents (individually and in groups) pursuing their own particular projects and plans. In some communities, the pursuit of these various projects and plans might be relatively harmonious but, in others, it might be highly competitive and conflictual. In the latter scenario, some agents might find that their plans and purposes are frustrated or not prioritised and, in consequence, there might be both some loss and disappointment. However, to the extent that we treat this as a 'harm', it clearly does not touch and concern the generic infrastructural conditions.

What, though, about acts that are harmful in a more obvious way, such as those involving the loss of life? Recall the case of A intentionally cutting the life-line to B. Is this act by A as harmful (relative to the generic infrastructural conditions) as A intentionally killing both B and C, or killing dozens of people, terrorising an area, and generating a climate of fear and concern? In the latter cases, not only are there many more direct victims of A's acts, there may also be many victims who are indirectly affected as the impact of A's killing radiates beyond the immediate area of the acts themselves. It is tempting to think that, while the environment for agency is degraded in the latter case, this is not usually so where there is an isolated killing. This, however, is a temptation to be avoided. The critical point is that, in all these scenarios (single killing or multiple killing), A has broken a rule (demanding respect for the lives of one's fellow agents) that is constituent of the generic infrastructural (the commons') conditions. The question is not how damaging

---

46 For discussion, see e.g., Iris H-Y Chiu, 'A new era in fintech payment innovations? A perspective from the institutions and regulation of payment systems' (2017) 9 *Law, Innovation and Technology* 190; Douglas W. Arner, Jànos Nathan Barberis, Ross P. Buckley, 'The Evolution of Fintech: A New Post-Crisis Paradigm?' [2015] University of Hong Kong Faculty of Law Research Paper No 2015/047, University of New South Wales Law Research Series [2016] UNSWLRS 62: available at papers.ssrn.com/sol3/papers.cfm?abstract_id=2676553 (last accessed 2 November 2018); and Aaron Wright and Primavera De Filippi, 'Decentralized Blockchain Technology and the Rise of *Lex Cryptographia*', papers.ssrn.com/sol3/papers.cfm?abstract_id=2580664 (March 10, 2015) (last accessed 2 November 2018)..

A's killing is; the question is whether a community of human agents could reasonably think that a zone lacking a rule against intentional and unwilled killing would be conducive to agency. A Hobbesian state of nature is no place for human agents.

That said, there will be a point at which we will judge that our physical and psychological integrity is sufficiently protected by the rules that specify the generic infrastructural context. However, some agents will be more robust and resilient than others and this may mean that a community is not entirely at one in agreeing which rule should be adopted and which rules speak to the generic infrastructural conditions. Communities will need to take a position on such questions and do so without allowing any partisan interests of agents to shape the commons' rules that are supposed to be pre-competitive and pre-conflictual.

(iv) A special regulatory jurisdiction

Arguably, four major regulatory implications follow from the analysis of the generic infrastructure that I am proposing.

First, while the deepest layer of the regulatory environment should be concerned with securing the generic infrastructure for agency itself, the remainder should set the ground rules for agents' on-stage interactions and transactions. While the former is foundational and universal, the latter can be more tuned to local cultural commitments and preferences. To put this in cosmopolitan terms, while all regulators share a responsibility for securing the essential infrastructural conditions (so to speak, 'the commons'), within each community there is room for some (legitimate) variation in the regulation of local (on-stage) activities.[47]

Secondly, if the infrastructure is to be secured, this implies—even if this is against the grain of current experience—a considerable degree of international co-ordination and shared responsibility.[48] Moreover, because politics tends to operate with short-term horizons, it also implies that the regulatory stewards should have some independence from the political branch, but not of course that they should be exempt from the Rule of Law's culture of accountability and justification.[49]

Thirdly, although (after its initial honeymoon period) the precautionary principle has been severely criticised—for example, it has been accused of being 'an overly-simplistic and under-defined concept that seeks to circumvent the

---

47 For further discussion, see Roger Brownsword, 'Regulatory Cosmopolitanism: Clubs, Commons, and Questions of Coherence' (TILT Working Paper series, University of Tilburg, 2010).
48 See Wirth (n 31), esp. 430–436.
49 See, too, Roger Brownsword, 'Responsible Regulation: Prudence, Precaution and Stewardship' (2011) 62 *Northern Ireland Legal Quarterly* 573; and Davies (n 31).

hard choices that must be faced in making any risk management decision'[50]—
a form of precautionary reasoning might well be acceptable in defence of the
infrastructure.[51] According to such reasoning, where regulators cannot rule
out the possibility that some activity threatens the infrastructure (which, on
any view, is potentially 'catastrophic'), then they should certainly engage a
precautionary approach.[52] This reasoning, it should be emphasised, assumes
an active employment of precaution. It is not simply that a lack of full scientific certainty is no reason (or excuse) for inaction—which puts one reason
for inaction out of play but still has no tilt towards action; rather, where the
harm concerns the infrastructure, there is a need to initiate preventative and
protective action.[53]

The range of precautionary measures is quite broad. At minimum, regulators should consider withdrawing any IP encouragement (notably patents)
for the relevant technology[54] and they may in good faith apply protective
measures or prohibitions even though such measures involve some sacrifice

50 Gary E. Marchant and Douglas J. Sylvester, 'Transnational Models for Regulation of Nanotechnology' (2006) 34 *Journal of Law, Medicine and Ethics* 714, 722. For an extended critique, see Cass R. Sunstein, *Laws of Fear* (Cambridge: Cambridge University Press, 2005).

51 Compare Deryck Beyleveld and Roger Brownsword, 'Complex Technology, Complex Cal-
culations: Uses and Abuses of Precautionary Reasoning in Law' in Marcus Duwell and Paul
Sollie (eds), *Evaluating New Technologies: Methodological Problems for the Ethical Assessment of
Technological Developments* (Dordrecht: Springer, 2009) 175; and 'Emerging Technologies,
Extreme Uncertainty, and the Principle of Rational Precautionary Reasoning' (2012) 4 *Law
Innovation and Technology* 35.

52 However, Sunstein changes tack when catastrophic harms are in contemplation. See Cass
R. Sunstein, *Worst-Case Scenarios* (Cambridge, Mass.: Harvard University Press, 2007), esp.
Ch.3, 167–168, where Sunstein develops the following precautionary approach:

> In deciding whether to eliminate the worst-case scenario under circumstances of uncer-
> tainty, regulators should consider the losses imposed by eliminating that scenario, and the
> size of the difference between the worst-case scenario under one course of action and the
> worst-case scenario under alternative courses of action. If the worst-case scenario under
> one course of action is much worse than the worst-case scenario under another course
> of action, and if it is not extraordinarily burdensome to take the course of action that
> eliminates the worst-case scenario, regulators should take that course of action. But if the
> worst-case scenario under one course of action is not much worse than the worst-case
> scenario under another course of action, and if it is extraordinarily burdensome to take the
> course of action that eliminates the worst-case scenario, regulators should not take that
> course of action.

53 Compare Elizabeth Fisher, Judith Jones, and René von Schomberg, 'Implementing the Precautionary Principle: Perspectives and Prospects' in Elizabeth Fisher, Judith Jones, and René
von Schomberg (eds), *Implementing the Precautionary Principle: Perspectives and Prospects*
(Cheltenham: Edward Elgar, 2006); and Elizabeth Fisher, *Risk Regulation and Administrative Constitutionalism* (Oxford: Hart, 2007).

54 Compare Estelle Derclaye, 'Should Patent Law Help Cool the Plant? An Inquiry from the
Point of View of Environmental Law—Part I' (2009) 31 *European Intellectual Property
Review* 168, and 'Part II' (2009) 31 *European Intellectual Property Review* 227.

of a valued activity (actual or anticipated).[55] It is true that, with the benefit of hindsight, it might be apparent that a precautionary sacrifice was actually unnecessary. However, the alternative is to decline to make the sacrifice even when this was necessary to defend the generic conditions. If regulators gamble with the generic infrastructure, and if they get it wrong, it is not just the particular valued activity, but all human activities, that will be affected adversely.

Fourthly, we come to the key feature for present purposes. I have made much of the point that, for communities that have moral aspirations, it is important that the regulatory environment does not design out (non-normatively) the opportunities for acting freely in trying to do the right thing. In other words, we need to be concerned about the impact of technological management on the conditions for moral community. Nevertheless, where the regulatory stewards are acting to protect the generic infrastructural resources, a resort to designed-in solutions may be more readily justified. For example, if humans will not comply with normative regulatory requirements that are designed to tackle global warming, a non-normative geo-engineering technical fix might be a legitimate way of dealing with the problem.[56] Of course, the conditions on which technological management is licensed for the purpose of protecting the generic infrastructure need to be qualified. It makes no sense to trade one catastrophe for another. So, bilateral precaution needs to be applied: if interventions that are containable and reversible are available, they should be preferred. As Gareth Davies rightly puts it, the rational approach is to balance 'the costs and risks of geoengineering against the costs and risks of global warming'.[57] If in doubt, perhaps regulators should give rules a fair chance to work before resorting to experimental technological management.

I have also suggested that the use of technological management to eliminate intentional wrongdoing might be more problematic, morally speaking, than it is to eliminate unintentional wrongdoing and harm. Yet, how can it be sensible to propose that, while it is perfectly fine to rely on measures of technological management to prevent unintended harm to the commons, we should not use such measures to eliminate intentionally harmful acts of this kind? After all, it is plausible to think that communities of human agents

---

55 For some insightful remarks on the difference between dealing with technologies that are already in use and newly emerging technologies, see Alan Randall, *Risk and Precaution* (Cambridge: Cambridge University Press, 2011).

56 For an overview of such possible fixes, see Jesse Reynolds, 'The Regulation of Climate Engineering' (2011) 3 *Law, Innovation and Technology* 113; and, for an indication of the burgeoning literature as well as possible responses to transboundary liability issues, see Barbara Saxler, Jule Siegfried and Alexander Proelss, 'International Liability for Transboundary Damage Arising from Stratospheric Aerosol Injections' (2015) 7 *Law Innovation and Technology* 112.

57 Davies (n 31) 5.

need to have secured the commons before they even begin to debate their moral commitments. Accordingly, if we are to develop a more coherent view, it seems that we should not categorically rule out the option of using technological management to protect the commons irrespective of whether the harmful acts at which our measures are aimed are intentional or unintentional. This does not mean that there are no acts of intentional wrongdoing in relation to which we should be slow to apply measures of technological management. For, with the commons secured, each community will articulate a criminal code that expresses its distinctive sense of serious wrongdoing, albeit that the wrongs in question will not be acts that touch and concern the essential generic infrastructure.

This leaves one other point. Although I have taken a particular approach to what counts as a generic infrastructural condition, I would not want to lose the benefits of Frischmann's insights into the productive value of infrastructural resources that are generic in his sense. Political communities might well decide that modern infrastructures, such as the Internet, are so valuable that they need to be protected by measures of technological management. Provided that these communities have a clear understanding not only the value of such infrastructural resources but also of any downsides to technological management, then they should be entitled to regulate the resource in this way.

VII Conclusion

In this chapter, we have suggested that the idea of the 'complexion' of the regulatory environment provides an important focus for tracking the shift from reliance on the moral regulatory register to the prudential register, and then from normative to non-normative registers and strategies. For those who are concerned that the use of emerging technologies as regulatory instruments might corrode the conditions for flourishing human agency and, especially, for moral community, this gives a lens through which to monitor regulatory action and the changing nature of the regulatory environment.

To the extent that technological instruments replace rules, and even if the former are generally more effective, economical and efficient than the latter, there is a major question about the acceptability of regulators employing measures of technological management.

We have tentatively suggested that, where technological management is employed for crime control purposes, this is prima facie problematic because it tends to corrode the conditions for moral community. The corrosion can occur where the State employs such a strategy but also where individuals take acts of otherwise justified self-defence. By contrast, where technological management is used for reasons of health and safety, it is less obviously morally problematic, particularly where it is adopted by an agent for the sake of their own health and safety and without any negative impact on others.

However, in other cases, there may be conflicts to be dealt with—for example, when A adopts measures of technological management in a paternalistic way for the sake of B's health and safety; and the general adoption of such measures may well present a conflict of interests between those who wish to retain lifestyle or working options in preference to increased levels of health and safety and those who wish to prioritise health and safety.

Finally, we have also identified the use of technological management in order to protect the critical infrastructure of any form of human social existence as a possible special case, with dedicated regulatory stewards entrusted with the use of instruments that might not otherwise be acceptable. Arguably, the concern that moral communities will have about privileging deliberation together with free and intentional action (which the deliberating agent judges to be the right thing to do) are not fully engaged until the commons' conditions are secured.

While these first thoughts about the use of measures of technological management in a community that has moral aspirations indicate that the context in which such measures are employed is important, we need to elaborate a more systematic view of the range of regulatory responsibilities. It is to this task that we turn in the next chapter.

4

THREE REGULATORY RESPONSIBILITIES

Red lines, reasonableness, and technological management

I Introduction

Building on our discussion in the last chapter, we can say that nation State regulators have three principal responsibilities. The first and paramount responsibility is to ensure that the preconditions (the commons) for any form of human social existence, for any kind of human community, are maintained and protected; the second is to articulate and respect the distinctive fundamental values of the community; and, the third is to maintain an acceptable balance between the legitimate (but potentially competing and conflicting) interests of members of the community (for example, an acceptable balance between the interest in innovation and technological progress and the interest in human health and safety, between economic growth and the preservation of employment options, between safety on the roads and the freedom to drive, and so on). These responsibilities, it should be emphasised, are lexically ordered: the first responsibility takes categorical priority over any other responsibilities; and the second responsibility takes categorical priority over the third. Relative to these responsibilities, we can begin to understand more about the challenges faced by regulators, where the regulatory priorities lie, and whether it is acceptable to use technological management in the discharge of these responsibilities.

Traditionally, regulators have sought to discharge their responsibilities by legislating various rules and standards; but, from now on, they also have the option of resorting to increasingly sophisticated measures of technological management. Of course, the latter option will not appeal unless it promises to be more effective than the former. However, measures of technological management, albeit effective, may come at a price and the question—a question that we have been engaged with throughout this first part of the book—is where it is appropriate to rely on such measures. In other words, when is the use of technological management consistent with the responsibilities that regulators have?

In this chapter, we start by restating the red lines that ring-fence the commons and that should guide regulators in relation to their first responsibility. Then, we can draw out the challenges that face regulators in relation to their other responsibilities (to respect the fundamental values of the community and to strike an acceptable balance between competing and conflicting legitimate interests). This leads to some reflections on the way in which the discharge of these regulatory responsibilities guides our assessment of the fitness of the regulatory environment, following which we draw together the threads by considering how the use of technological management sits with these responsibilities. The chapter concludes by highlighting the key points—the acts of re-imagination in relation to the nature and complexion of the regulatory environment, and the range of regulatory responsibilities—that emerge from the discussion in the first Part of the book.

II Regulatory red lines and the first responsibility

According to Neil Walker, 'the idea of a law that is not bound to a particular territorial jurisdiction is a recurrent theme in the history of the legal imagination and its efforts to shape and reshape the regulatory environment.'[1] Moreover, this is an idea that 'claims a global warrant ... in the sense of claiming or assuming a universal or globally pervasive justification for its application'.[2] In other words, the assumption, the promise, and the ambition of global law is that it 'is, or should be, applicable to all who might be covered by its material terms regardless of location or association with a particular polity, because it is justifiable to all who might be covered by those material terms ...'[3] So understood, global law re-opens 'minds to the idea of law as something more than a patchwork of difference and a collection of particulars'.[4]

In the spirit of global law so imagined, we can start by restating a simple but fundamental idea. This is that it is characteristic of human agents—understood in a thin sense akin to that presupposed by the criminal law[5]—that they have the capacity to pursue various projects and plans whether as

---

1 Neil Walker, *Intimations of Global Law* (Cambridge: Cambridge University Press, 2015) 19.
2 *Ibid.*, 22.
3 *Ibid.*
4 *Ibid.*, 198.
5 Compare, Stephen J. Morse, 'Uncontrollable Urges and Irrational People' (2002) 88 *Virginia Law Review* 1025, 1065–1066:

> Law guides human conduct by giving citizens prudential and moral reasons for conduct. Law would be powerless to achieve its primary goal of regulating human interaction if it did not operate through the practical reason of the agents it addresses and if agents were

individuals, in partnerships, in groups, or in whole communities. Sometimes, the various projects and plans that they pursue will be harmonious; but, often, human agents will find themselves in conflict or competition with one another as their preferences, projects and plans clash. However, before we get to particular projects or plans, before we get to conflict or competition, there needs to be a context in which the exercise of agency is possible. This context is not one that privileges a particular articulation of agency; it is prior to, and entirely neutral between, the particular plans and projects that agents individually favour; the conditions that make up this context are generic to agency itself. In other words, as we put it in the last chapter, there is a deep and fundamental critical infrastructure, a commons, for any community of agents. It follows that any agent, reflecting on the antecedent and essential nature of the commons, must regard the critical infrastructural conditions as special. Indeed, from any practical viewpoint, prudential or moral, the protection of the commons must be the highest priority.

The commons comprises three sets of pre-conditions for viable human social existence—that is, for viable human life together with the exercise of individual and collective agency. Each of these pre-conditions sets a red line for regulators, whose primary responsibility is to defend and maintain the commons.

These red lines relate, first, to the existence conditions for humans; secondly, to the conditions for self-development; and, thirdly, where agents are committed to values that demand respect for the legitimate interests of others, to the conditions for the moral development of agents. These conditions yield a triple bottom line for regulators. Quite simply, the terms and conditions of any regulatory (or social) licence for new technologies should be such as to protect, preserve, and promote:

- the essential conditions for human existence (given human biological needs);
- the generic conditions for the self-development of human agents; and,
- the essential conditions for the development and practice of moral agency.

---

not capable of rationally understanding the rules and their application under the circumstances in which the agent acts.

Responsibility is a normative condition that law and morality attribute only to human beings. We do not ascribe responsibility to inanimate natural forces or to other species. Holding an agent responsible means simply that it is fair to require the agent to satisfy moral and legal expectations and to bear the consequences if he or she does not do so; holding an agent non-responsible means simply that we do not believe the agent was capable in the context in question of satisfying moral and legal expectations. The central reason why an agent might not be able to be guided by moral and legal expectations is that the agent was not capable of being guided by reason.

Moreover, these are imperatives for regulators in all regulatory spaces, whether international or national, public or private. From the responsibilities for the commons, or the essential infrastructure, there are no exemptions or exceptions; we are dealing here with principles that are truly cosmopolitan.[6] These are the conditions that give shape to global law.[7]

Of course, determining the nature of these conditions will not be a mechanical process and I do not assume that it will be without its points of controversy (including, as I indicated in the previous chapter, how closely these conditions can be assimilated to ideas such as Martha Nussbaum's threshold level for human capabilities or that of global public goods).[8] Nevertheless, let me build on my remarks in the previous chapter by giving an indication of how I would understand the distinctive contribution of each segment of the commons.

In the first instance, regulators should take steps to protect, preserve and promote the natural ecosystem for human life.[9] At minimum, this entails that the physical well-being of humans must be secured; humans need oxygen, they need food and water, they need shelter, they need protection against contagious diseases, if they are sick they need whatever medical treatment is available, and they need to be protected against assaults by other humans or non-human beings. It follows that the intentional violation of such conditions is a crime, not just against the individual humans who are directly affected, but against humanity itself.[10] Quite simply, as Stephen Riley puts the point so well, 'No legal system should threaten the continued existence of the society it is intended to serve'.[11] For, in Riley's terms, we are dealing here

---

6 Compare Roger Brownsword, *Rights, Regulation and the Technological Revolution* (Oxford: Oxford University Press, 2008) Ch.7 (on the challenge of regulatory cosmopolitanism); and 'Regulatory Cosmopolitanism: Clubs, Commons, and Questions of Coherence' TILT Working Papers No 18 (2010).

7 Neil Walker (n 1), it should be said, sketches two conceptions (or visions) of global law, one 'convergence-promoting' and the other 'divergence-accommodating' (at 56). While the former approaches 'involve one or both of the attributes of hierarchy and normative singularity, [the latter] are characterised by the twin themes of heterarchy and normative plurality' (*ibid.*).

8 Moreover, even if it is agreed where the bottom lines are to be drawn, a community still has to decide how to handle proposals for uses of technologies that do not present a threat to any of the bottom line conditions.

9 Compare J. Rockström, et al, 'Planetary Boundaries: Exploring the Safe Operating Space for Humanity' (2009) 14 *Ecology and Society* 32, available at www.ecologyandsociety.org/vol14/iss2/art32/ (last accessed 2 November 2018); and Kate Raworth, *Doughnut Economics* (London: Random House Business Books, 2017) 43–53.

10 Compare Roger Brownsword, 'Crimes Against Humanity, Simple Crime, and Human Dignity' in Britta van Beers, Luigi Corrias, and Wouter Werner (eds), *Humanity across International Law and Biolaw* (Cambridge: Cambridge University Press, 2014) 87.

11 Stephen Riley, 'Architectures of Intergenerational Justice: Human Dignity, International Law, and Duties to Future Generations' (2016) 15 *Journal of Human Rights* 272, 272. In this light, Riley suggests that, while we might accept the 'legal coherence' of the Advisory

with conditions that are 'constitutive' of the legal enterprise and upon which all 'regulative' requirements are predicated.[12]

Secondly, the conditions for meaningful self-development and agency need to be constructed (largely in the form of positive support and negative restriction): there needs to be a sense of self-esteem and respect; agents need to feel that their agency matters; and there needs to be sufficient trust and confidence in one's fellow agents, together with sufficient predictability to plan, so as to operate in a way that is interactive and purposeful rather than merely defensive. Let me suggest that the distinctive capacities of prospective agents include being able:

- to freely choose one's own ends, goals, purposes and so on (to make one's own choices and 'to do one's own thing')
- to reason both prudentially and instrumentally rationally
- to prescribe rules (for oneself and for others) and to be guided by rules (set by oneself or by others)
- to form a sense of one's own identity ('to be one's own person').

Accordingly, the essential conditions are those that support the exercise of these capacities.[13] With existence secured, and under the right conditions, human life becomes an opportunity for agents to be who they want to be, to have the projects that they want to have, to form the relationships that they want, to pursue the interests that they choose to have and so on. In the twenty-first century, no other view of human potential and aspiration is plausible; in the twenty-first century, it is axiomatic that humans are prospective agents and that agents need to be free.

Thirdly, the commons must secure the conditions for an aspirant moral community, whether the particular community is guided by teleological or deontological standards, by rights or by duties, by communitarian or liberal or libertarian values, by virtue ethics, and so on. The generic context for moral community is impartial between competing moral visions, values, and ideals; but it must be conducive to 'moral' development and 'moral' agency in a formal sense. So, to add to the several articulations of this point that we have

---

Opinion of the ICJ on the *Legality of the Threat or Use of Nuclear Weapons* (1996)—to the effect that it might be permissible to use nuclear weapons under the prevailing rules of international law—we should surely 'question the intelligibility of the international legal system as a whole if it permits any such ruling' (at 272).

12 Riley (n 11); and, for further elaboration and application of this crucial idea, see Stephen Riley, *Human Dignity and the Law: Legal and Philosophical Investigations* (Abingdon: Routledge, 2018).

13 Compare the insightful analysis of the importance of such conditions in Maria Brincker, 'Privacy in Public and the Contextual Conditions of Agency' in Tjerk Timan, Bryce Clayton Newell, and Bert-Jaap Koops (eds), *Privacy in Public Space* (Cheltenham: Edward Elgar, 2017) 64.

already given, in her discussion of techno-moral virtues, (sous)surveillance, and moral nudges, Shannon Vallor is rightly concerned that any employment of digital technologies to foster prosocial behaviour should be consistent with conduct 'remain[ing] our *own conscious activity and achievement* rather than passive, unthinking submission.'[14] She then invites readers to join her in imagining that Aristotle's Athens had been ruled by laws that 'operated in such an unobtrusive and frictionless manner that the citizens largely remained unaware of their content, their aims, or even their specific behavioral effects.'[15] In this regulatory environment, we are asked to imagine that Athenians 'almost never erred in moral life, either in individual or collective action.'[16] However, while these fictional Athenians are reliably prosocial, 'they cannot begin to explain *why* they act in good ways, why the ways they act *are* good, or *what* the good life for a human being or community might be.'[17] Without answers to these questions, we cannot treat these model citizens as moral beings. Quite simply, their moral agency is compromised by technologies that do too much regulatory work.

Agents who reason impartially will understand that each human agent is a stakeholder in the commons where this represents the essential conditions for human existence together with the generic conditions of both self-regarding and other-regarding agency; and, it will be understood that these conditions must, therefore, be respected. While respect for the commons' conditions is binding on all human agents, it should be emphasised that these conditions do not rule out the possibility of prudential or moral pluralism. Rather, the commons represents the pre-conditions for both individual self-development and community debate, giving each agent the opportunity to develop his or her own view of what is prudent as well as what should be morally prohibited, permitted, or required. However, the articulation and contestation of both individual and collective perspectives (like all other human social acts, activities and practices) are predicated on the existence of the commons.

In this light, we can readily appreciate that—unlike, say, Margaret Atwood's post-apocalyptic dystopia, *Oryx and Crake*[18]—what is dystopian about George Orwell's *1984*[19] and Aldous Huxley's *Brave New World*[20] is not that human *existence* is compromised but that human *agency* is compromised;[21] that today's dataveillance practices, as much as *1984*'s surveillance, 'may be

14 Shannon Vallor, *Technology and the Virtues* (New York: Oxford University Press, 2016) 203 (emphasis in original).
15 *Ibid.*
16 *Ibid.*
17 *Ibid.*, (emphasis in the original).
18 (London: Bloomsbury, 2003).
19 (London: Penguin Books, 1954) (first published 1949).
20 (New York: Vintage Books, 2007) (first published 1932).
21 To be sure, there might be some doubt about whether the regulation of particular acts should be treated as a matter of the existence conditions or the agency conditions. For

doing less to deter destructive acts than [slowly to narrow] the range of tolerable thought and behaviour';[22] and that what is dystopian about Anthony Burgess' *A Clockwork Orange*,[23] and what is at least worrying about the supposed utopia of B.F. Skinner's *Walden Two*,[24] is not that human *existence* is compromised but that human *moral* agency is compromised to the extent that those who do the right thing—whether reformed criminals or young children—do not seem to have any choice in the matter. We will also appreciate that, while technological management might not kill us, it may not be our preferred mode of governance and it may crowd out the possibility of moral development and debate.[25]

To return to Neil Walker, the prospects for global law turn on 'our ability to persuade ourselves and each other of what we hold in common and of the value of holding that in common'.[26] Regulators should need no persuading: what all human agents have in common is a fundamental reliance on the commons; if the commons is compromised, the prospects for any kind of legal or regulatory activity, or any kind of persuasive or communicative activity, indeed for any kind of human social activity, are diminished. If we value anything, if we are positively disposed towards anything, we must value the commons.

III Fundamental values and the second responsibility

Beyond the fundamental stewardship responsibilities, regulators are also responsible for ensuring that the fundamental values of their particular community are respected. Just as each individual human agent has the capacity to develop their own distinctive identity, the same is true if we scale this up to communities of human agents. All communities have common needs (the need for a supportive commons), but each community may treat the commons

---

present purposes, however, resolving such a doubt is not a high priority. The important question is whether we are dealing with a bottom-line condition.

22 Frank Pasquale, *The Black Box Society* (Cambridge, Mass.: Harvard University Press, 2015) 52. Compare, too, the embryonic Chinese social credit system: see, e.g., Didi Tang, 'Big Brother Xi bids to turn 1.3bn Chinese into perfect citizens' *The Times*, November 4, 2017, p.48; Rogier Creemers, 'China's Social Credit System: An Evolving Practice of Control' (May 9, 2018) available at SSRN: ssrn.com/abstract=3175792 (last accessed 2 November 2018); and, for incisive commentary, see Yongxi Chen and Anne S.Y. Cheung, 'The Transparent Self under Big Data Profiling: Privacy and Chinese Legislation on the Social Credit System' (2017) 12 *Journal of Comparative Law* 356, 357 ('Individuals risk being reduced to transparent selves before the state in this uneven battle. They are uncertain what contributes to their social credit scores, how those scores are combined with the state system, and how their data is interpreted and used.').

23 (London: Penguin Books, 1972) (first published 1962).

24 (Indianapolis: Hackett Publishing Company Inc, reprinted 2005) (first published 1948).

25 Again, see the discussion in Vallor (n 14).

26 Neil Walker (n 1) 199.

as a platform for developing its own distinctive identity, articulating the values that make it the particular community that it is.[27] Such values represent the distinctive peaks in the normative landscape of each particular community.

From the middle of the twentieth century, many nation states have expressed their fundamental (constitutional) values in terms of respect for human rights and human dignity.[28] These values (most obviously the human right to life) clearly intersect with the commons conditions and there is much to debate about the nature of this relationship and the extent of any overlap. For example, if we understand the root idea of human dignity in terms of humans having the capacity freely to do the right thing for the right reason,[29] then human dignity reaches directly to the commons' conditions for moral agency;[30] and, similarly, if we understand privacy (whether derived from human rights or human dignity) as ring-fencing spaces for the self-development of agents, then it is more than a fundamental value recognised by a particular community, it reaches through to the commons' conditions themselves. However, those nation states that articulate their particular identities by the way in which they interpret their commitment to respect for human dignity are far from homogeneous. Whereas, in some communities, the emphasis of human dignity is on individual empowerment and autonomy, in others it is on constraints relating to the sanctity, non-commercialisation, non-commodification, and non-instrumentalisation of human life.[31] These differences in emphasis mean that communities articulate in very different ways on a range of beginning of life and end of life questions as well as questions of lifestyle and human enhancement, and so on.

---

27 Insofar as we can map Neil Walker's (nn 1 and 7) contrast between convergence-promoting and divergence-accommodating conceptions of global law onto my conception of the regulatory responsibilities, the former relates to what I am terming the first tier responsibilities and the latter to the second and third tier responsibilities. In a similar way, the distinction that Walker highlights between 'an understanding of law as possessing a general and presumptively universal rationality of its own, and law as an expression and instrument of human will' (n1, 196–197) might be related to the first tier (universal rational) responsibilities and the other tiers of (human willed) responsibilities.

28 See Roger Brownsword, 'Human Dignity from a Legal Perspective' in M.Duwell, J. Braavig, R. Brownsword, and D. Mieth (eds), *Cambridge Handbook of Human Dignity* (Cambridge: Cambridge University Press, 2014) 1.

29 For such a view, see Roger Brownsword, 'Human Dignity, Human Rights, and Simply Trying to Do the Right Thing' in Christopher McCrudden (ed), *Understanding Human Dignity* (Proceedings of the British Academy 192) (Oxford: The British Academy and Oxford University Press, 2013) 345.

30 See, Roger Brownsword, 'From Erewhon to Alpha Go: For the Sake of Human Dignity Should We Destroy the Machines?' (2017) 9 *Law, Innovation and Technology* 117.

31 See Deryck Beyleveld and Roger Brownsword, *Human Dignity in Bioethics and Biolaw* (Oxford: Oxford University Press, 2001); Tim Caulfield and Roger Brownsword, 'Human Dignity: A Guide to Policy Making in the Biotechnology Era' (2006) 7 *Nature Reviews Genetics* 72; and Roger Brownsword, *Rights, Regulation and the Technological Revolution* (Oxford: Oxford University Press, 2008).

To anticipate our discussion in Part Two of the book, a question that will become increasingly important is whether, and if so how far, a community sees itself as distinguished by its commitment to regulation by rule rather than by technological management. In some smaller scale communities or self-regulating groups, there might be resistance to a technocratic approach because compliance that is guaranteed by technological means compromises the context for trust—this might be the position, for example, in some business communities (where self-enforcing transactional technologies, such as blockchain, are rejected).[32] Or, again, a community might prefer to stick with regulation by rules because it values not only the flexibility of rules but also public participation in setting standards; and, it is worried that technological management might prove inflexible as well as that regulatory debates might become less accessible if the debate were to become technocratic.

If a community decides that it is generally happy with a regulatory approach that relies on technological features rather than rules, it then has to decide whether it is also happy for humans to be out of the loop. Where the technologies involve AI (as in anything from steering public buses to decisions made by the tax authorities), the 'computer loop' might be the only loop that there is. As Shawn Bayern and his co-authors note, this raises an urgent question, namely: 'do we need to define essential tasks of the state that must be fulfilled by human beings under all circumstances?'[33] Furthermore, once a community is asking itself such questions, it will need to clarify its understanding of the relationship between humans and robots—in particular, whether it treats robots as having moral status, or legal personality, and the like.[34]

It is, of course, essential that the fundamental values to which a particular community commits itself are consistent with (or cohere with) the commons conditions; and, if we are to talk about a new form of coherentism—as, in due course, I will suggest that we might—it should be focused in the first instance on ensuring that regulatory operations are so consistent.[35]

---

32 See, the excellent discussion in Karen E.C. Levy, 'Book-Smart, Not Street-Smart: Blockchain-Based Smart Contracts and The Social Workings of Law' (2017) 3 *Engaging Science, Technology, and Society* 1; and, see Roger Brownsword, 'Regulatory Fitness: Fintech, Funny Money, and Smart Contracts' (2019) *European Business Organization Law Review* (forthcoming).

33 Shawn Bayern, Thomas Burri, Thomas D. Grant, Daniel M. Häusermann, Florian Möslein, and Richard Williams, 'Company Law and Autonomous Systems: A Blueprint for Lawyers, Entrepreneurs, and Regulators' (2017) 9 *Hastings Science and Technology Law Journal* 135, 156.

34 See, e.g., Bert-Jaap Koops, Mireille Hildebrandt, and David-Olivier Jaquet-Chiffelle, 'Bridging the Accountability Gap: Rights for New Entities in the Information Society?' (2010) 11 *Minnesota Journal of Law, Science and Technology* 497; and Joanna J. Bryson, Mihailis E. Diamantis, and Thomas D. Grant, 'Of, for, and by the people: the legal lacuna of synthetic persons' (2017) 25 *Artif Intell Law* 273.

35 The values that are fundamental to the community and which should guide 'new coherentism' might include various ideas of justice and equality that are characteristic of traditional

IV The third regulatory responsibility: the quest for an acceptable balance of interests

This takes us to the third tier of regulatory responsibility, that of finding an acceptable balance between the legitimate interests that make up the pluralistic politics of the community. Given that different balances will appeal to different interest groups, finding an acceptable balance is a major challenge for regulators.

Today, we have the perfect example of this challenge in the debate about the liability (both criminal and civil) of Internet intermediaries for the unlawful content that they carry or host.[36] Should intermediaries be required to monitor content or simply act after the event by taking down offending content? In principle, we might argue that such intermediaries should be held strictly liable for any or some classes of illegal content—for example, for imminently harmful, or dangerous, or manifestly illegal content;[37] or that they should be liable if they fail to take reasonable care; or that they should be immunised against liability even though the content is illegal. If we take a position at the strict liability end of the range, we might worry that the liability regime is too burdensome to intermediaries and that online services will not expand in the way that we hope; but, if we take a position at the immunity end of the range, we might worry that this treats the Internet as an exception to the Rule of Law and is an open invitation for the illegal activities of copyright infringers, paedophiles, terrorists and so on. In practice, most legal systems balance these interests by taking a position that confers an immunity but only so long as the

---

legal thinking. For example, it has been argued that the decisive reason for treating autonomous software agents as responsible for the transactional decisions made is not 'efficiency, transaction cost savings, utilitarian considerations, issues of sociological jurisprudence or policy-questions'. Rather, it is the principle of equal treatment that demands liability here. If the execution of the contract is delegated to a human actor, the principal is liable for his breach of duty, consequently, he cannot be exempted from liability if a software agent is used for the identical task ... If legal doctrine unswervingly adheres to traditional legal terms and refuses to assign legal action capacity to software agents, it would have to be accused of treating equal cases unequally.

See, Gunther Teubner, 'Digital Personhood: The Status of Autonomous Software Agents in Private Law' *Ancilla Juris* (2018) 107, 135.

36 Almost by the day, the media carry pieces that further fuel and contribute to this debate: see, e.g., David Aaronovitch, 'Bringing law and order to digital Wild West' *The Times*, January 4, 2018, p.25; and Edward Munn, 'YouTube severs (some of) its ties with Logan Paul' available at www.alphr.com/life-culture/1008081/youtube-severs-some-of-its-ties-with-logan-paul (last accessed 2 November 2018).

37 Compare the views of the public reported by the European Commission in its *Synopsis Report on the Public Consultation on the Regulatory Environment for Platforms Online Intermediaries and the Collaborative Economy* (May 25, 2016) pp 16–17: available at ec.europa.eu/digital-single-market/en/news/full-report-results-public-consultation-regulatory-environment-platforms-online-intermediaries (last accessed 5 November 2018); and, for public views on specific duties of care (for particular kinds of content) see *ibid* at p.19.

intermediaries do not have knowledge or notice of the illegal content. Predictably, now that the leading intermediaries are large US corporations with deep pockets, and not fledgling start-ups, many think that the time is ripe for the balance to be reviewed.[38] However, finding a balance that is generally acceptable, in both principle and practice, is another matter.[39]

Where the content that is carried or hosted is perfectly lawful, we might think that there is no interest to set against its online presence. Indeed, we might think that, in a community that is fundamentally committed to freedom of expression, there are strong reasons for keeping such content available. However, there may be an interest, not in relation to the removal of the content, but in relation to the way in which search engines 'advertise' or 'signpost' or 'direct towards' the content at issue. In other words, there may be a 'right to be forgotten' of the kind upheld by the Court of Justice of the European Union (the CJEU) in the much-debated *Google Spain* case.[40]

Here, the CJEU accepted that a right to be forgotten is implicit in the conjunction of Articles 7 (respect for private life) and 8 (protection of personal data) of the EU Charter of Fundamental Rights[41] together with Articles 12(b) and 14(a) of the Data Protection Directive[42]—these provisions of the Directive concerning, respectively, the data subject's right to obtain rectification, erasure or blocking where the processing of the data is not compliant with the Directive and the data subject's right to object on 'compelling legitimate grounds' to the processing of the data which itself is ostensibly justified by reference to the legitimate interests of the controller or third parties. The significance of the newly recognised right to be forgotten is that a data subject who objects to certain personal data being flagged up where a search is made under that data subject's name may require the link to be erased—in the *Google Spain* case itself, the information in question was an announcement made some 16 years earlier in a Spanish newspaper that identified the data subject in connection with a real estate auction that was related to

---

38 For a particularly compelling analysis, see Marcelo Thompson, 'Beyond Gatekeeping: the Normative Responsibility of Internet Intermediaries' (2016) 18 *Vanderbilt Journal of Entertainment and Technology Law* 783.

39 In the EU, there is also the question whether national legislative initiatives—such as the recent German NetzDG, which is designed to encourage social networks to process complaints about hate speech and other criminal content more quickly and comprehensively—are compatible with the provisions of Directive 2000/31/EC on e-commerce: see, for discussion of this particular question, Gerald Spindler, 'Internet Intermediary Liability Reloaded—The New German Act on Responsibility of Social Networks and its (In-) Compatibility With European Law', available at www.jipitec.eu/issues/jipitec-8-2-2017/4567 (last accessed 3 November 2018)

40 Case C-131/12, *Google Spain SL, Google Inc v Agencia Española de Protection de Datos (AEPD), Mario Costeja González* [2014] available at curia.europa.eu/juris/document/document_print.jsf?doclang=EN&docid=152065 (last accessed 3 November 2018).

41 Charter of Fundamental Rights of the European Union (2000/C 364/01) (18.12.2000).

42 Directive 95/46/EC.

attachment proceedings for the recovery of social security debts. Moreover, this right may be exercised even if the data to be forgotten is perfectly lawful and accurate and even if there is no evidence of prejudice to the data subject.

However, the judgment is riddled with references to the 'balancing of interests' leaving the precise basis of the right unclear. If the right is derived from Articles 7 and 8 of the Charter then, as the Court observes, it belongs to a privileged class of rights that 'override, as a rule, not only the economic interests of the operator of the search engine but also the interest of the general public in finding that information upon a search relating to the data subject's name'.[43] In other words, it would only be other, conflicting, fundamental rights (such as the fundamental right to freedom of expression that is recognised by Article 11 of the Charter) that could be pleaded against such an overriding effect. Immediately after saying this, though, the court muddies the waters by suggesting that the right to be forgotten would not have overriding effect if 'it appeared, *for particular reasons*, such as the role played by the data subject in public life, that the interference with his fundamental rights is justified by the preponderant interest of the general public in having, on account of inclusion in the list of results, access to the information in question'.[44] Clearly, care needs to be taken that the only reasons that qualify as 'particular reasons' here are that fundamental rights are implicated. If, on the other hand, the right to be forgotten rests on the rights in Articles 12(b) and 14(a) of the Directive, it would not be privileged in the way that fundamental rights are and a general balancing of interests (seeking an acceptable or reasonable accommodation of relevant interests) would be appropriate. On this analysis, the particular reasons relied on against the right to be forgotten could be much broader—or, at any rate, this would be so unless we read the more particular provisions of Article 8 of the Charter as elevating the specific rights of the Directive to the status of fundamental rights.

Applying its principles to the case at hand, the Court held as follows:

> As regards a situation such as that at issue in the main proceedings … it should be held that, having regard to the sensitivity for the data subject's private life of the information contained in those announcements and to the fact that its initial publication had taken place 16 years earlier, the data subject establishes a right that that information should no longer be linked to his name by means of such a list. Accordingly, since in the case in point there do not appear to be particular reasons substantiating a preponderant interest of the public in having, in the context of such a search, access to that information, a matter which is, however, for the referring court to establish, the data subject may, by

---

43 (n 40) [97].
44 (n 40) [97] (emphasis added).

virtue of Article 12(b) and subparagraph (a) of the first paragraph of Article 14 of Directive 95/46, require those links to be removed from the list of results.[45]

What is puzzling here is that fundamental rights (to privacy) are being mixed with rights (in the Directive) that are subject to balancing and that belong to a different class of interests.[46] Whereas, from a fundamental rights perspective, it makes no sense to think that the passage of 16 years is a relevant consideration, from a balancing perspective, the privacy-sensitive nature of the data has no privileged status.[47] Arguably, the Court is trying to strike some intermediate position between fundamental rights and simple balancing. What might this be?

In principle, a community might treat a right to be forgotten as: (i) a fundamental right that is necessarily privileged and overriding in relation to all non-fundamental rights (as a right that is constitutive of this particular community); or (ii) as an interest that is not protected as a fundamental right but which, in the general balancing of interests, has more weight (although still susceptible to being outweighed by the preponderance of interests); or (iii) as a simple legitimate interest to be balanced against other such interests. Arguably, *Google Spain* is an example of a community that, being uncertain about its priority of informational rights and interests, needs to place the right to be forgotten in category (ii). While such an ad hoc approach might offend lawyers who reason in a traditional way, it fits well enough with a regulatory-instrumental mind-set where there is uncertainty about the most acceptable balancing point. That said, if this is the nature of the exercise, we might think that it is better undertaken by the legislative rather than the judicial branch.[48]

---

45 (n 40) [98].

46 Compare the insightful critique in Eleni Frantziou, 'Further Developments in the Right to be Forgotten: The European Court of Justice's Judgment in Case C-131/12, *Google Spain SL, Google Inc v Agencia Española de Proteccion de Datos*' (2014) 14 *Human Rights Law Review* 761, esp 768–769. See, too, Juliane Kokott and Christoph Sobotta, 'The Distinction between Privacy and Data Protection in the Jurisprudence of the CJEU and the ECtHR' in Hielke Hijmans and Herke Krannenborg (eds), *Data Protection Anno 2014: How to Restore Trust?* (Cambridge: Intersentia, 2014) 83.

47 Whether or not the elapse of time is a relevant consideration seems to depend on the particular facts of the case: see Article 29 Data Protection Working Party, *Guidelines on the Implementation of the Court of Justice of the European Union Judgment on "Google Spain and Inc v Agencia Española de Protección de Datos (AEPD) and Mario Costeja González"* C-131/32 (November 26, 2014) at 15–16 ('Depending on the facts of the case, information that was published a long time ago ... might be less relevant [than] information that was published 1 year ago.').

48 In Europe, Article 17 of the GDPR, Regulation (EU) 2016/679, now provides for a right to erasure with 'a right to be forgotten' placed alongside this in the heading to the Article. Whether this provision helps to clarify the law after the *Google Spain* case remains to be seen. See, further, Chapter Eight.

V The regulatory responsibilities and the fitness of the regulatory environment

The fitness of the regulatory environment in relation to emerging technologies will be tested (or audited) in a number of respects—for example, for its effectiveness, for its connectedness, and for its acceptability.[49] However, these criteria tend to assume that regulators are discharging their primary responsibility to maintain the commons' conditions. Just like debates about what might be prudent or moral for the community, debates about the nicer matters of regulatory fitness presuppose that the generic infrastructural conditions are secure. If there has been a catastrophic regulatory failure in relation to these conditions, not only will will there be nothing to debate (the regulatory environment is clearly unfit), there will be difficulty in debating anything.

If we focus on the second and third of the regulatory responsibilities, regulators will be expected to be responsive to the community's fundamental values and to its preferences and priorities. In modern communities, where citizens are neither technophiles nor technophobes, the expectation will be that regulators act in ways that are designed:

- to support rather than stifle beneficial innovation;
- to provide for an acceptable management of risks to human health and safety, the environment, and the economy; and
- to respect fundamental community values (such as privacy and confidentiality, freedom of expression, liberty, justice, equality, human rights and human dignity).

In this context, the regulatory environment will be judged to be adequate when fundamental values are respected and when a reasonable balance of interests is struck. However, this presents a major challenge because the demands are characterised by a double layer of tensions—with tensions existing both *between* the demands and *within* each of them.[50]

Between the first demand, on the one side, and the second demand, on the other, there is an obvious tension. The first demand fears over-regulation, the second fears under-regulation. The first prefers rapid access to potentially beneficial innovation (even during a developmental period), the second takes a more precautionary view. Similarly, there can be a tension between, on the

---

49 See, Roger Brownsword, *Rights, Regulation and the Technological Revolution* (n 2); and Roger Brownsword and Morag Goodwin, *Law and the Technologies of the Twenty-First Century* (Cambridge: Cambridge University Press, 2012).
50 See Roger Brownsword, 'Law, Regulation, and Technology: Supporting Innovation, Managing Risk and Respecting Values' in Todd Pittinsky (ed), *Science, Technology and Society: Perspectives and Directions* (Cambridge: Cambridge University Press, forthcoming).

one side, the first and second demands, and, on the other, the third demand. Here, robust commitment to values such as human rights and human dignity might oppose innovation that appeals to some as beneficial and that is not a risk to human health and safety. Provided that there is a clear commitment to these values, regulators should simply treat respect for these values as an overriding reason and resolve the tension accordingly. However, as we have seen (in cases such as *Google Spain*), it may not be so straightforward in practice and the regulatory waters become muddied by a discourse of balancing and proportionality.[51]

Within each of the demands, there is a central contestable concept. We should not assume that everyone will agree about which innovations are 'beneficial' (and, of course, all technologies invite both use and abuse); we should not assume that everyone will agree about the level and nature of risk that they judge to be 'acceptable'; and, we certainly should not assume that everyone will agree about the relevant values or about the criterion of doing the right thing.[52]

Faced with these potentially conflicting demands, regulators need to act with integrity and reasonably—which implies that their processes for decision-making are transparent and inclusive and that the decisions themselves can be explained in a way that plausibly relates to these processes. If regulators cannot readily satisfy this expectation—because, for example, they are reliant on 'intelligent' machines that operate in a way that 'works' but that is nonetheless alien to humans—then something will have to give: either regulatory reliance on the machines or regulatee expectations. Taking a stand on this, the House of Lords Select Committee on Artificial Intelligence has recommended that where, as with deep neural networks, 'it is not yet possible to generate thorough explanations for the decisions that are made, this may mean delaying their deployment for particular uses until alternative solutions are found'.[53]

Given these contestable and (in pluralistic communities) contested responsibilities, there is unlikely to be immediate agreement that regulators

---

51 Compare the extended discussion of Case C-34/10, *Oliver Brüstle v Greenpeace e.V.* (Grand Chamber, 18 October 2011) in Chapter Six. There, the Court held that the innovative products and processes of pioneering medical research were not patentable to the extent that they relied on materials derived from human embryos. Human dignity dictated that human embryos should not be so 'instrumentalised'.

52 Compare, Roger Brownsword, 'Bioethics Today, Bioethics Tomorrow: Stem Cell Research and the "Dignitarian Alliance"' (2003) 17 *University of Notre Dame Journal of Law, Ethics and Public Policy* 15; and 'Stem Cells and Cloning: Where the Regulatory Consensus Fails' (2005) 39 *New England Law Review* 535.

53 Report on *AI in the UK; ready, willing and able?* (Report of Session 2017–19, published 16 April 2017, HL Paper 100) at para 105: available at publications.parliament. uk/pa/ld201719/ldselect/ldai/100/10007.htm#_idTextAnchor025 (last accessed 3 November 2018).

are backing the right values or striking the most reasonable balances. Into this mix, we must now add the question of how regulators should pursue their (contested) purposes. Given that regulators have traditionally relied on rules and standards, the question is whether it is acceptable for them to rely on measures of technological management? Again, there is unlikely to be an easy consensus.

So, we have at least three potential sites for intense debate and disagreement: in relation to the interpretation of each of the demands that we make of regulators, in relation to the priorities as between these demands, and in relation to the way in which regulators should operationalise their responses to these demands. That said, once we take into account the lexical ordering of regulators' responsibilities, there will be some views that are clearly unreasonable. In particular, it will be unreasonable to make any demand or proposal that would involve compromising the commons' conditions; and it will be unreasonable to argue that some legitimate interest (such as economic gain) should override the community's fundamental values (the priority is the other way round).

VI Technological management and the three responsibilities

In the previous chapter, it was suggested that it might be acceptable to employ measures of technological management, rather than rules, to protect the commons. Or, to put this another way, it has been suggested that the case for relying on technological management rather than rules might be stronger where regulators are seeking to discharge their first and paramount responsibility. For example, if rules do not suffice to protect the planetary boundaries[54] or to ensure a reasonable degree of human physical security such that a risk is presented to the conditions for human existence, then perhaps technological management should be used. However, maintaining the conditions for human existence may come at a price with regard to the conditions for both self-development and moral development. This suggests that regulators, guided by their communities, will need to make a judgment about trade-offs between the conditions that comprise the commons. In general, we can suppose that the existence conditions will be prioritised (life is better than no life) but, as with the second and third responsibilities, regulators will find that there are tensions in discharging their first responsibility.[55]

---

54 See, J. Rockström, et al (n 9); and Kate Raworth (n 9).
55 By way of analogy, consider the case of a university library. It is a 'commons' for scholars and students. If the rules requiring respect for the library books and imposing silence in the library do not work, a technological response might be considered if it promises to be more effective. However, a technological response will change the complexion of the regulatory environment and, in the university community, this might be viewed by many as a change

Moreover, we can imagine that the pattern of tensions might again be not only between each of the commons' demands but within each of them.

To illustrate a possible internal tension—one that takes us back to a difficulty met in the previous chapter—we may find that in some communities there is a division between (i) 'unrestricted moralists' who hold that human agents should aspire to do the right thing for the right reason in relation to all acts, including acts that concern the commons' conditions themselves, and (ii) 'restricted moralists' who hold that this moral aspiration is *fully* engaged only in relation to acts that, while relying on the commons, do not directly impact on the commons' conditions themselves. For example, let us assume that, in a particular community, the core rules of the criminal law are treated as instruments that are designed to secure the commons' (existence and agency) conditions. Now, when modern technologies of surveillance and DNA-identification are adopted in support of these laws, unrestricted moralists will be concerned that the amplification of prudential reasons for compliance might result in the crowding out of moral reasons. When the use of technology is taken a stage further, such that 'non-compliance' is prevented and 'compliance' is technologically forced, agents no longer reason in terms of what they ought to do, either prudentially or morally: at this point, both prudential and moral reason are crowded out. For unrestricted moralists, this is a move in the wrong direction; as some might put it, human dignity is compromised.[56] However, for those who take a restricted view, a technological fix for the sake of the existence or agency pre-conditions is not necessarily a problem.[57]

Having drawn this distinction, two questions arise. First, is there any rational basis for the (unrestricted moralist) view that, even where rules do not work in protecting the commons' conditions and where technological management might fix the problem, the priority is still to try freely to do the right thing? Secondly, does the restricted view leave much significant space for agents to debate what is morally required and for agents to do the right thing for the right reason? The first question is difficult. Regulators might judge that unrestricted moralism sounds more like a credo—or a mere 'ideology' as some like to label views with which they disagree—than an aspect of the commons conditions. If so, we can say that, unless a rational basis for such a view can be articulated, regulators will be guided by the expectation

---

for the worse. Still, if the commons is seriously compromised by theft of books or by noise, it might be thought to be the lesser of two evils: a trade-off has to be made.

56 See Roger Brownsword (n 30).

57 For these changes in the regulatory registers, see Roger Brownsword, 'Lost in Translation: Legality, Regulatory Margins, and Technological Management' (2011) 26 *Berkeley Technology Law Journal* 1321. And, for the general question whether we should try to eliminate the possibility of offending, compare Michael L. Rich, 'Should We Make Crime Impossible?' (2013) 36 *Harvard Journal of Law and Public Policy* 795.

that protection of the commons is more important than privileging some particular moral dogma. However, if unrestricted moralism is widespread in a particular community, regulators may not be able to dismiss it as misguided dogma or moral heroism. So far as the second question is concerned, modern communities find plenty of matters to debate where the community is divided as to the morality of particular lifestyles or practices, such as the respect that should be accorded to non-human animals, to embryonic and foetal human life, to future generations, and so on, none of which relates directly to the commons conditions to the extent that the debates concern the status of such actual, potential, and prospective beings. Similarly, the value and weight to be accorded by a community to values such as freedom of expression and religious conviction can be matters of intense moral discussion and difference without this necessarily reaching back to the commons' conditions.[58] There is a difference between what kind of community a group of humans want their community to be and the conditions that must be in place for any kind of human community to be viable.

Taking stock, we can anticipate that regulators will find a similar pattern of contestation in relation to each of their responsibilities. In relation to each tier of responsibility, the expectation is that regulators, faced with competing interpretations and conflicting demands, will take up reasonable and accept-able positions. Moreover, we can anticipate that there will be disagreements about where the boundary lies between the tiers of responsibility and about which tier is engaged in a particular case. All this said, it seems that the case for the use of technological management might be more compelling (although not costless) in relation to the first sphere of responsibility; and that measures of technological management that compromise the commons' conditions should not be employed where regulators are seeking to discharge either their second or third level of responsibility.

VII Part One: Concluding remarks

Drawing together the threads of this and the previous chapters, the main points that arise from this initial re-imagining of the regulatory environment are the following.

First, jurists should take an interest in the field not only of legal and non-legal norms but also of technologies that function as regulatory instruments. In order to frame this field of inquiry, I have suggested that an extended concept of the regulatory environment, encompassing both normative and non-normative channelling strategies, is an appropriate device. Thus, the first step in re-imagining the regulatory environment is to conceive of regulatory

---

58 For one such complex and contested debate, see Dominic McGoldrick, *Human Rights and Religion: The Islamic Headscarf Debate in Europe* (Oxford: Hart, 2006).

instruments comprising not only rules and standards but also technologies (running from soft assistive or advisory technologies to hard technological management). While rules engage with the normative realm of paper liberty, hard technological measures target the realm of real, or practical, liberty.[59]

Secondly, the idea of the complexion of the regulatory environment provides an important focus for tracking the shift from reliance on the moral regulatory register to the prudential register, and then from normative to non-normative registers and strategies. For those who are concerned that emerging technologies might corrode the conditions for flourishing human agency and, especially, for moral community, this gives a lens through which to monitor regulatory action and the changing nature of the regulatory environment..

Thirdly, to the extent that technological instruments replace rules, and even if the former are generally more effective, economical and efficient than the latter, there is a major question about the acceptability of regulators employing measures of technological management. Schematically, it has been suggested that, in responding to this question, we should conceive of regulators as having three tiers of responsibility: first, to protect the global commons; secondly, to construct on the commons the kind of community that citizens distinctively value; and, thirdly, to undertake routine adjustments to the balance of competing and conflicting legitimate interests in the community. It has been suggested that, other things being equal, the case for resorting to technological management is likely to be more compelling where regulators are seeking to discharge the first of these responsibilities; and, looking ahead, one of the major questions for each community will be to determine whether the use of technological management rather than rules raises issues that speak to its distinctive commitments and values (thereby engaging the second tier of regulatory responsibility).

Fourthly, where technological management is employed for purposes of crime control, this is prima facie problematic because it tends to corrode the conditions for moral community. The corrosion can occur where the State employs such a strategy but also where individuals take acts of otherwise justified self-defence. To the extent that such technological measures are intended to protect the commons' conditions relating to human existence and/or self-development, the question for regulators is whether they can justify making the trade-off against the conditions for moral development.

Fifthly, where technological management is used for reasons of health and safety, it is less obviously morally problematic, particularly where it is adopted by an agent for the sake of their own health and safety and without any negative impact on others. However, in other cases, there might be conflicts to be dealt with—for example, when A adopts measures of technological

---

59 See, further, Chapter Seven.

management in a paternalistic way for the sake of B's health and safety; and the general adoption of such measures may well present a conflict between those who wish to retain lifestyle or working options in preference to increased levels of health and safety and those who wish to prioritise health and safety. To the extent that the use of technological management for reasons of health and safety impacts on the commons' conditions, regulators may face difficult choices in assessing the acceptability of trade-offs within those conditions. To the extent that such use does not impact on the commons' conditions, regulators may again face difficult choices as regulatees present them with conflicting demands. In this latter case, the question for regulators is whether they can defend their accommodation of these demands as consistent with the fundamental values of the community and as otherwise reasonable and acceptable relative to whatever legitimate interests are at stake.

PART TWO
Re-imagining legal values

5

THE IDEAL OF LEGALITY AND THE RULE OF LAW

I Introduction

According to Mireille Hildebrandt, the 'challenge facing modern law is to reinvent itself in an environment of pre-emptive computing without giving up on the core achievements of the Rule of Law'.[1] If we do not rise to this challenge, Hildebrandt leaves readers in no doubt about the seriousness of the consequences:

> If we do not learn how to uphold and extend the legality that protects individual persons against arbitrary or unfair state interventions, the law will lose its hold on our imagination. It may fold back into a tool to train, discipline or influence people whose behaviours are measured and calculated to be nudged into compliance, or, the law will be replaced by techno-regulation, whether or not that is labelled as law.[2]

In other words, it is the ideal of legality together with the Rule of Law that stands between us and a disempowering techno-managed future. How, though, are we to understand these protective ideals?

I take it that, even though there are many conceptions of the Rule of Law, it is agreed that this is an ideal that sets its face against both arbitrary governance and irresponsible citizenship. Advocates of particular conceptions of

---

1 Mireille Hildebrandt, *Smart Technologies and the End(s) of Law* (Cheltenham: Edward Elgar, 2015) 17.
2 Hildebrandt (n 1) xiii; and 226, Hildebrandt concludes by emphasising that it is for *lawyers* to involve themselves with such matters as 'legal protection by design'. See, too, Mireille Hildebrandt and Bert-Jaap Koops, 'The Challenges of Ambient Law and Legal Protection in the Profiling Era' (2010) 73 *Modern Law Review* 428.

the ideal will specify their own favoured set of conditions (procedural and substantive, thin or thick) for the Rule of Law which, in turn, will shape how we interpret the line between arbitrary and non-arbitrary governance as well as whether we judge citizens to be acting responsibly or irresponsibly in their response to acts of governance.[3] Viewed in this way, the Rule of Law represents a compact between, on the one hand, lawmakers, law-enforcers, law-interpreters, and law-appliers and, on the other hand, the citizenry. The understanding is that the actions of those who are in the position of the former should always be in accordance with the authorising constitutive rules (with whatever procedural and substantive conditions are specified); and that, provided that the relevant actions are in accordance with the constitutive rules, then citizens (including lawmakers, law-enforcers, law-interpreters, and law-appliers in their capacity as citizens) should respect the legal rules and decisions so made. In this sense, no one—whether acting offline or online—is above the law;[4] and the Rule of Law signifies that the law rules.

Similarly, if we apply this ideal to the acts of regulators—whether these are acts that set standards, or that monitor compliance, or that take corrective steps in response to non-compliance—then those acts should respect the constitutive limits and, in turn, they should be respected by regulatees provided that the constitutive rules are observed.[5]

In principle, we might also—and, indeed, I believe that we should—apply the ideal of the Rule of Law to technological management. The fact that regulators who employ technological management resort to a non-normative instrument does not mean that the compact is no longer relevant. On the one side, it remains important that the exercise of power through technological management is properly authorised and limited; and, on the other, although citizens might have less opportunity for 'non-compliance', it is important that the constraints imposed by technological management are respected. To be sure, the context of regulation by technological management is very different from that of a normative legal environment but the spirit and intent of the compact remains relevant.

The compact represented by the Rule of Law hinges on reciprocal constraints—first, constraints on the way in which the power of the Law is exercised by its institutions, its officers, and its agents; and, secondly, constraints on citizens who are expected to respect laws that are properly made. However, in at least two respects, this compact is unstable. First, if there is defection on either side, and if this escalates, then there can be a downward

---

3   Generally, see Joseph Raz, 'The Rule of Law and its Virtues' (1977) 93 LQR 195; and David Dyzenhaus, 'Recrafting the Rule of Law' in David Dyzenhaus (ed), *Recrafting the Rule of Law* (Oxford: Hart, 1999) 1.

4   Compare Joel R. Reidenberg, 'Technology and Internet Jurisdiction' (2005) 153 *University of Pennsylvania Law Review* 1951, resisting the claims of the 'Internet separatists' and defending the application of the Rule of Law to on-line environments.

5   Compare Karen Yeung, *Securing Compliance* (Oxford: Hart, 2004).

spiral that leads to a breakdown of trust and even a breakdown in social order. Secondly, the constitutive rules may be more or less constraining. In a well-known thick substantive conception, the International Commission of Jurists' Declaration of Delhi recognises the Rule of Law as

> a dynamic concept for the expansion and fulfilment of which jurists are primarily responsible and which should be employed not only to safeguard and advance the civil and political rights of the individual in a free society, but also to establish social, economic, educational and cultural conditions under which his legitimate aspirations and dignity may be realized.[6]

By contrast, the Rule of Law might demand only that acts of governance should be capable of guiding the behaviour of citizens, in which case, as Joseph Raz points out, this does not speak to 'how the law is to be made: by tyrants, democratic majorities or any other way'; and it 'says nothing about fundamental rights, about equality or justice'.[7] It follows that, if the only constraint on the power of the Law is whatever set of constitutive rules is adopted locally, and if these rules constrain only marginally, then the commitment to the Rule of Law might be some, but no great, improvement on unbridled power.[8] Although there may be some political contexts—for example, when a State is being rebuilt—where such an undemanding requirement might be sufficient to create the expectation that citizens will respect the law, in general, this will be an unreasonable demand. As a rule, where the constraints on the power of the Law are weak, the demand for respect by citizens will be correspondingly weak. The risk is that, before long, there will be no real commitment to the compact and the Rule of Law will be deployed rhetorically in what becomes a contest of power and persuasion.

There is also the possibility of internal tensions within the Rule of Law. For example, the 'negative' dimension of legality, holding that 'the highest concern of a legal system should be to protect the citizenry against an aggressive state' and insisting that the State should give citizens a fair warning that they are breaking the rules, might come into tension with the 'positive' dimension of legality 'which stands for consistency and completeness in the application of the law' and which emphasises the importance of the guilty being punished.[9] While the former (the 'shield' of the Rule of Law) tends to encourage restrictive (and literal) interpretation of penal statutes, the latter

---

6 Available at www.icj.org/wp-content/uploads/1959/01/Rule-of-law-in-a-free-society-conference-report-1959-eng.pdf (last accessed 3 November 2018).

7 Raz (n 3) 198.

8 For a compelling discussion, see Judith N. Shklar, *Legalism* (Cambridge, Mass.: Harvard University Press, 1964).

9 See George P. Fletcher, *Basic Concepts of Criminal Law* (Oxford: Oxford University Press, 1998), both quotes at 207.

(the 'sword') encourages a broad reading of criminal statutes 'with a view to capturing within their scope all those who are guilty or who can usefully be regarded as guilty'.[10]

In this chapter, we can consider three questions relating to the ideal of the Rule of Law when technological management is a significant part of the regulatory environment. The first question draws on the concern expressed by Brian Tamanaha that, in some legal orders, there has been a rise in 'instru-mentalism'—for example, rules of law being brokered in political institu-tions by powerful interest groups and elites, without any concern for the larger public interest or common good—all with deleterious consequences for the Rule of Law.[11] Given that technological management might seem to be an example of instrumentalism writ large, we need to ask whether, for this reason, it is antithetical to the Rule of Law. The second question considers how we might judge whether the use of technological management, by both public and private actors, involves an abuse of power. The third question is whether, and if so how, the generic Fullerian ideal of legality that nests within the Rule of Law can be applied, and perhaps reinvented, where non-normative technological management displaces normative regulation.

Before we broach these three questions, it should be said that my discus-sion of questions about the ideals of principled adjudication and, concomi-tantly, the coherence of the law—ideals that, again, are associated with the Rule of Law—will be largely reserved for the next chapter.

II Instrumentalism and technological management

How does respect for the Rule of Law conduce to good governance? Brian Tamanaha suggests that four ideals are fundamental to this project. These are:

> that the law is a principled preserver of justice, that the law serves the public good, that legal rules are binding on government officials (not only the public), and that judges must render decisions in an objective fashion based upon the law.[12]

Potentially, instrumentalist politics and adjudication are a threat to legal-ity and good governance. However, provided that these fundamental ideals are respected, Tamanaha has no problem about instrumentalist reasoning per se—indeed, he says that the idea of law as a means to an end 'would be a positive component if integrated within a broader system with strong

---

10 *Ibid.*, 209.
11 See, e.g., Brian Z. Tamanaha, *Law as a Means to an End* (Cambridge: Cambridge University Press, 2006).
12 *Ibid.*, 249.

commitments to these four ideals'.[13] The pathology of instrumentalism arises where the use of legislatures and courts in pursuit of (partisan) ends together with the acts of officials within those institutions is no longer constrained by the Rule of Law. It follows, argues Tamanaha, that 'legislators must be genuinely oriented toward enacting laws that are in the *common* good or *public* interest'; that 'government officials must see it as their solemn duty to abide by the law in good faith; this duty is not satisfied by the manipulation of law and legal processes to achieve objectives'; and 'that judges, when rendering their decisions, must be committed to searching for the strongest, most *correct* legal answer; they must resist the temptation to succumb to the power they have to exploit the inherent indeterminacy of law to produce results they desire'.[14]

If the use of technological management, its overt instrumentalism notwithstanding, is to be compatible with the Rule of Law, what lessons should we take from Tamanaha? Clearly, Tamanaha's thesis is not that instrumentalism (or means-to-an-end reasoning) in itself is objectionable; rather, it is unconstrained instrumentalism (rather like the instrumentalisation of others, treating them only as a means and not also as an end) that is the problem. From Tamanaha's perspective, if there is a problem with technological management, it is that its use is not sufficiently or appropriately constrained.

Quite which constraints are sufficient and appropriate, quite which conception of the common good or the public interest, quite which conception of justice is to be adopted, are matters that remain to be debated. However, in the light of our discussion in Part One of the three tiers of regulatory responsibilities, it will be appreciated that we should not regard debates about the common good or the public interest as entirely at large. At all events, if the constraints set by the Rule of Law are so weak as to tolerate an instrumentalism that turns the institutions of (so-called) law into arenas for pure power politics, then there is a problem because the underlying compact is likely to break down. To expect citizens to respect the use of technological management, or the particular purposes served by technological management, in such circumstances is manifestly unreasonable; it puts an intolerable strain on the compact. If the Rule of Law is to have any chance of prospering, the constitutive rules need to be acceptable to all sections of the community and then regulators need to operate in accordance with those rules.

III Technological management and abuse of power

Building on the closing remarks of the previous section, if technological management is to be compatible with the Rule of Law, then a minimum

---

13 *Ibid.*
14 *Ibid.*, 250 (emphases in original).

requirement is that such a strategy is used only in ways that accord with the constitutive rules, or any authorising or limiting rules made thereunder. This applies to the use of technological management by both public and private regulators; and, any failure to respect the Rule of Law compact will be an abuse of power.

To start with *private* use of technological management, it was some years ago that Lawrence Lessig flagged up the risk that digital rights management technologies might be used in ways that overreach the proprietors' lawful interests—for example, by denying some uses that would fall within one of the exceptions recognised by copyright law.[15] No doubt, we could generalise this concern. For example, in an earlier chapter, we mentioned the potential use of the 'Mosquito' device as a means of preventing young people from congregating in certain places where they might interfere with private interests. The principle that governs such cases is clear: private power must be exercised in ways that are compatible with the protected legal interests of others as well as the larger public interest. So, whether copyright holders try to protect their interests by contract or by technology, they should do so only to the extent that this is consistent with the protected legal interests of others; and the same applies to Mosquito-using shopkeepers who would prefer teenagers to congregate somewhere else. However, the novelty of these issues together with the lack of transparency in some cases means that there is a practical challenge even if the general principles are clear.

When we turn to the use of regulating technologies by *public* bodies, the principle is again clear. Public bodies must act within their lawful powers and in a way that is compatible with respect for the constitutive rules of their community (for example, respect for human rights and human dignity).[16] Currently, when the criminal justice agencies are increasingly relying on tracking, monitoring, identifying, and locating technologies (such as CCTV, body-worn cameras, GPS, RFID, and DNA profiling as well as the latest AI-assisted technologies), respect for the right to 'privacy' is the main restriction on unbridled use. In the *Marper* case,[17] the European Court of Human Rights has emphasised that the impingement on privacy must be 'proportionate'; and, in *Jones*, the United States' Supreme Court has affirmed the relevance of constitutional limits, specifically the Fourth Amendment, to chipping and tracking vehicles on public roads.[18] It is not altogether clear how public use of the 'Mosquito' might be limited; possibly this would be by

---

15 See Lawrence Lessig, *The Future of Ideas: The Fate of the Commons in a Connected World* (New York: Knopf Doubleday, 2002).
16 See Ben Bowling, Amber Marks, and Cian Murphy, 'Crime Control Technologies: Toward an Analytical Framework and Research Agenda' in Roger Brownsword and Karen Yeung (eds), *Regulating Technologies* (Oxford: Hart, 2008) 51.
17 *S and Marper v United Kingdom* [2008] ECHR 1581.
18 *United States v Jones* 565 US 400, 132 S.Ct 945 (2012).

appealing to freedom of association or, again, the elastic concept of 'privacy' might be adequate; and, presumably, it would be 'privacy' that would be the basis for challenging a disproportionate future use of brain imaging or other scanning (e.g. thermal imaging) technologies.[19] The problem, however, is that new technologies and new applications are being developed very quickly and our constitutional repertoires (which were largely articulated for low-technology societies) may not be fit for purpose.

What would we say, for example, in response to the proposal that human prison officers should be replaced by robots—an idea that, apparently, is already being piloted in South Korea?[20] Assuming that the robots function reliably, would this use of technology violate any constitutional principles? If privacy is not the relevant principle, then how would the provisions of the constitution be engaged? In many constitutions, the German Basic Law perhaps being the clearest example, respect for human dignity is the most fundamental principle;[21] and, certainly, one of the possible applications of the principle is in relation to prison conditions that are judged to be demeaning or degrading.[22] Whether or not robotic prison officers would fall foul of this principle is a moot point. Recalling the views of 'Richard'—one of Sherry Turkle's interviewees whom we met in the opening chapter of the book—it might be argued that, if prisoners are denied contact with human prison officers, then this treats them as less than human—in the language of the German constitutional jurisprudence, the prisoners are 'degraded to a mere object of state action'.[23] Do the robot prison officers really understand the difference between mere objects/things and human beings? On the other hand, those humans who design the prison management regimes do understand this difference and, if their designs reflect this understanding, then perhaps this suffices. Moreover, if cars and trains operated by robots do not offend the principle of human dignity, why should robotic prisoner officers be any different?

The problem with appeals to human dignity is that, as with appeals to the Rule of Law, the concept is contested and it is subject to many competing interpretations. In fact, some see human dignity as underlying the Rule of Law because, by following the publicly declared rules, government and legal

---

19 See *Kyllo v United States* 533 US 27 (2001) (police use of heat-sensing technology to detect domestic cannabis growing).

20 Lena Kim, 'Meet South Korea's New Robotic Prison Guards' Digital Trends (April 21, 2012): available at http://www.digitaltrends.com/international/meet-south-koreas-new-robotic-prison-guards/ (last accessed 3 November 2018)

21 See, Roger Brownsword, 'Human Dignity from a Legal Perspective' in M. Duwell, J. Braavig, R. Brownsword, and D. Mieth (eds), *Cambridge Handbook of Human Dignity* (Cambridge: Cambridge University Press, 2014) 1.

22 See, Deryck Beyleveld and Roger Brownsword, *Human Dignity in Bioethics and Biolaw* (Oxford: Oxford University Press, 2001) 14–15.

23 See, the *Honecker* Decision BerlVerfGH NJW 1993, 515, 517.

officials act in ways that are reasonably predictable which, in turn, enables citizens to plan their lives—that is to say, the Rule of Law is in line with respect for human dignity because it respects human autonomy.[24] However, more 'conservative' readings of human dignity, possibly of the kind reflected in Richard's sentiments—and certainly reflected in the view that we should not transfer our distinctively human responsibilities to robots—tend to emphasise the limits on human autonomy. Faced with the all too familiar stand-off between those who appeal to 'human dignity as empowerment' in order to support access to new technologies and those who appeal to 'human dignity as constraint' (or simply to dignity) in order to resist access,[25] we might retreat to a position in which human dignity is equated with the capacity for moral reason. On this view, the distinctive dignity of humans resides in their capacity freely to do the right thing even when they have the opportunity to do the wrong thing.[26] From this it follows, as we argued in Part One, that we need to monitor the changing complexion of the regulatory environment lest, through the use of technological management, the conditions for moral community are compromised; and, if we treat it as a defining characteristic of the Rule of Law that it serves human dignity in this sense, then we will judge that government abuses its powers where it resorts to technological management in this fundamentally compromising way.

IV Technological management and the generic ideals of legality

In (what I have termed) a first generation regulatory environment, regulators will rely entirely on normative signals, the standard set by the law being one such signal. This, of course, is the context in which we have crafted our understanding of the Rule of Law. However, as regulators begin to rely on non-normative design and coding, our understanding of legality will surely need some modification.

In a classic discussion of the ideal of 'legalism', understood as a set of procedural requirements, Lon Fuller proposed that the standards set should be general, promulgated, prospective, clear, non-contradictory, (reasonably) constant, and possible (of compliance).[27] He also suggested that it was of

---

24 For example, Raz (n 3) explicitly takes the view.

25 See, Beyleveld and Brownsword (n 22).

26 See, Roger Brownsword, 'Human Dignity, Human Rights, and Simply Trying to Do the Right Thing', in Christopher McCrudden (ed), *Understanding Human Dignity* (Oxford: Proceedings of the British Academy and Oxford University Press, 2013) 470; and 'Developing a Modern Understanding of Human Dignity' in Dietmar Grimm, Alexandra Kemmerer, and Christoph Moller (eds), *Human Dignity in Context* (Berlin: Hart/Nomos, 2018) 295.

27 Lon L. Fuller, *The Morality of Law* (New Haven: Yale University Press, 1969). For an application of the Fullerian principles to particular instances of cyberlaw, see Chris Reed, 'How to Make Bad Law: Lessons from Cyberspace' (2010) 73 MLR 903, esp 914–916. As Reed

the essence of the Rule of Law that enforcement should be congruent with the standards so promulgated. Where the standards are not promulgated, prospective, clear, non-contradictory, and (reasonably) constant, regulatees will simply not know where they stand; even if they wish to comply with the regulatory standard, they will not know what it is. If the standard set requires impossible acts of compliance, then ex hypothesi regulatees cannot comply. Reliance on highly specific regulations will drain most regulatory resource and, again, it will leave many regulatees unclear about their position. And, if there is a disconnect between the standards set and the enforcement practice, not only will regulatees be unclear about their position, they will lose respect for the regulatory regime.

For many years, jurists have debated whether the Fullerian principles speak only to the conditions for effective regulation or whether, as Fuller insists, they go to the heart of distinctively *legal* forms of regulation.[28] According to Fuller, there is a critical distinction between legal direction and mere managerial direction. As he puts it, 'law is not, like management, a matter of directing other persons how to accomplish tasks set by a superior, but is basically a matter of providing the citizenry with a sound and stable framework for their interactions with one another, the role of the government being that of standing as a guardian of the integrity of this system.'[29] Although, in the context of debates concerning the essential nature (or concept) of law, there is a fundamental choice between a moralised idea of law (evincing a necessary connection between law and morals) and an idea of law as a by and large effective institution for the direction of social life, in the larger regulatory picture other distinctions loom large. In particular, as I have said, there is the choice between normative and non-normative ordering, between rules (signalling ought and ought not) and design (signalling can and cannot). For Fuller, as for his critics, law and management alike operate with normative registers; they are traditional regulatory environments. However, what happens when we move into second and third generation regulatory environments, where non-normative signals (speaking only to what is practicable or possible) predominate? In such environments, there are many concerns for the virtues of the Rule of Law; but what would we make of the Fullerian criteria in such settings?

Rather than dismissing out of hand the relevance of the Fullerian principles—on the ground that they relate only to normative regulatory strategies and not at all to non-normative strategies—we can consider briefly each of the principles, trying to capture its spirit and intent, with a view to

---

summarises it (at 927): 'Complexity makes laws hard to understand, contradictory rules make compliance impossible and frequent change compounds these difficulties.'

28 See, e.g., HLA Hart's review of *The Morality of Law* (1964–65) 78 *Harvard Law Review* 1281.

29 Fuller (n 27) 210; and, see further, text below at 126–128.

understanding what it might signify for the acceptable use of technological management.[30]

(i) Laws should be prospective rather than retrospective

We can start with the injunction against the use of retrospective rules. No one surely would gainsay Fuller's remark that a regime 'composed exclusively of retrospective rules could exist only as a grotesque conceit worthy of Lewis Carroll or Franz Kafka'.[31] Certainly, where penalties are applied for breach of rules that have retrospective effect, this represents a paradigm of unfairness; regulatees not only do not know where they stand, they are denied a fair warning that they are non-compliant and that they are liable to be penalised. Even if the rules are themselves extremely unfair or difficult to justify, it is an independent requirement that they should be applied only with prospective effect. This is a basic requirement of due process.

In a non-normative context, it is very difficult to imagine how techno-logical management could itself operate retrospectively (although, no doubt, digital records can be wiped clean and amended). Quite simply, where tech-nological management is introduced to make a particular act impossible, or to remove what was previously a practical option, it takes effect as of then. No doubt, as in the Fullerian world of rules, it would be good practice to give regulatees fair warning that such technological measures are to be intro-duced; and, if regulatees operate on the assumption that what is possible in technologically managed environments is permissible, then it would be unfair to penalise by retrospective rule or decree those regulatees who, in good faith, have acted on that assumption. However, on the face of it, technologi-cal management does not in itself introduce new risks of unfair retrospective penalisation of conduct.

(ii) Laws should not require the impossible

The injunction against requiring the impossible responds not only to the irrationality of requiring persons to defy the law of gravity or to be in two places at one and the same time, but also to the unfairness of penalising

---

30 Compare Lodewijk Asscher, '"Code" as Law. Using Fuller to Assess Code Rules' in E. Dommering and L. Asscher (eds), *Coding Regulation: Essays on the Normative Role of Infor-mation Technology* (The Hague: TMC Asser, 2006) 61, 86:

> Code can present constraints on human behaviour that can be compared with constraints by traditional laws. We have argued that even though code is not law, in some instances it can be useful to ask the same questions about code regulation as we do about traditional regulation. Code as law must be assessed by looking at the results of regulation in terms of freedom and individual autonomy and compared to the balance struck in traditional law.

31 Fuller (n 27) 74.

persons for failing to comply with rules that require the literally impossible or that impose penalties on persons who, through no fault of their own, find themselves in circumstances where compliance is simply not possible.[32] On the face of it, because technological management operates in a mode that redefines what is possible and what is impossible within a particular regulatory space, it should not fall foul of requiring the impossible. To be sure, in some dystopian world, regulators might introduce a rule that requires regulatees to defy the restrictions imposed by technological management and penalises those who fail to succeed; but we can discount this bizarre possibility.

Rather, if the injunction against requiring the impossible has any continuing relevance, its thrust will be to avoid unfairly penalising regulatees in their various encounters with technological management. For example, would it be unfair to penalise an agent who attempts (but fails) to perform an act which that agent knows is impossible because it is prevented by technological management? For example, would a one-time and now frustrated vandal be fairly punished for persistently throwing stones at bus shelters equipped with shatterproof glass? Would it be unfair if the defendant responded, not by pleading the Fullerian principle that he should not be punished for failing to satisfy an impossible 'legal' requirement, but by arguing that his act was innocuous and that he should not be penalised for failing to do something that, because of technological measures, was no longer possible? This is a question to which we will return briefly in Chapter Seven. Conversely, suppose that some measure of technological management is prone to malfunction from time to time, such that an agent might in some circumstances be unclear whether a particular act is or is not possible (and by implication permissible). Should it be open to such an agent to plead that, where technological management is in place, 'can implies may'? Or, is this a defence more appropriately placed under the requirement that the regulatory signals should be clear and non-contradictory?

(iii) Laws should be clear

The Fullerian principle relating to clarity insists that regulatees should not be left uncertain about what the rules require. If such uncertainty exists, the law is likely to be less effective than it might be; and, where penalties are attached to non-compliance, it is unfair to punish regulatees who are unclear about what the law requires—or, as it is commonly expressed, due process demands that there should be a fair warning that a particular act will breach the law.

---

32 For Fuller's critical remarks on strict liability criminal offences, see esp. *ibid.*, 77–78. This is a matter to which we will return in Chapters Eight and Nine.

When we switch to a context of technological management, the clarity of the regulatory signal might be somewhat less important but it is not entirely redundant. Regulators still need to communicate with their regulatees; and, crucially, they need to signal that only certain options are practically available. For example, in the smart cities and places of the future, where the available options are heavily technologically managed, there probably needs to be clear signalling to regulatees that they are entering or acting in such a zone. To this extent, clarity of transmission is still something that matters. Of course, if the regulatory environment, even a regulatory environment in which the signals are not clear, is designed in such a way that regulatees have no option other than to do *x*, they will eventually do *x*. Even so, *x* should be done with less friction and confusion where the regulatory signal is clearly and decisively transmitted.

(iv) Laws should be relatively constant

While laws need to be revised from time to time, there is a problem if they are changed so frequently that regulatees are uncertain of their legal position. Just as a lack of clarity in the law breaches the fair warning principle, the same applies to a lack of constancy. Similarly, we might think that constancy has some value even in a non-normative regulatory environment. Indeed, as we have already noted, we might imagine that if some application of technological management sometimes prevents an act but at other times permits it—whether this arises from a technological malfunction or by a deliberate change made to the regulatory coding—this can leave regulatees uncertain of their position. This invites some confusion, which is undesirable; but perhaps the real sting in this requirement is if penalties or other forms of detriment are suffered as a result of too many technological modifications.

(v) Laws should not be contradictory

Legality, Fuller argues, aspires to the avoidance of contradiction—that is to say, contradiction between rules and/or precedents, within legislation, and so on. To recall one of Fuller's examples, we might detect a contradiction in section 704 of the Federal Food, Drug, and Cosmetic Act when it seemingly prohibited an inspector from entering a factory without the owner's permission but, at the same time, prohibited the owner from refusing permission for the purpose of entering or inspecting the factory as authorised by section 704.[33] No doubt, there are ways of resolving whatever contradiction is thought to arise from coupling (i) the inspector's needing the owner's permission to enter (implying the owner's having the right to withhold

---

33 Fuller (n 27) 67–68.

permission) with (ii) the owner's being required to grant permission—although, as Fuller points out, not necessarily in ways that make any sense relative to the legislative purposes. However, in *United States v Cardiff*,[34] the Supreme Court treated this contradiction as fatal, holding that the clash between these provisions meant that, for the purposes of a conviction under the criminal law, this was simply not good enough; defendants were entitled to be given a fair warning that their conduct was subject to criminal penalty.

In the next chapter, we will look quite carefully at the circumstances in which rules or rulings will offend the principle of non-contradiction—in particular, we will consider whether it is contradictory for one rule to permit the doing of x in conjunction with another rule or ruling that declines to encourage or incentivise the doing of x. Clearly, the paradigmatic case of contradiction is where one rule prohibits the doing of x while another permits or requires the doing of x. However, what does non-contradiction imply in a context of technological management? Perhaps the obvious implication is that, in a particular situation, the relevant technologies should be consistent in allowing or disallowing a certain 'act'. Where the technologies are simply talking to one another, some inconsistency might be inconvenient. However, if humans are misled by the inconsistency, and if there are penalties for doing some act that should have been prevented, but where the technology has failed, then it would seem to be unfair to apply the penalty—or, at any rate, it would be unfair if the agent acted in the good faith belief that, because the signal was set to 'possible', this implied that the act was permitted.

(vi) The administration of the law should be congruent with its published rules

As is well known, Fuller attaches a very high importance to the principle of congruence, to officials administering the rules in accordance with the declared rules, rather than operating with a secret rule book. However, because technological management promises to close the normative gap (the possible gap between the standard as declared and as administered), congruence takes on a different significance.

Where rules are administered by automated systems, congruence demands that the technology should faithfully follow the rules as intended. This presents a considerable challenge to the coding of rules.[35] However, this is still recognisably an issue of legality within a Fullerian universe of rules. The question is whether congruence, or at any rate the spirit of congruence, has an application to a context of technological management.

---

34 344 US 174 (1952).
35 See Danielle Keats Citron, 'Technological Due Process' (2008) 85 *Washington University Law Review* 1249.

The spirit of congruence is that regulators and their enforcement agents should operate in a way that accords with the expectations of regulatees as reasonably formed on the basis of the regulatory signals. In a context of technological management, as we have remarked already, regulatees might reasonably expect that where an act is possible, then regulators treat it as optional and no negative regulatory reaction should ensue where the act is done—at least, this is so unless regulatees clearly know that there has been a malfunction or something of that kind (analogous to regulatees looting shops during a police strike or bringing in excess tobacco and alcohol during a strike by the customs and excise people). So, congruence, along with clarity, constancy, and the like, demands that regulators and their agents do not penalise regulatees who, in good faith, have misunderstood the regulatory position.

It is also within the spirit of congruence that the articulation of technological management should be within the limits that have been published for its particular use as well as coherent with background limiting principles. On any understanding of the Rule of Law, powers should be operationalised in a way that is intra vires; and, as we have seen already in the previous section of this chapter, the rules and principles that set the limits for the use of technological management are a key reference point for the purpose of determining whether there has been an abuse of power.

(vii) Laws should be general

Fuller, recalling the historic abuse of prerogative power, also identifies ad hominem 'rules' as contrary to the very idea of legality. Of course, courts hand down rulings that apply only to the particular parties, but legislative acts should be of general application. By contrast, some articulations of technological management might be specific to particular individuals—for example, as profiling becomes more fine-grained and the management of access is more personalised, the targeting of particular individuals will be even more pronounced.[36] Instead of focusing on dangerous acts or dangerous classes, precision profiling is likely to identify and isolate dangerous individuals. If this form of technological management offends a community's ideas of fairness or legality but is more accurate than more general profiling and targeting strategies, some hard choices will need to be made. This is a matter to which we will return in Chapter Nine.

---

36 Compare, too, the possibility of tailored guidance by 'micro-directives': see, Anthony Casey and Anthony Niblett 'Self-Driving Laws' (2016) 66 *University of Toronto Law Journal* 429 and 'The Death of Rules and Standards' (2017) 92 *Indiana Law Journal* 1401.

(viii) *Laws should be promulgated*

Finally, there is the requirement that the rules should be promulgated. Regulatees need to know where they stand; and, particularly where there are penalties for non-compliance, regulatees need to be given a fair warning if they are at risk. This implies that transparency is an important aspect of legality and this continues to be a critical ideal in a context of technological management. There should be openness in authorising the use of technological management, in knowing that it is in operation, and arguably in knowing how it operates (otherwise there might be difficulties in challenging decisions made by the technology). As we have seen already, the question of transparency and the right to an explanation are becoming major debating points in relation to the leading-edge developments in AI and machine learning.[37]

(ix) Overview

Picking up this last point, Bert-Jaap Koops highlights the importance of this kind of transparency by hypothesising the case of a street food seller who is denied a licence to operate in a zone that security services require to be risk-free.[38] The seller does not understand why he is judged to be a safety risk; and, if there is to be due process, he needs to know on what basis the automated decision was made. Where one piece of data is determinative (such as, in Koops' example, a criminal conviction twenty years earlier for being in possession of drugs), it should be possible for this to be given as the reason and then the seller might challenge the accuracy of, or weight given to, this data point. In other kinds of cases, where 'advanced self-learning algorithms calculate risks based on complex combinations of factors', it might be necessary to bring in independent third-party auditors, thereby providing 'another type of checks and balances on the fairness of profiling-based decisions'.[39] Summing up, Koops says that decision transparency in such cases could be effected,

> first, by a legal obligation to inform the applicant that the decision was based on profiling and allowing the applicant to request information about the logic involved in the profiling and, second, by architectural

---

37 See, e.g., House of Lords Select Committee on Artificial Intelligence, Report on *AI in the UK; ready, willing and able?* (Report of Session 2017–19, published 16 April 2017, HL Paper 100) at para 105: available at publications.parliament.uk/pa/ld201719/ldselect/ldai/100/10007.htm#_idTextAnchor025 (last accessed 3 November 2018).
38 Bert-Jaap Koops, 'On Decision Transparency, or How to Enhance Data Protection after the Computational Turn' in Mireille Hildebrandt and Katja de Vries (eds), *Privacy, Due Process and the Computational Turn* (Abingdon: Routledge, 2013) 196, 212–213.
39 *Ibid.*, 212.

safeguards in risk assessment systems that aim at making the profiling more transparent, for example by marking in which proportion the outcome was influenced by data from each data source fed into the system and marking data in those sources that were used in consecutive steps when the profiling algorithm was run. Periodic independent audits could supplement the accountability of the decision-making process.[40]

This seems to me to be consistent with the spirit of the Rule of Law. Accordingly, I suggest that, in second generation regulatory environments, where technological measures are widely used, openness, or transparency—in authorising the use of as well as in operationalising technological management—supported by ideals of fairness and due process, continue to be key principles of legality.[41] So far as public regulators are concerned, there needs to be an authorising rule framework setting out the process for adopting measures of technological management, with particular proposed uses being openly debated (for example, in the legislative assembly or by administrative notice and comment procedure). As Danielle Keats Citron, has recommended:[42]

> [A]gencies should explore ways to allow the public to participate in the building of automated decision systems …
>
> In the same vein, agencies could establish information technology review boards that would provide opportunities for stakeholders and the public at large to comment on a system's design and testing. Although finding the ideal makeup and duties of such boards would require some experimentation, they would secure opportunities for interested groups to comment on the construction of automated systems that would have an enormous impact on their communities once operational.

Moreover, private use of technological management should be permitted only within publicly agreed limits and, if new uses are proposed, they should be approved by open special procedures (possibly akin to applications for planning permission). In all cases, ideals of fairness should support the process by insisting that tricks or traps should be avoided.

There is a huge amount of unfinished business here. However, the guiding spirit of legality is captured in Fuller's attempts to trace his differences with

---

40 *Ibid.*

41 Compare Ian Kerr, 'Prediction, Pre-emption, Presumption' in Hildebrandt and de Vries (n 38) 91, 109:

> At its core—whether in the public or private sector, online or off—the due process concept requires that individuals have an ability to observe, understand, participate in and respond to important decisions or actions that implicate them.

42 Danielle Keats Citron (n 35) 1312.

his critics. This led him to identify the following two key assumptions made by the legal positivists:

> The *first* of these is a belief that the existence or non-existence of law is, from a moral point of view, a matter of indifference. The *second* is an assumption … that law should be viewed not as the product of an inter-play of purposive orientations between the citizen and his government but as a one-way projection of authority, originating with government and imposing itself upon the citizen.[43]

The second of these assumptions is elaborated in a crucial contrast that Fuller draws between a legal form of order and simple managerial direction. He sketches the distinction between the two forms of order in the following terms:

> The directives issued in a managerial context are applied by the subor-dinate in order to serve a purpose set by his superior. The law-abiding citizen, on the other hand, does not apply legal rules to serve specific ends set by the lawgiver, but rather follows them in the conduct of his own affairs, the interests he is presumed to serve in following legal rules being those of society generally. The directives of a managerial system regulate primarily the relations between the subordinate and his supe-rior and only collaterally the relations of the subordinate with third per-sons. The rules of the legal system, on the other hand, normally serve the primary purpose of setting the citizen's relations with other citizens and only in a collateral manner his relations with the seat of authority from which the rules proceed. (Though we sometimes think of the criminal law as defining the citizen's duties towards his government, its primary function is to provide a sound and stable framework for the interactions of citizens with one another.)[44]

As Fuller concedes, these remarks need 'much expansion and qualification';[45] and he tries to give more substance to them by characterising the relation-ship, in a legal order, between government and citizens in terms of 'reciproc-ity' and 'intendment'.[46] Perhaps, Fuller's most evocative observation is that 'the functioning of a legal system depends upon a cooperative effort—an effective and responsible interaction—between lawgiver and subject'.[47]

---

43 Fuller (n 27) 204.
44 Fuller (n 27) 207–208.
45 Fuller (n 27) 208.
46 Fuller (n 27) 209 et seq.
47 Fuller (n 27) 219.

From this clutch of ideas, it is the association of legal ordering with an open and two-way reciprocal process that is most fruitful. For, in the larger context of the regulatory environment, it implies that the legal approach—an approach to be valued—is one that embeds participation, transparency, due process and the like. Hence, if we take our lead from Fuller, we will reason that, whether we are dealing with a regulatory enterprise that subjects human conduct to the governance of rules (in the way that both Fuller and his critics agreed was the pre-theoretical nature of law) or that relies on technological control to design-in or design-out conduct, we should hold on to the idea that what we value is a reciprocal enterprise, not just a case of management, let alone technological management, by some regulatory elite.

V Technological management and a re-imagined understanding of legality

Although the spirit of the Fullerian ideals of legality can be brought to bear on the adoption of technological management, we might still feel that, if not quite a fresh start, some supplementation needs to be made. To be sure, the Fullerian ideals continue to be applicable to the normative dimensions of the regulatory environment; but, once technological management is employed, these ideals as specified no longer apply in an entirely adequate way. Accordingly, the question that we should start with is this: if the use of technological management is not to be arbitrary, if its use is to be 'respected' by citizens, what terms and conditions should be set for its use? I suggest that the following eight conditions merit serious consideration.[48]

First, for any community, it is imperative that technological management (just as with rules and standards) does not compromise the essential conditions for human social existence (the commons). The Rule of Law should open by emphasising that the protection and maintenance of the commons is always the primary responsibility of regulators.

Yet, in counterpoint to this imperative, we have suggested that technological management might be seriously considered as a response to threats to the commons (particularly, where rules that are intended to protect the commons are not being observed); but, of course, no such measures should be adopted unless regulators believe that they will be effective and are confident that they will not make things worse. It bears emphasising that this is no proposal to treat regulators as 'technocratic Leviathans', mandated to act in whatever way they judge to be in the prudential interests of the community. Human agents have compelling reasons, both prudential and moral, for protecting the existence conditions for human life and the generic conditions

---

48 Compare Roger Brownsword, 'The Rule of Law, Rules of Law, and Technological Management' Amsterdam Law School Research Paper No. 2017–35 (2017) 9–17. Available at: ssrn. com/abstract=3005914 (last accessed 3 November 2018).

for agency; but the Rule of Law would underline that the jurisdiction of the regulatory stewards (and, concomitantly, whatever use they make of technological management) is strictly limited to maintaining the essential infrastructural conditions. Moreover, anticipating ongoing debates about where precisely the boundary lies between these conditions and the activities of agents who are supported by this infrastructure, the community might insist that the regulatory stewards deploy technological management only when they are comfortably intra vires.

It is worth repeating that if the Rule of Law requires that the use of measures of technological management should be consistent with the maintenance of the commons, then such measures must be compatible with the aspiration for moral community. Stated minimally, this aspiration is that all members of the community, whether regulators or regulatees, should try to do the right thing (relative to the legitimate interests of both themselves and others). As we have explained, technological management, by compelling acts that regulators judge to be right (or by excluding acts that they judge to be wrong), might compromise the conditions for moral community. Where the moral judgments made by regulators are contested, technological management might compel some regulatees to act against their conscience; and even where regulatees accept the moral judgments made by regulators, there might be a concern that this interferes with the cultivation of moral virtue, with an understanding of what it is to respect others and with authentic moral action (namely, freely doing the right thing for the right reason). Precisely how these essential conditions are articulated and operationalised is a matter for each community to determine. However, if the community is concerned to avoid compelling members to act against their conscience, then this suggests that regulators should eschew the use of technological management where there are significant moral disagreements about the regulatory purpose or provision. Similarly, if the community is concerned to maintain an active and engaged moral membership, with members taking personal moral responsibility, then regulators should eschew the use of technological management where they sense that it is either doing too much work for regulatees or disrupting permissible group or sectoral standards—for example, the standards of 'neighbourliness' and of 'cooperation and compromise' that characterise the self-regulating groups that we meet, respectively, in Robert Ellickson's study of the ranchers and farmers of Shasta County, California,[49] and in Stewart Macaulay's study of the transactional practices of Wisconsin business people.[50]

Secondly, where the aspiration is not simply to be a moral community (a community committed to the primacy of moral reason) but a particular kind of moral

49 Robert C. Ellickson, *Order Without Law* (Cambridge, Mass.: Harvard University Press, 1991).
50 Stewart Macaulay, 'Non Contractual Relations in Business: A Preliminary Study' (1963) 28 *American Sociological Review* 55.

community, then it will be a condition of the Rule of Law that the use of tech-nological management (just as with rules and standards) should be consistent with its particular constitutive features—whether those features are, for instance, liberal or communitarian in nature, rights-based or utilitarian, and so on.

Recalling earlier discussions in the book, one thought is that a community might attach particular value to preserving both human officials (rather than machines) and rules (rather than technological measures) in the core areas of the criminal justice system. Indeed, it might be suggested that core crime should be ring-fenced against technological management. However, while this proposal might appeal to liberals who fear that technological manage-ment will impede due process and to 'unrestricted' moralists who hold that it is always important to have the opportunity freely to do the right thing, it is likely to be opposed by those who are 'restricted' moralists and who fear that such a strategy is a hostage to fortune.[51]

Thirdly, where the use of technological management is proposed as part of a risk management package, so long as the community is committed to the ideals of deliberative democracy, it will be a condition of the Rule of Law that there needs to be a transparent and inclusive public debate about the terms of the package. It will be a condition that all views should be heard with regard to whether the package amounts to both an acceptable balance of benefit and risk as well as representing a fair distribution of such risk and benefit (including adequate compensatory provisions). Before the particular package can command respect, it needs to be somewhere on the spectrum of reasonableness. This is not to suggest that all regulatees must agree that the package is optimal; but it must at least be reasonable in the weak sense that it is not a package that is so unreasonable that no rational regulator could, in good faith, adopt it.

For example, where technologically managed places or products operate dynamically, making decisions case-by-case or situation-by-situation, then one of the outcomes of the public debate might be that the possibility of a human override is reserved. In the case of driverless cars, for instance, we might want to give agents the opportunity to take control of the vehicle in order to deal with some hard moral choice (whether of a 'trolley' or a 'tunnel' nature) or to respond to an emergency (perhaps involving a 'rescue' of some kind). Beyond this, we might want to reserve the possibility of an appeal to humans against a decision that triggers an application of technological man-agement that forces or precludes a particular act or that excludes a particular person or class of persons. Indeed, the concern for last resort human inter-vention might be such a pervasive feature of the community's thinking that it is explicitly embedded as a default condition in the Rule of Law.

---

51 For the distinction between 'restricted' and 'unrestricted' moralists, see Chapter Four, Part VI.

Similarly, there might be a condition that interventions involving techno-logical management should be reversible—a condition that might be particu-larly important if measures of this kind are designed not only into products and places but also into people, as might be the case if regulators contemplate making interventions in not only the coding of product software but also the genomic coding of particular individuals.

Fourthly, where following community debate or public deliberation, par-ticular limits on the use of technological management have been agreed, those limits should be respected. As we explained in an earlier section of the chapter, it would be an abuse of power to exceed such limits. In this sense, the use of technological management should be congruent with the particu-lar rules agreed for its use, as well as being coherent with the community's constitutive rules.

Fifthly, the community will want to be satisfied that the use of techno-logical measures is accompanied by proper mechanisms for accountability. When there are problems, or when things go wrong, there need to be clear, accessible, and intelligible lines of accountability. It needs to be clear who is to be held to account as well as how they are to be held to account; and, the accounting itself must be meaningful.[52]

Sixthly, a community might be concerned that the use of technological management will encourage some mission creep. If so, it might stipulate that the restrictive scope of measures of technological management or their forc-ing range should be no greater than would be the case were a rule to be used for the particular purpose. In this sense, the restrictive sweep of technologi-cal management should be, at most, co-extensive with that of the equivalent (shadow) rule.

Seventhly, as we have seen in the previous section, it is implicit in the Fullerian principles of legality that regulators should not try to trick or trap regulatees; and this is a principle that is applicable whether the instrument of regulation is the use of rules or the use of technological management. Accordingly, it should be a condition of the Rule of Law that technological management should not be used in ways that trick or trap regulatees and that, in this sense, the administration of a regime of technological man-agement should be in line with the reasonable expectations of regulatees. Crucially, if the default position in a technologically managed regulatory environment is that, where an act is found to be available, it should be treated as permissible, then regulatees should not be penalised for doing the act on the good faith basis that, because it is available, it is a lawful option.

---

52 See Joshua A. Kroll, Joanna Huey, Solon Barocas, Edward W. Felten, Joel R. Reidenberg, David G. Robinson, and Harlan Yu, 'Accountable Algorithms' (2017) 165 *University of Pennsylvania Law Review* 633, 702–704.

Eighthly, regulatees might also expect there to be a measure of public scrutiny of the private use of technological management. Even if public regulators respect the conditions set by regulatees, it will not suffice if private regulators are left free to use technological management in ways that compromise the community's moral aspirations, or violate its constitutive principles, or exceed the agreed and authorised limits for its use. Accordingly, it should be a condition of the Rule of Law that the *private* use of technological management should be compatible with the general principles for its use.

Unlike the Fullerian principles of legality, which focus on formal and procedural desiderata, these conditions for the lawful use of technological management bring into play a thicker morally substantive set of requirements. To claim that the Fullerian principles were only about *effective* regulation was never convincing; once they are set in the context of legal ordering, the principles are clearly about fairness. To claim that the conditions for the use of technological management are simply about effectiveness is manifestly absurd. Technological management appeals because it promises to be more effective than rules; but, taking a leaf from Tamanaha's book, its brute instrumentalism demands that its use be conditioned by principles that give it legitimacy—otherwise, there is no reason why regulatees should at least acquiesce in its use.

VI Conclusion

The compact that underlies the Rule of Law is the fulcrum of normative legal orders. It constrains against arbitrary governance and, where governance satisfies the relevant conditions, it demands responsible citizenship (paradigmatically in calling for respect for the law). With the introduction of technological management, the compact needs to be updated and its emphasis modified, but its spirit persists and its importance is greater than ever. In a context of technological management, those laws that authorise the use of technological management must be promulgated and their administration needs to be congruent with the terms of the authorisation. However, the key point is that there needs to be systematic openness about the use of technological management. Regulatees need to be part of the process of adopting measures of technological management and there must be an ongoing transparency about their use and about how the technology works.

Returning to the Hartian conceptualisation of legal systems, while the idea that legal systems distinctively feature an intersection between primary and secondary rules fails to anticipate the impact of technological management on primary rules, it rightly tries to correct the Austinian failure to take account of public law by developing its layer of secondary rules. However, in the Twenty-First Century, the challenge for the secondary rules is not so much to empower those officials who legislate, enforce and apply the primary rules but to authorise the use of technological management provided that it is within acceptable limits. It is in these limits that we find the new focus for the Rule of Law and the fulcrum for legal orders that employ technological management.

6

THE IDEAL OF COHERENCE

I Introduction

Famously, Ronald Dworkin argued that one of the characteristic aspirations of legal (especially judicial) practice is that it should be principled in its approach, displaying a certain kind of integrity,[1] and concomitantly that the body of legal doctrine (together with its underlying jurisprudence) should be coherent.[2] Arguably, such concerns are closely linked to the values of the Rule of Law.[3] In one sense, of course, it is axiomatic that observance of the Rule of Law entails that the judges should 'apply the law'; but we know that, where the jurisprudence is open to more than one reading, judicial practice—most conspicuously the practice of the Appeal Courts—in 'applying the law' is more complex. Dworkinian integrity demands that, in such hard cases, there should be a dual coherence: first, the reading that is adopted must maintain a thread of continuity with the jurisprudence; and, secondly, the reading must cohere with the constitutive (moral) values of the particular legal order. Of course, it bears repetition that, to be comprehensively coherent, the reading must also be consistent with respect for the essential conditions of any kind of legal order (the commons).

---

1 Ronald Dworkin, *Law's Empire* (Cambridge: Harvard University Press, 1986).
2 For an extended discussion of the particular issue introduced in this section, see Roger Brownsword, 'Regulatory Coherence—A European Challenge' in Kai Purnhagen and Peter Rott (eds), *Varieties of European Economic Law and Regulation: Essays in Honour of Hans Micklitz* (New York: Springer, 2014) 235.
3 Compare Brian Z. Tamanaha, *Law as a Means to an End* (Cambridge: Cambridge University Press, 2006).

One of the most obvious and worrying examples of doctrinal incoherence is a straightforward contradiction between two precedents, where the courts deciding the cases are at the same level. Without knowing which precedent to follow, the legal position is unclear and the law falls short of the standards set by the Rule of Law. Sometimes, however, the incoherence in the law can be rather more subtle. For example, for some time, English contract lawyers have been wrestling with the relationship between two streams of doctrine, one dealing with the contextual interpretation (or construction) of contracts and the other with implied terms. Controversially, in *Attorney-General of Belize v Belize Telecom Ltd*,[4] the Privy Council (led by Lord Hoffmann) sought to assimilate these two doctrinal streams and instate a single unifying test based on the reasonable expectations of the parties in the particular context of their transaction.[5] One of the points made by supporters of this assimilation is that it 'promotes the internal coherence of the law'; and this, it is said, is important 'because it enables the courts to identify the aims and values that underpin the law and to pursue those values and aims so as to achieve consistency in the structure of the law'.[6] However, in a trio of recent Supreme Court decisions—*Marks and Spencer plc v BNP Paribas Services Trust Company (Jersey) Limited*[7] (on implied terms); and *Arnold v Britton*[8] and *Wood v Capita Insurance Services Ltd*[9] (on interpretation)—we find a reaction against expansive implication and interpretation of terms, particularly in carefully drafted commercial contracts. If we value doctrinal coherence, the question now is how to square contextualist with non-contextualist approaches.[10]

In this chapter, we can explore the ideal of coherence by focusing, not on contracts, but on the regulatory environment for patents in Europe. This is an instructive test case because the challenge is to achieve coherence between a number of strands of regional law, namely: the European Patent Convention 1973 (the EPC); Directive 98/44/EC on the Legal Protection of Biotechnological Inventions (the Directive); and the European Convention on Human Rights. Technological management is not yet in the mix to complicate this issue; but aligning the various elements of the normative dimension of the regulatory environment for patents in Europe is plenty to be going on with.

---

4 [2009] UKPC 10.

5 See Richard Hooley, 'Implied Terms after *Belize Telecom*' (2014) 73 CLJ 315.

6 Per Arden LJ in *Stena Line Ltd v Merchant Navy Ratings Pension Fund Trustees Ltd* [2011] EWCA Civ 543 at [36].

7 [2015] UKSC 72.

8 [2015] UKSC 36.

9 [2017] UKSC 24.

10 See, further, Roger Brownsword, 'After Brexit: Regulatory-Instrumentalism, Coherentism, and the English Law of Contract' (2018) 35 *Journal of Contract Law* 139; and, for an attempt to restore coherence, see Zhong Xing Tan, 'Beyond the Real and the Paper Deal: The Quest for Contextual Coherence in Contractual Interpretation' (2016) 79 *Modern Law Review* 623.

In both the EPC and the Directive, provision is made for exclusion against patentability on moral grounds. Compared to other patent regimes worldwide, this kind of provision is unusual[11] and it has given rise to major questions concerning the patentability of innovative stem cell research—questions that have been raised for determination both at the European Patent Office (in relation to the EPC) and before the European Court of Justice (CJEU). The two leading cases, *Wisconsin Alumni Research Foundation (WARF)* at the European Patent Office,[12] and *Brüstle* at the CJEU,[13] have generated a storm of controversy about both the substance of the decisions and the coherence of the patent regime.

Briefly, in October 2011, the CJEU, responding to a reference from the German Federal Court of Justice, ruled that the processes and products associated with the innovative stem cell research conducted by the neuroscientist Oliver Brüstle were excluded from patentability. The basis of the decision was that Brüstle's research crossed one of the moral red lines provided for by the Directive—in this case, a 'dignitarian' line,[14] set out in Article 6(2)(c), that protects human embryos against instrumentalisation, commodification,

---

11 See, e.g., Gregory N. Mandel, 'Regulating Nanotechnology through Intellectual Property Rights' in G. Hodge, D.M. Bowman, and A.D. Maynard (eds), *International Handbook on Regulating Nanotechnologies* (Cheltenham: Edward Elgar, 2010) 388 at 395, reporting that 'no [US] court has held an invention invalid for being immoral, deceptive, or illegal for almost 100 years'.

12 Case G 0002/06, November 25, 2008. Here, the Enlarged Board of Appeal at the EPO was asked by the Technical Board of Appeal (T 1374/04 (OJ EPO 2007, 313)) to rule on four questions of law—one of which was whether Article 6(2)(c), as incorporated in the EPC Rules, forbids the patenting of a human embryonic stem cell culture which, at the time of filing, could be prepared only by a method that necessarily involved the destruction of human embryos (even though the method in question is not part of the claim). Treating this as an exercise in the interpretation of a particular rule, rather than a more general essay in European morality, the EBA said (at [18]):

> On its face, the provision ... is straightforward and prohibits the patenting if a human embryo is used for industrial or commercial purposes. Such a reading is also in line with the concern of the legislator to prevent a misuse in the sense of a commodification of human embryos ... and with one of the essential objectives of the whole Directive to protect human dignity ...

Rejecting the argument that human embryos were not actually being used for commercial or industrial purposes, the EBA held that, where the method of producing the claimed product necessarily involved the destruction of human embryos, then such destruction was 'an integral and essential part of the industrial or commercial exploitation of the claimed invention' (para.25); and, thus, the prohibition applied and precluded the patent.

13 Case C-34/10, *Oliver Brüstle v Greenpeace e.V.* (Grand Chamber, 18 October 2011).

14 Here, I use the term 'dignitarian' to capture a range of duty-based ethics, both secular and non-secular, that highlight the importance of not compromising human dignity. See, further, Roger Brownsword, 'Bioethics Today, Bioethics Tomorrow: Stem Cell Research and the "Dignitarian Alliance"' (2003) 17 *University of Notre Dame Journal of Law, Ethics and Public Policy* 15, and 'Stem Cells and Cloning: Where the Regulatory Consensus Fails' (2005) 39 *New England Law Review* 535.

and commercialisation by researchers (or, in the words, of Article 6(2)(c), that protects human embryos against use for 'industrial or commercial purposes'). Although the Court's ruling was, broadly speaking, in line with the decision of the Enlarged Board of Appeals at the European Patent Office in the *WARF* case (where, once again, the products of pioneering human embryonic stem cell research were excluded from patentability), and even though Article 27(2) of the GATT/TRIPs Agreement permits members to exclude inventions from patentability on grounds of *ordre public* or morality, the decision in *Brüstle* (like the decision in *WARF*) has attracted widespread criticism.[15]

Amongst the criticisms is the charge that *Brüstle* lacks coherence, both *formal* and *substantive*. With regard to the former, the question is: how can it be coherent to exclude Brüstle's research from patentability when it was perfectly lawful to conduct it in Germany, a jurisdiction that has some of the strongest embryo protection laws in Europe? As for the latter, the thrust of the charge is that the CJEU, being bound by the provisions of the European Convention on Human Rights, should always decide in terms that are compatible with the jurisprudence of the European Court of Human Rights;[16] in that jurisprudence, it is clear that human embryos are not treated as bearers of human rights; and, in particular, it is clear that the treatment (including destruction) of human embryos does not directly engage the Convention right to life. It follows, so the objection runs, that there is no support in European human rights law for the dignitarian moral concerns that underpin Article 6(2)(c) of the Directive; and, thus, it was incoherent for the Court to exclude patentability in relation to Brüstle's research on this ground.

While it might be possible to defend the *Brüstle* decision against both these charges, further charges of incoherence are conceivable—for example, that the CJEU violates a Rule of Law standard of impartiality, that its prescriptive approach fails to cohere with the margin of appreciation typically accorded

---

15   On the *WARF* case, see e.g., Paul L.C. Torremans, 'The Construction of the Directive's Moral Exclusions under the EPC', in Aurora Plomer and Paul Torremans (eds), *Embryonic Stem Cell Patents* (Oxford: Oxford University Press, 2009), 141. There is also the question of how coherently the *WARF* reading of the exclusion fits with other EC legal measures that license human embryonic stem cell research: see, Aurora Plomer, 'Towards Systemic Legal Conflict: Article 6(2)(c) of the EU Directive on Biotechnological Inventions', in Plomer and Torremans, above, 173. Nevertheless, this broad interpretation was foreshadowed in EDINBURGH/Animal Transgenic Stem Cells (Patent App. No. 94 913 174.2, July 21, 2002, Opposition Division), on which, see Shawn H.E. Harmon, 'From Engagement to Re-engagement: the Expression of Moral Values in European Patent Proceedings, Past and Future' (2006) 31 E.L. Rev 642.

16   See, most explicitly, Article 6 (ex Article 6 TEU) of the Consolidated Version of the Treaty on European Union, *Official Journal of the European Union*, C 115/13, 9.5.2008. Generally, see Sionaidh Douglas-Scott, 'The European Union and Human Rights after the Treaty of Lisbon' (2011) 11 *Human Rights Law Review* 645.

by the Strasbourg court to Contracting States where moral consensus is lacking, and that it fails to cohere with a certain vision of a regional association of communities of rights. It also invites some clarification of the very idea that legal and regulatory decision-making should be 'coherent'. After all, one very obvious thought is that the best defence of the CJEU is that the Court was simply articulating the agreed terms of the Directive and that such incoherence as there might be resides in the Directive itself. In the real world, we know that the political branch will generate legislation that will often involve compromises and accommodations that indicate a lack of coherent purpose. Indeed, in the case of the Directive, we know that it was all but lost in 1995 and that it was only because of compromise and accommodation that it was rescued—from a political perspective, perhaps an incoherent agreed Directive would seem better than no Directive at all. In any event, if we are to assess the activities of courts and legislators by reference to a standard of coherence, in the context of the pluralistic nation state democracies that collectively comprise the European Union, we need to work out just how much, and what kind of coherence, we can insist upon.

Given such a 'reality check', it is tempting to argue that, once a regulatory mind-set is adopted (such as that of the European Commission in its approach to creating a harmonised European marketplace), then instrumental reasoning takes over and the only question is whether the chosen means serve the regulatory ends.[17] Or, again, we might argue that we simply have to accept that, as legislation overtakes the case law, there will be some incoherence. After all, this is politics; compromises and accommodations have to be made; inevitably, there can be a lack of coherence. However, these concessions to instrumental rationality and to politics do not just give up on coherence; for many, these concessions involve giving up on the Rule of Law.[18] Moreover, if we are giving up on the ideal of coherence in relation to the normative parts of the regulatory environment, then we surely have no chance of maintaining it with the onset of technological management. So long as we are sticking with the Rule of Law, these are concessions to resist.

This chapter takes up these issues in six main parts. First the central provisions of the Directive and the ruling in *Brüstle* are laid out. Secondly, the charge made against the *Brüstle* decision, that it is formally incoherent, is reviewed and rejected. Thirdly, the charge that *Brüstle* is substantively incoherent relative to the European jurisprudence of human rights is considered; and one particular version of the charge is again rejected. Fourthly, three further charges of incoherence are suggested and very briefly reviewed. Fifthly, the aspiration of 'coherence' in legal and regulatory decision-making is itself

---

17 Compare Hugh Collins, *Regulating Contracts* (Oxford: Oxford University Press, 2002) 8: 'The trajectory of legal evolution alters from the private law discourse of seeking the better coherence for its scheme of principles to one of learning about the need for fresh regulation by observations of the consequences of present regulation.'

18 See Brian Z. Tamanaha (n 3).

analysed in an attempt to identify the types of coherence that we can plausibly use to evaluate the work of legislators, regulators and judges. Finally, the relevance of regulatory coherence in a context of technological management is outlined.

So far as the CJEU's decision in *Brüstle* is concerned, my conclusions are twofold: the first is that, while some charges of incoherence do not stick, there are others that might be more telling; and, the second is that, if there is a serious incoherence in European patent law, it is in the Directive itself rather than in the Court's ruling. More generally, the lesson for coherence and technological management is that all dimensions of the regulatory environment, both normative and non-normative, need to cohere with the constitutive values that inform the ideal of the Rule of Law and that underpin any sustainable compact between regulators and regulatees.

II The Directive and the ruling in *Brüstle*

For present purposes, we need not review the full range of the Directive.[19] So far as the dispute in *Brüstle* is concerned, the core provisions of the Directive are those in Article 6—comprising, in Article 6(1), a general moral exclusion and then, in Article 6(2), four specific exclusions—together with the underlying guidance given by Recital 38.

Article 6(1) of the Directive (in language that very closely resembles that of Article 53(a) of the European Patent Convention) provides:

> Inventions shall be considered unpatentable where their commercial exploitation would be contrary to *ordre public* or morality; however, exploitation shall not be deemed to be so contrary merely because it is prohibited by law or regulation.

Article 6(2) then provides for four specific exclusions that follow from the general exclusion in Article 6(1). Thus:

> On the basis of paragraph 1 [i.e. Article 6(1)], the following, in particular, shall be considered unpatentable:
>
> (a) processes for cloning human beings;
> (b) processes for modifying the germ line genetic identity of human beings;
> (c) uses of human embryos for industrial or commercial purposes;

---

19 For such a review, see Deryck Beyleveld, Roger Brownsword, and Margaret Llewelyn, 'The Morality Clauses of the Directive on the Legal Protection of Biotechnological Inventions: Conflict, Compromise, and the Patent Community' in Richard Goldberg and Julian Lonbay (eds), *Pharmaceutical Medicine, Biotechnology and European Law* (Cambridge: Cambridge University Press, 2000) 157.

> (d) processes for modifying the genetic identity of animals which are likely to cause them suffering without any substantial medical benefit to man or animal, and also animals resulting from such processes.

It is clear from the jurisprudence, and especially from *Commission v Italy*,[20] that there must be strict and unequivocal implementation of the exclusions in Article 6(2). It is also clear from Recital 38 of the Directive that the four particular exclusions listed are not exhaustive. According to Recital 38:

> Whereas the operative part of this Directive should also include an illustrative list of inventions excluded from patentability so as to provide referring courts and patent offices with a general guide to interpreting the reference to *ordre public* and morality; whereas this list obviously cannot presume to be exhaustive; whereas processes, the use of which offend against human dignity, such as processes to produce chimeras from germ cells or totipotent cells of humans and animals, are obviously also excluded from patentability.

Although the third clause of this Recital invites more than one interpretation, it certainly implies that human dignity is the key underlying value in both Article 6(1) and Article 6(2).

In response to the several questions referred to the CJEU, the key ruling handed down by the Court, expressed at the conclusion of the judgment, is as follows:

> Article 6(2)(c) of Directive 98/44 excludes an invention from patentability where the technical teaching which is the subject-matter of the patent application requires the prior destruction of human embryos or their use as base material, whatever the stage at which that takes place and even if the description of the technical teaching claimed does not refer to the use of human embryos.

---

20 Case C-456/03, [2005] ECR I-5335, [78] et seq. Interestingly, at [82], the Court draws on the obscure proviso in Article 6(1), first, to underline the point that prohibition of commercial exploitation by law or regulation does not entail exclusion from patentability, and then to insist that any possible uncertainty is removed by legislating for the Article 6(2) exclusions. However, the first of these points invites clarification because, on the face of it, it is formally incoherent for a regulator to prohibit the commercial exploitation of *x* but, at the same time, to permit the patenting of *x*: that is, in the ordinary way of things, prohibition of commercial exploitation of *x* does entail exclusion of *x* from patentability. Nevertheless, there might be contextual factors that resolve the apparent contradiction—for example, if the prohibition is in the nature of a moratorium for a limited period, then, in the particular setting, regulatees might understand that the red prohibitory signal in conjunction with the green patentability signal actually amounts to an amber signal (to proceed cautiously).

Accordingly, even though Brüstle was not responsible for the destruction of human embryos, even though there was some distance between the destruction of the embryos and Brüstle's use of the embryonic (base) materials, the invention was still seemingly excluded from patentability.[21]

While there was no doubt that Brüstle's base materials were derived from human embryos, the Bundesgerichtshof also sought guidance on the interpretation of 'human embryos' in Article 6(2)(c). Stated simply, how far does the definition of a 'human embryo' extend into the kinds of cells created and used by stem-cell researchers? Addressing this question, the CJEU ruled (at the conclusion of the judgment) that 'any human ovum after fertilisation, any non-fertilised human ovum into which the cell nucleus from a mature human cell has been transplanted, and any non-fertilised human ovum whose division and further development have been stimulated by parthenogenesis constitute a "human embryo".' Finally, in a further ruling, the CJEU referred back to the national court the question whether, in the light of scientific developments, 'a stem cell obtained from a human embryo at the blastocyst stage constitutes a "human embryo" within the meaning of Article 6(2)(c) of [the Directive]'.[22]

III Is the *Brüstle* decision formally incoherent?

The charge of formal incoherence starts by observing that the research into Parkinson's disease undertaken by Oliver Brüstle was perfectly lawful in Germany; and, we need no reminding that, in German law, the protection of human embryos is taken particularly seriously. Assuming that German law and EU law are sufficiently connected to be viewed as parts of one regional regulatory regime, the question posed by the objectors is this: how can it be coherent to permit Brüstle to carry out research that uses materials derived from human embryos and yet to deny a patent on the products of that research for just the reason that human embryonic materials were utilised? At first blush, it looks as though German regulators are showing Brüstle a green light while, at the same time, the CJEU is holding up a red light. Is this not a case of formal incoherence?

It is a good question but, as a first step, we need to specify the ways in which a regulatory or legal regime might suffer from formal incoherence. Then, we can cross-check this against the supposed incoherence of permitting research,

---

21 However, the Federal Court of Justice has subsequently ruled that Brüstle's patent does not involve the destruction of human embryos and that, in an amended form, the patent is valid. See Antony Blackburn-Starza, 'German court upholds Brüstle patent as valid' BioNews 684, available at www.bionews.org.uk/page_222080.asp (last accessed 4 November 2018).

22 The Federal Court of Justice has now ruled that, on their own, human embryonic stem cells are not capable of developing into a born human and, thus, should not be treated as a 'human embryo' (see Antony Blackburn-Starza, n 21).

the resulting processes or products of which are excluded from patentability; or, excluding patentability where the research is already permitted.

Minimally, the idea of a coherent legal or regulatory enterprise presupposes that the signals given to regulatees should not be formally contradictory.[23] If the signals take the form of prohibitions, permissions and requirements, then regulatory coherence entails that the signals should not indicate, for example, that x is both prohibited and permitted or that x is both prohibited and required. To be sure, at different levels within the hierarchies of a legal or regulatory order, ostensibly contradictory signals may be given—for example, a higher court may overrule an old precedent or reverse the decision of a lower court, or a legislature may enact a statute to repeal an earlier statute or to change the effect of a court decision. Within the legal order's own hierarchical rules, these apparent contradictions are resolved; the later signals are understood to supersede and replace the earlier ones. Similarly, the context might make it clear what a potentially conflicting combination of signals actually means. For example, in the particular context, regulatees might understand perfectly well that a combination of red and green lights is actually signalling a prohibition (the green light is to be ignored, being treated as misleading or mistaken or even a trap); or that the lights are signalling a permission (the red light is to be ignored, perhaps because it is attached to a law that is now, so to speak, a 'dead letter'); or that the conjunction of red and green lights amounts to some kind of cautionary amber signal.[24] What is clearly formally incoherent is the co-existence of red lights in competition with green lights.

That said, without contradiction, the law may signal that x is both permitted and encouraged; or that x is permitted but neither encouraged nor discouraged; or even that x is permitted but discouraged (for example, as with the regulation of smoking). For present purposes, the key point is that permission does not entail encouragement. In other words, it is not formally incoherent to signal permission but without also signalling encouragement.

While patent lawyers might pause over this proposition, it will surely seem plausible to Contract lawyers. In the common law, there are many transactions that are perfectly permissible under the background law but which are nevertheless not treated as enforceable on grounds of good morals or the public interest. For example, in English law, although for many years it has been perfectly lawful to place a bet on a horserace or some other sporting event in one of the thousands of betting offices that open their doors to the public, until a legislative change in 2005, gaming and wagering contracts

---

23 Compare the seminal analysis in Lon L. Fuller, *The Morality of Law* (New Haven: Yale University Press, 1969) (revised edition).
24 For one such amber signal, in the context of a moratorium, see n 20. For further combinations that signal both prohibition and encouragement but the formal incoherence of which is more apparent than real, see Maurice Schellekens and Petroula Vantsiouri, 'Patentability of Human Enhancements' (2013) 5 *Law, Innovation and Technology* 190.

were treated as unenforceable. In this way, Victorian morals cast a long shadow over the regulation of this class of transaction; but, the shadow did not amount to a prohibition so much as a refusal to encourage what in modern times had become permissible. The same story applies to transactions that violate Victorian sexual mores as well as more modern arrangements such as surrogacy and pre-nuptial agreements.[25] In all cases, background permissions are conjoined with a refusal to signal encouragement.[26]

Pulling this together, and mindful of the contextual considerations and caveats already mentioned, I suggest that the two general rules of formal coherence and incoherence are: (i) if x is prohibited, it is not coherent to signal that x is permitted or required, or that x is encouraged;[27] but (ii) if x is permitted, it is coherent to signal at the same time that x is (a) encouraged, or (b) neither encouraged nor discouraged, or (c) discouraged.

Applying this analysis to patent law, we must start by clarifying how patenting fits into the regulatory array. What is the function of making patents available for inventive work? Once patents have been granted, their function (as property entitlements) is to put the proprietor in a position to control the use of the invention (for a limited period of time). However, the pre-grant, and primary, function of patent law is not to control access to the invention so much as (i) to incentivise and encourage innovation that is in the public interest and (ii) to incentivise innovators putting their knowledge into the public domain.[28] It follows that, when patents are available, their intended (if not always their actual) role is to encourage innovators;[29] and when patents are excluded, as in *Brüstle*, the signal is one of either discouraging a particular kind of innovation or at least not encouraging it (neither encouraging nor discouraging it).[30]

---

25 For the latter, however, see now *Radmacher v Granatino* [2010] UKSC 42.

26 Away from the law, whether contracts or patents, we can find common examples of regulators signalling that an action is permitted but not encouraged, or even that it is discouraged. Moreover, I suspect that liberal-minded parents often signal something rather similar to their teenage children.

27 It is worth emphasising that this is subject to the various caveats in the text above and at n 24.

28 The extent to which patent and other IP rights operate in the public interest is, of course, a moot point: compare the critique in James Boyle, *The Public Domain* (New Haven: Yale University Press, 2008).

29 One of many problems with patents in practice is that a liberal policy on granting patents can lead to blockages for downstream researchers. Famously, see Michael A. Heller and Rebecca Eisenberg, 'Can Patents Deter Innovation? The Anticommons in Biomedical Research' (1998) 280 *Science* 698.

30 Yet is it not incoherent to permit research while prohibiting (on the grounds of violation of human rights or human dignity) IP *ownership* in relation to the products of that research (compare, e.g., Aurora Plomer, 'Patents, Human Dignity and Human Rights', in Christophe Geiger (ed), *Research Handbook on Human Rights and Intellectual Property* (Cheltenham: Edward Elgar, 2015) 479)? Certainly, to permit research without also incentivising

Where the background regulatory position is one of prohibition (if, for example, Brüstle's research work had been prohibited in Germany) it would be formally incoherent for regulators to treat the products of the prohibited activity as patentable. That would be signalling prohibition with encouragement; and, applying the first of the above rules of formal coherence, that is not formally coherent. However, it does not follow (as some critics seem to imply) that, unless the background regulatory position is one of prohibition, the only coherent position for patent law is to encourage the activity.[31] To repeat, applying the second of the above rules of formal coherence, it is formally coherent to conjoin a background permission with something other than encouragement. On this analysis, there is no formal incoherence between the background German legal permission in relation to Brüstle's research work and the CJEU's refusal to encourage such research.

IV Is the *Brüstle* decision substantively incoherent?

Even if *Brüstle* does not suffer from formal incoherence, can it withstand scrutiny substantively—and specifically in relation to the European jurisprudence of human rights?[32] After all, even if non-encouragement is formally coherent alongside a permission, it calls for explanation. Why not encourage potentially beneficial research that is permitted?

Defenders of *Brüstle* will have to concede that in the key cases at Strasbourg, the Court has held that human embryos (and fetuses) do not have rights under the Convention.[33] However, the inference that the European jurisprudence of human rights gives no support for discouraging research that makes use of human embryos is much more questionable. For example, defenders might point to Article 18 of the Convention on Human Rights and Biomedicine[34] as a signal that any use of human embryos as research tools should never be encouraged and that some uses should be positively discouraged. That said, as Aurora Plomer has pointed out, it is difficult 'to

---

it invites explanation. However, none of this affects the claim that *formal* coherence allows for permitting some act while declining to encourage it (by an IP incentive or otherwise).

31 See the range of criticisms of the Wisconsin Alumni Research Foundation case in Aurora Plomer and Paul Torremans (eds), *Embryonic Stem Cell Patents* (Oxford: Oxford University Press, 2009).

32 Compare the reading of moral exclusions in Deryck Beyleveld and Roger Brownsword, *Mice, Morality and Patents* (London: Common Law Institute of Intellectual Property, 1993).

33 *Evans v United Kingdom* (Application no. 6339/05) Grand Chamber, April 10, 2007; *Vo v France* (Application no. 53924/00) Grand Chamber, July 8, 2004.

34 Article 18 (1) provides that 'Where the law allows research on [human] embryos in vitro, it shall ensure adequate protection of the embryo'; and Article 18(2) prohibits the 'creation of human embryos for research purposes'.

read Article 18 as indicative of a European consensus that research destructive of human embryos is contrary to human dignity ...'.[35]

Be that as it may, the more important point is that the Strasbourg jurisprudence does no more than deny that human embryos hold rights *directly* under the Convention. This leaves open the possibility that Contracting States may grant *indirect* protection to human embryos, just as in many places protection is afforded to non-human animals. Provided that these indirect protections (motivated by the desire to do the right thing) are not incompatible with the rights directly recognised by the Convention—and, it should be noted, there is no Convention right that explicitly protects the interests of researchers in having access to human embryos for their base materials—then no regulatory incoherence arises.

We can put this point to the test in a straightforward way. Germany, because of its sensibilities about embryo protection, already places very considerable legal restrictions on researchers such as Oliver Brüstle. Notably, in German law, there is a general prohibition on the importation and use of human embryonic stem cells, from which derogation is permitted only where a number of restrictive conditions are met—for example, that the stem cells were sourced from embryos that were created for reproductive purposes but that have now become supernumerary. I take it that no one would suggest that these laws are incompatible with the European Convention on Human Rights. Suppose, though, that Germany took the restrictions a step further and prohibited the kind of research in which Brüstle was engaged. Would this violate any of Brüstle's rights under the Convention? If, despite heroic attempts to construct a supporting right,[36] the answer to this question is that it would not, then there is surely no way that we can accuse the CJEU of violating Brüstle's rights by merely excluding patents on his research. To repeat, the CJEU merely signals that it is contrary to the Directive to encourage this kind of research by use of the patent regime.

So, far from being substantively incoherent relative to the jurisprudence of European human rights, it is arguable that the decision in *Brüstle* is very much in line with Strasbourg, neither protecting human embryos against permissive national laws nor encouraging researchers to use human embryos. Moreover, if *Brüstle* violates some right of researchers that we construct out of the Convention, then a broad sweep of restrictive background law is likely to be even more seriously in violation.

This, however, might be thought to let the CJEU off too easily. To be substantively coherent, it might be insisted that the Court's position must

---

35 Aurora Plomer, 'After *Brüstle*: EU Accession to the ECHR and the Future of European Patent Law' (2012) 2 *Queen Mary Journal of Intellectual Property* 110, 132.

36 For an argument tapping into Article 1 of the 2001 Protocol to the Convention, which concerns the protection of property (and, by implication, intellectual property) rights, see Plomer (n 35) at 130–131.

not only be human rights compatible but compatible for sound human rights reasons. We can sketch two ways in which such comprehensive coherence might be argued for, one line of argument relying on precautionary reasoning, the other on considerations of comity.

First, human rights considerations might indicate that a precautionary approach should be taken with regard to the treatment of human embryos. This, in turn, might be argued for in two ways. One argument is that we simply cannot be confident about the moral status of the human embryo. We are confident that a human embryo is distinguishable from, say, a table and chairs, but how confident can we be that it is distinguishable from born humans? In the human rights jurisprudence a line is drawn between unborn and born humans; but why should we draw the line there? And, whatever characteristics that we think born humans have that are the basis for recognising them as direct holders of rights, can we be sure that unborn humans do not also have these characteristics? If we have got this wrong in relation to human embryos, we do them a terrible wrong when we use them for research purposes.[37]

The other precautionary argument is that instrumentalising human embryos might indirectly corrode respect for the rights of born humans.[38] In the different context of the well-known *Omega Spielhallen* case,[39] we find an example of this kind of precautionary reasoning. In paragraph 12 of the judgment, we read that:

> The referring court states that human dignity is a constitutional principle which may be infringed either by the degrading treatment of an adversary, which is not the case here, or by the awakening or strengthening in the player of an attitude denying the fundamental right of each person to be acknowledged and respected, such as the representation, as in this case, of fictitious acts of violence for the purposes of a game. It states that a cardinal constitutional principle such as human dignity cannot be waived in the context of an entertainment, and that, in national law, the fundamental rights invoked by Omega cannot alter that assessment.

---

37 Compare Deryck Beyleveld and Roger Brownsword, 'Emerging Technologies, Extreme Uncertainty, and the Principle of Rational Precautionary Reasoning' (2012) 4 *Law Innovation and Technology* 35.

38 Similar arguments might be offered for the protection of non-human animals: see Peter Carruthers, *The Animals Issue* (Cambridge: Cambridge University Press, 1992). And, for a succinct expression of the concern, see Sherry Turkle, *Alone Together* (New York: Basic Books, 2011) at 47: 'This is, of course, how we now train people for war. First we learn to kill the virtual. Then, desensitized, we are sent to kill the real.'

39 Case C-36/02 *Omega Spielhallen- und Automatenaufstellungs-GmbH v Oberbürgermeisterin der Bundesstadt Bonn* 2004 ECR 1–9609 (October 14, 2004).

No doubt, claims of this kind—to the effect that 'permitting x causes y', or 'permitting x increases the likelihood of y', or 'permitting x encourages y'—are highly contentious. However, where 'y' is of high value in a scheme of human rights thinking, there are precautionary reasons for at least taking a hard look at such claims.

Secondly, there is the idea of comity: where communities are morally divided, there is an argument that respect for different views justifies some finessing of the regulatory position in order to cause members the least moral distress. Arguably, this is a plausible reason for declining to enforce contracts that still deeply offend some moral sensibilities. Similarly, might it be argued that the CJEU in *Brüstle* was declining to encourage human embryonic stem cell research for reasons of comity? The basis of this decision would then be that the CJEU judged that those members of the human rights community who believe that a precautionary approach should be taken would be more offended by the encouragement of patentability than those members who do not take such a (precautionary) view would be offended by the Court's unwillingness to signal such encouragement.

Although such human rights arguments might have guided the thinking in the *Brüstle* case, all the surface indications are that the thinking of the CJEU is much more dogmatically dignitarian. Quite simply, where human embryos are used as research tools, human dignity is compromised and such research activities are not to be encouraged.[40] If the better interpretation of the European jurisprudence is that it is predicated on a liberal articulation of human dignity, then the former (human rights inspired) accounts have the better credentials; for, on this reading, the latter account would involve a deeper kind of regulatory incoherence.[41] It would mean that, although a human rights-compatible defence of the decision in *Brüstle* can be mounted, the CJEU is actually operating with a dignitarian ethic that is antithetical to human rights—and, relative to the more demanding test, it follows that the decision in *Brüstle* lacks comprehensive coherence.

V Three further charges of incoherence

The two charges of incoherence that I have considered do not exhaust the options available to those who question the coherence of the *Brüstle* decision. Let me sketch three further charges of incoherence. These are as follows: first, that the CJEU violates a Rule of Law standard of impartiality;

---

40 Similarly, two accounts might be given of the decision of the ECJ First Chamber in *Omega Spielhallen* (n 39).

41 In support of such a liberal rights-driven interpretation, I would rely on arguments derived from Alan Gewirth, *Reason and Morality* (Chicago: University of Chicago Press, 1978).

secondly, that its prescriptive approach fails to cohere with the margin of appreciation typically accorded to Contracting States where moral consensus is lacking; and, thirdly, that it fails to cohere with a certain vision of a regional association of communities of rights.

(i) The CJEU violates a Rule of Law standard of impartiality

There is no gainsaying that the CJEU should decide in accordance with the values of the Rule of Law. What this requires (and prohibits) in relation to adjudication depends, of course, on how one articulates the Rule of Law. I take it, though, that, on anyone's view, the Rule of Law demands that judges should stay neutral and impartial, in the sense that they should not side with a particular political view as such. Drawing on this minimal requirement, critics might detect a lack of impartiality—and, hence, incoherence—in the decision in *Brüstle*, because the Court (as the critics would have it) sides with the prohibitionists.

This charge of incoherence strikes me as particularly weak. To be sure, the outcome of the case is that the CJEU upholds the view of those who oppose patentability; and, insofar as the court 'takes sides', it is with those who argue for non-encouragement rather than those who argue for encouragement of this kind of research. Yet, given that a decision has to be made, the CJEU cannot avoid making a choice; if it had ruled *in favour of* patentability, it would have been open to just the same kind of charge (although it would now be the prohibitionists raising the objection). Even if we allow that excluding patents on Brüstle's research is generally in line with the prohibitionist view, the CJEU is not actually prohibiting anything; and, perhaps more importantly, its decision on patentability is not aligned with the views of the prohibitionists for that reason (that is, for the reason that it so aligns). Moreover, there is no suggestion that the CJEU takes into account considerations that are improper, in the sense that they compromise its impartiality and independence; and there is surely no way that such a charge can be substantiated.

We can compare comparable criticisms of the European Patent Office's (EBA's) ruling in the *WARF* case: If we side with the critics, we will say that the EBA should have noted that there is no agreement amongst members as to the morality of using human embryos for research purposes; that to exclude the patent on moral grounds would be to privilege the dignitarian views of those members who already have domestic prohibitions against the destruction of human embryos for research purposes; and that the EBA has no warrant for such partiality. But, of course, we might turn this argument on its head. If, in the absence of a common moral position amongst members, the EBA declines to exclude the patent on moral grounds, then it privileges the liberal view of those members that already have permissive domestic regimes with regard to human embryo research. Whichever way the EBA decides, its ruling will align with one or other of the rival constituencies; there is no third option which will enable it to stay neutral. However, the fact that the outcome is not neutral as between the positions advocated by rival political constituencies is not the same as saying that the EBA (or the CJEU) defected from the impartiality requirements of the Rule of Law.

Courts, especially appellate courts, make decisions all the time that are welcomed by some political groups and denounced by others. But, so long as the courts are applying themselves in good faith to the question of what the law means or requires in such cases, so long as the courts have not been corrupted, captured, or otherwise compromised by the political branch or by political interests, there is no breach of the Rule of Law.[42]

(ii) The CJEU's approach is out of line with the margin of appreciation typically given to Member States

In the jurisprudence of the ECHR, the doctrine of the margin of appreciation allows some room for Contracting States to interpret and apply the provisions of the Convention in their own way. Where questions, such as the moral status of the human embryo, are deeply contested, where members take many different views on the matter, the Strasbourg jurisprudence treats this pluralism as, so to speak, a sleeping dog that is best left to lie. If a Contracting State appears as a serious outlier, it is likely to be pulled back into the main band of difference;[43] but, in general, Strasbourg does not insist that all members take the same line.

By contrast, so the objection runs, in the *Brüstle* case, the CJEU demands that all members treat Article 6(2)(c) as excluding patentability on research products or processes that involve, or rely on, the proximate or remote destruction of human embryos. Accordingly, there is no room for manoeuvre in Article 6(2)(c) and this, the objectors maintain, fails to cohere with the general approach of giving some margin.

Is this a good objection? As I have said before, *Brüstle* does not in any way impinge on the background regulatory options that are available to Member States. Members states may prohibit, permit, or require the use of human embryos for research, indeed even require such use for industrial or commercial purposes. Patent exclusion notwithstanding, Member States may also find ways of supporting and encouraging human embryonic stem cell research through, for example, their science funding programmes or tax policies. After *Brüstle*, the impingement is purely and simply on the patentability of this

---

42 Compare the defence of appellate court judges against the accusation that they act inappropriately as 'legislators' in 'hard cases'; seminally, see Ronald Dworkin, *Taking Rights Seriously* (revised edition) (London: Duckworth, 1978).
43 For a case in point, see *S and Marper v United Kingdom* [2008] ECHR 1581.

kind of research; but that is the extent of the intervention. In the larger regulatory picture, where the margin of appreciation for Member States remains wide and where there are ways of incentivising stem cell research outside the patent regime, the Court's intervention might seem like something of an empty gesture. Accordingly, with such a minor impingement on the margin, this charge of incoherence does not look like a particularly strong objection.

There is, however, another angle to this objection. The point here is not so much that the CJEU requires all Member States to come unequivocally into line with the exclusion in Article 6(2)(c) but that it interprets the concept of a 'human embryo' very broadly. Following the advice of Advocate-General Bot,[44] the CJEU in the *Brüstle* case insists that, for the purposes of the Directive, there must be a common understanding of the term 'human embryo'; and, it will be recalled, (according to the CJEU) that understanding is to be specified broadly as covering 'any human ovum after fertilisation, any non-fertilised human ovum into which the cell nucleus from a mature human cell has been transplanted, and any non-fertilised human ovum whose division and further development have been stimulated by parthenogenesis ...'. Now, the objection here is that such a broad and inclusive reading, denying to Member States any leeway in their interpretation of 'human embryo', lacks coherence alongside the margin of discretion that is often given in Europe.

This objection might resonate with, among others, European consumer lawyers. To make a short detour, since the mid-1980s, the Community's principal strategy for harmonising the law of the consumer marketplace has been to rely on Directives. Until recently, the Directives have been measures of minimum harmonisation—that is, Directives leaving the Member States some discretion (or margin) to provide for stronger measures of consumer protection going beyond the minimum.[45] As a result, the landscape of

---

44 Fully aware of the many different views as to both the meaning of a human embryo and the degree to which such embryos should be protected. Advocate General Bot insists that the legal position as settled by the Directive is actually perfectly clear. Within the terms of the Directive, the concept of a human embryo must be taken as applying from 'the fertilisation stage to the initial totipotent cells and to the entire ensuing process of the development and formation of the human body' ([115]). *In themselves*, isolated pluripotent stem cells would not fall within this definition of a human embryo (because they could not go on to form a whole human body). For the Advocate General's Opinion see: http://curia.europa.eu/jurisp/cgi-bin/form.pl?lang=en&alljur=alljur&jurcdj=jurcdj&jurtpi=jurtpi&jurtfp=jurtfp&numaff=C-34/10&nomusuel=&docnodecision=docnodecision&allcommjo=allcommjo&affint=affint&affclose=affclose&alldocrec=alldocrec&docor=docor&docav=docav&docsom=docsom&docinf=docinf&alldocnorec=alldocnorec&docnoor=docnoor&docppoag=docppoag&radtypeord=on&newform=newform&docj=docj&docop=docop&docnoj=docnoj&typeord=ALL&domaine=&mots=&resmax=100&Submit=Rechercher (last accessed on 4 November 2018).

45 Examples include the earlier Directives on doorstep selling (85/577/EEC), consumer credit (87/102/EEC), package holidays (90/314/EEC), unfair terms (93/13/EC), and distance selling (97/7/EC).

consumer protection law across Europe is still quite variable; the minimum standards apply in all consumer markets but there are different degrees of national 'gold-plating' above the minimum. For suppliers who wish to extend their business across borders, the variation in national laws can operate as an obstacle to trade.[46] The margin of discretion, in other words, damages the unity of the market; instead of a single market, suppliers are faced with, so to speak, '27 [or more] mini-markets'.[47] In response, the Commission's more recent interventions—such as the Directive on Unfair Commercial Practices[48] and the Directive on Consumer Rights[49]—apply maximum harmonisation measures; the minimum standards now become the maximum; and the margin of discretion is eliminated.[50] For many consumer lawyers, this is a step too far, impinging on local control in respect of the fine-tuning of the balance of interests between suppliers and consumers.

In the context of consumer contract law, the question of how far the Commission needs to go in securing the unity of the market hinges on the importance that one attaches to cross-border trading. Generally, consumers prefer to shop close to home (although on-line consumption can change this); and critics of maximum harmonisation will ask whether such deep intervention into local control is proportionate to the regulatory aims. Why not, critics might ask, limit these maximum harmonising measures to cross-border contracts, leaving local consumer law within the discretion of the Member States?[51]

To return to patentability and the definition of a 'human embryo', we might ask in a similar fashion whether the view in *Brüstle* is appropriate relative to the aims of the Directive. According to Recital 5, the raison d'être for Directive 98/44/EC is that differences between the laws and practices of each member state 'could create barriers to trade and hence impede the proper functioning of the internal market'. Again, in *Commission v Italy*, the aim of the Directive is expressed as being 'to prevent damage to the unity of the internal market'.[52] Prima facie, this has some plausibility because venture

---

46 For evidence, see Stefan Vogenauer and Stephen Weatherill, 'The European Community's Competence to Pursue the Harmonisation of Contract Law—An Empirical Contribution to the Debate' in Stefan Vogenauer and Stephen Weatherill (eds), *The Harmonisation of European Contract Law* (Oxford: Hart, 2006) 105; and the supporting case for a proposed Regulation on a Common European Sales Law, COM/2011/0635 final.

47 Per Commissioner Kuneva, giving a public lecture on 'Transformation of European Consumer Policy' at Humboldt University, Berlin, 28 March 2008.

48 2005/29/EC.

49 2011/83/EU.

50 See, Roger Brownsword, 'Regulating Transactions: Good Faith and Fair Dealing' in Geraint Howells and Reiner Schulze (eds), *Modernising and Harmonising Consumer Contract Law* (Munich: Sellier, 2009) 87.

51 For one such example, see Christian Twigg-Flesner, *A Cross-Border-Only Regulation for Consumer Transactions in the EU: A Fresh Approach to EU Consumer Law* (New York: Springer, 2011).

52 *Commission v Italy* (n 20), [58].

capitalists will surely prefer to invest in small biotech start-ups where innovative and commercially exploitable processes or products are patentable. If Europe has many different patent regimes, investment (and the companies in which investments are made) will tend to be concentrated in those regulatory areas where patents are available. However, on closer inspection, it is apparent that there are two major flaws in this argument.

First, harmonising patent law falls a long way short of creating a level playing field. To be sure, patentability is one element in the regulatory environment for research and development in biotechnology; but the more fundamental elements are the background prohibitions and permissions. Even if patents are not excluded, investors will prefer to back research and development in those areas where the regulatory environment overall is most congenial. Accordingly, harmonising patent law prevents some damage to the unity of the internal market but does nothing to correct the most important differences between national laws. Within Europe, there will still be room for a significant element of regulatory arbitrage.

Secondly, the idea that investment in biotechnology should not be impeded by regulatory barriers that might vary from one part of the market to another makes good sense where (as was the Commission's overall intention) the policy is to encourage investment in biotechnology. Hence, the harmonising of patent law to improve the chances of investment crossing borders makes some sense where the thrust of the Directive is to affirm the patentability of new biotechnology, as is the case in relation to sequencing work around the human genome. However, those who argue that *Brüstle* is incoherent in disallowing some definitional margin are surely right in thinking that it is far from obvious that insisting upon a broad harmonised definition of an embryo makes the same kind of sense where the provisions in question are exclusions. Having the same laws excluding patents where human embryos are used by researchers simply means that investment will be inhibited across the region; and, defining human embryos broadly for exclusionary purposes aggravates matters by extending the scope of the inhibition. The problem with the unity of the market argument (as applied to the exclusions from patentability) is that it simply does not fit with the view that the regulatory tilt of the Directive is to encourage research and development in modern biotechnologies (which surely was the Commission's dominant regulatory purpose).[53]

---

53 In the post-*Brüstle* case of *International Stem Cell Corporation v Comptroller General of Patents, Designs and Trade Marks* (Case C-364/13), the referring court (the High Court of England and Wales) remarked that the broad definition of a human embryo, excluding parthenotes from patentability 'does not strike [an acceptable] balance between, on the one hand, research in the field of biotechnology which is to be encouraged by means of patent law and, on the other hand, respect for the fundamental principles safeguarding the dignity and integrity of the person' (see [19]). Responding to this view, the Court modified and clarified its position in *Brüstle*, ruling (at [38]) that 'Article 6(2)(c) of Directive 98/44

By a somewhat meandering route, therefore, we arrive at the conclusion that the elimination of the margin of discretion with regard to the definition of 'human embryo' might be out of line with the general strategy of minimum harmonisation; but, more tellingly perhaps, we conclude that broadening the scope of the exclusion, far from securing the desired unity of the market, fails to cohere with the fundamental purpose of the Directive (namely, enabling investment in modern biotechnology).

(iii) The decision fails to cohere with a certain vision of a regional association of communities of rights.

No doubt, there are many visions of communities that take rights seriously. However, as we have argued in Part One, there is a rationally defensible view that starts with building and protecting a secure platform for agency (this platform being constituted by the generic conditions that enable agents freely to choose and to pursue their own plans and projects) and then licenses communities to articulate their rights commitments in their own way.[54] This is a vision of each community of rights giving its best interpretation of its rights commitments. Within such communities there are many potential points of disagreement—for example, about the interpretation, scope, and ranking of particular rights, about the priorities where rights conflict, and about the qualifying conditions for membership of the community of rights (and, concomitantly, how marginal or prospective members are to be treated). In all these cases, a moral community needs to have very good reasons to surrender local control over the decisions that need to be made. For good reason, a community of rights will be reluctant to cede control over the best interpretation of its rights commitments; members of such communities need to believe that they are doing the right thing.

In this context, how strong a reason is given by the need to harmonise trade rules? Without doubt, communities cannot expect to trade in regional or global clubs without some loss of control. As Chief Justice Burger put it in *Bremen v Zapata Off-Shore Co*,[55] the expansion of business and industry is unlikely to be encouraged if nation states 'insist on a parochial concept that all disputes must be resolved under [their local] laws and in [their] courts'; it is simply not possible to 'have trade and commerce in world markets and

---

must be interpreted as meaning that an unfertilised human ovum whose division and further development have been stimulated by parthenogenesis does not constitute a 'human embryo', within the meaning of that provision, if, in the light of current scientific knowledge, that ovum does not, in itself, have the inherent capacity of developing into a human being, this being a matter for the national court to determine.'

54 See Roger Brownsword, *Rights, Regulation and the Technological Revolution* (Oxford: Oxford University Press, 2008).
55 407 US 1 (1972).

international waters exclusively on [local] terms ...'.[56] However, it is one thing agreeing to harmonise trade rules that are largely morally neutral, it is quite another to surrender control over important moral decisions. To insist on parochialism in such important matters is entirely appropriate for a community that takes its rights commitments seriously. From this perspective, the (economic) unity of the European market looks like an inadequate reason for a community of rights to hand over the right to decide on such an important matter as the moral status of a human embryo.

Granted, as I have emphasised several times, the *Brüstle* decision touches and concerns only patentability, not the background regulatory prohibitions or permissions. For moral communities to cede decision-making power on patentability is nowhere near as serious as ceding control over the background prohibitions or permissions. The fact remains, though, that there is no good reason for making even such a small concession. Accordingly, if we track back to the seat of the concession, we will say that the Directive should have been drafted in such a way that reserved to each member state the right to apply its best moral judgment in setting its own regulatory environment for the use of human embryos by researchers. And, relative to the aspirations of communities of rights, we will conclude that the incoherence ultimately lies, not so much in the *Brüstle* decision, as in the deal done that underpinned the Directive itself.

VI Regulatory coherence and legal coherence

For private lawyers, it goes without saying that the body of doctrine should be coherent. For example, we expect the rules of contract law to form a coherent code. So, for many years, contract lawyers of my generation asked: if 'freedom of contract' is the governing principle, how do we explain the protection of consumer contractors, the hard look at standard forms, and the policing of 'unfair' terms? To satisfy the ideal of coherence, it eventually had to be acknowledged that consumer transactions were now regulated by a separate scheme.[57] In the same way, we might wonder how well the *WARF* case and *Brüstle* cohere with the previous jurisprudence which, for the most part, holds that the moral exclusion is triggered only where it would be inconceivable to grant a patent or where there is an overwhelming consensus that it would be immoral to do so.[58] In both *WARF* and *Brüstle* the fabric of patent

---

56 *Ibid.*, at 9.

57 Generally, see Roger Brownsword, *Contract Law: Themes for the Twenty-First Century* (Oxford: Oxford University Press, 2006). See, further, Chapter Eleven.

58 See, e.g., Plant Cells/PLANT GENETIC SYSTEMS Case T 0356/93; and Edward Armitage and Ivor Davis, *Patents and Morality in Perspective* (London: Common Law Institute of Intellectual Property, 1994).

law seems to be badly torn; and, for the sake of coherence, lawyers need to work hard at stitching the pieces together again.

Anticipating the argument in Part Three of the book, we can say that, in relation to both contracts and patents, the premise of doctrinal coherence is severely disrupted once the Commission adopts 'a regulatory mind-set' focused on harmonising the marketplace. From such a regulatory perspective, the only idea of coherence is an instrumental one: provided that a regulatory intervention works relative to the regulator's purposes, then coherence is satisfied. Crucially, a regulator need not cross-check that the principle under-lying the latest intervention is coherent with the principle underlying some other intervention; regulators are simply directing regulatees.[59] Of course, it is important that regulatory interventions do not interfere with one another's purposes (in the way, for example, that there are concerns about competition law interfering with the incentives for innovation offered by patent law);[60] but, there is no problem of incoherence if the regulators take a pragmatic approach to the effective pursuit of their policies.

Moreover, even without the Brussels agenda, private lawyers must accept that, as legislation overtakes the case law, there will be some incoherence. Rival political constituencies seek to influence the policy and text of the law; and the outcome sometimes falls short relative to the ideal of coherence.[61] Directive 98/44/EC is surely a perfect example. On the one hand, the Directive secures the interests of the scientific research community by declar-ing that innovative work around the human genome (including work that replicates naturally occurring sequences) is in principle patentable.[62] On the other hand, Article 6 of the Directive sets limits on patentability that reflect some of the concerns of those various constituencies that oppose develop-ments in modern biotechnologies (especially the use of human embryos as research tools).[63]

In the light of arguments sketched in this chapter, what should we make of the apparent (and possibly inevitable) decline of coherence in regulatory Europe?

First, we should hold on to the European vision of a regional associa-tion of communities of rights. Perhaps we do not need to cross-check for coherence but we certainly need to check upwards to the fundamental rights

---

59 This is drawn out very clearly in Collins (n 17). See, too, Edward L. Rubin, 'From Coher-ence to Effectiveness' in Rob van Gestel, Hans-W Micklitz, and Edward L. Rubin (eds), *Rethinking Legal Scholarship* (New York: Cambridge University Press, 2017) 310.
60 For an extensive jurisprudence, see Richard Whish and David Bailey, *Competition Law* 7th edn (Oxford: Oxford University Press, 2012) Ch.19.
61 For an excellent case-study, albeit in a very different context, see Andrew Murray, 'The Reclassification of Extreme Pornographic Images' (2009) 72 *Modern Law Review* 73.
62 See, e.g., Recitals 20 and 21, and Article 5(2).
63 See, e.g., Recitals 37, 38 and 42.

commitments of the community. In all its phases, the regulatory environment should be compatible with respect for human rights and underlying human dignity.

Secondly, trade imperatives should not be permitted to 'collateralise' or displace or compromise moral imperatives (whether these are imperatives that relate directly to human rights or indirectly so as in the case of environmental protection).[64] Communities of rights, in a regional association of such communities, should value their (relative) moral autonomy. Moral subsidiarity is not simply a recognition that, on some matters, local decision-making is more efficient; the point is that moral communities need to give their commitments their own best interpretation. Any weakening on these ideals is again a worrying case of incoherence.

Thirdly, the values that seem to underpin Recital 38 and Article 6 of the Directive reflect a conservative conception of human dignity. With the development of modern biotechnologies, such dignitarian values have attracted considerable support.[65] However, there is a tension between this conception of human dignity and the autonomy-based conception that underlies the modern articulation of human rights.[66] For a community of rights to permit a conservative, duty-based, conception of human dignity to intrude through human rights gateways in the law is clearly incoherent. In this light, the moral exclusions of the Directive and the decision in *Brüstle* seem to be motivated by a value that does not cohere with human rights commitments.[67]

Finally, to underscore the previous point, it should be appreciated that there is an important difference between Europe as a region (and project) of 'closed' pluralism, where human rights (together with 'human dignity as empowerment') represent the fundamental values, and Europe as a region (and project) of 'open' pluralism, where rights-based values are in competition with conservative dignitarianism.[68] In the latter, the moral bonds are

---

64 Compare, Sheldon Leader, 'Collateralism' in Roger Brownsword (ed), *Human Rights* (Oxford: Hart, 2004) 53.

65 See, Roger Brownsword, 'Human Dignity, Human Rights, and Simply Trying to Do the Right Thing' in Christopher McCrudden (ed), *Understanding Human Dignity* (Oxford: Proceedings of the British Academy and Oxford University Press, 2013) 470.

66 Deryck Beyleveld and Roger Brownsword, *Human Dignity in Bioethics and Biolaw* (Oxford: Oxford University Press, 2001).

67 For an even more worrying example, see the judgment of the Grand Chamber at the European Court of Human Rights in the case of *SH v Austria* (application no. 57813/00) November 3, 2011. There, the Court seems to permit Austria to rely on conservative dignitarian values to justify its restrictive IVF laws (prohibiting third-party donation of gametes). For critique, see Rosamund Scott, 'Reproductive Health: Morals, Margins and Rights' (2018) 81 *Modern Law Review* 422.

68 For the implications of the difference between 'open' and 'closed' value pluralism, see Roger Brownsword, 'Regulating the Life Sciences, Pluralism, and the Limits of Deliberative Democracy' (2010) 22 *Singapore Academy of Law Journal* 80; and 'Framers and Problematisers: Getting to Grips with Global Governance' (2010) 1 *Transnational Legal Theory* 287.

relaxed; tolerance implies more than acceptance of margins of difference—there is an acceptable heterogeneity; and the demand for regulatory coherence is correspondingly weaker.

VII Regulatory coherence and technological management

Where technological management is employed, the courts will have an important role to play in reviewing the legality of any measure that is challenged relative to the authorising and constitutive rules. Granted, if there are apparent contradictions (horizontally) within the body of rules, the courts might also have to engage with questions of doctrinal coherence. However, the principal work for the courts will be to assess the coherence of particular instances of technological management relative to the ideal of the Rule of Law. In other words, the ideal of coherence is not monopolised by private law, it is also material to public and constitutional law, and beyond that 'new coherence' reaches through to the maintenance of the essential conditions for any community of human agents.

Recalling our discussion in Chapter Four, a renewed ideal of coherence should start with the paramount responsibility of regulators, namely, the protection and preservation of the commons. All regulatory interventions should cohere with that responsibility. This means that the conditions for human existence and the context for flourishing agency should be respected. Moreover, as we have emphasised, it also means that technological management should not be employed in ways that compromise the context for moral community. Hence, the Courts should always ask whether a challenged act of technological management coheres with this limiting principle. In order to do this, the Courts will develop a jurisprudence that identifies the key considerations that regulators should take into account before they adopt technological management. They might also develop some presumptions (for example, relating to the purposes for which technological management is being used) that affect the intensity of the review.

Next, measures of technological management should cohere with the particular constitutive values of the community—such as respect for human rights and human dignity, the way that non-human agents are to be treated, and so on—and its particular articulation of the Rule of Law. For example, the House of Lords Select Committee on Artificial Intelligence has recommended that a Code for the use of AI should be developed around the following five overarching principles:

(1) Artificial intelligence should be developed for the common good and benefit of humanity.
(2) Artificial intelligence should operate on principles of intelligibility and fairness.

(3) Artificial intelligence should not be used to diminish the data rights or privacy of individuals, families or communities.

(4) All citizens have the right to be educated to enable them to flourish mentally, emotionally and economically alongside artificial intelligence.

(5) The autonomous power to hurt, destroy or deceive human beings should never be vested in artificial intelligence.[69]

Although these principles are not presented as constitutive, they do speak to the kind of relationship that is envisaged between smart machines and humans. No doubt, the courts will face many challenges in developing a coherent account of these principles but their principal task will be to ensure that particular instances of technological management cohere with, rather than abuse, these or similar principles.

There will also be challenges to technological management on procedural grounds. Once again, there will be work for the courts. Where explicit procedures are laid out for the adoption of technological management, the courts will be involved in a familiar reviewing role. However, there might also be some doctrinal issues of coherence that arise—for example, where it is argued that the explicit procedural requirements have some further procedural entailments; or where the courts, having developed their own implicit procedural laws (such as a practice raising a legitimate expectation of consultation), find that the body of doctrine is not internally coherent.

Coherence might be an ideal that is dear to the hearts of private lawyers but, in an era of technological management, it is once coherence is brought into the body of public law that we see its full regulatory significance. Regulation, whether normative or non-normative, will lack coherence if the procedures or purposes that accompany it are out of line with the authorising or constitutive rules that take us back to the Rule of Law itself; and, regulation will be fundamentally incoherent if it is out of line with the responsibility for maintaining the commons. In short, we can continue to treat coherence as an ideal that checks backwards, sideways, and upwards; but, the re-imagination of this ideal necessitates its engagement with both the full range of regulatory responsibilities and the full repertoire of regulatory instruments.

VIII Conclusion

The CJEU's decision in the *Brüstle* case finds itself at the centre of a more general debate about the coherence of the European regulatory environment. Although several of the charges of incoherence that have been made in relation to the CJEU's decision do not stick, there may well be others that

---

69 Report on *AI in the UK; ready, willing and able?* (Report of Session 2017–19, published 16 April 2017, HL Paper 100) at para 417: available at publications.parliament. uk/pa/ld201719/ldselect/ldai/100/10007.htm#_idTextAnchor025 (last accessed 4 November 2018).

have better prospects. However, it is tempting to think that the most serious examples of incoherence are to be found in the Directive itself—in the ceding of moral judgment for the sake of regional trade and the opening of the door to a conservative dignitarian ideology. From the perspective of the many communities of rights that have signed to the European Convention on Human Rights and that populate the European Union, these are acts that lack coherence.

We can place these concluding remarks within the context of a more general concern, especially amongst private lawyers, about the prospects for legal and regulatory coherence. Traditionally, private lawyers set the bar for coherence rather high. As transactions become more heavily 'regulated', coherence seems to be less of a desideratum; what matters is that regulatory interventions are effective relative to their purposes. With the development of a multi-level regime of law and regulation in Europe, coherence seems even more elusive. Nevertheless, even though pragmatists might want to immunise legislation and regulation against anything more than charges of instrumental incoherence, that is not rational where there are fundamental governing values—it is simply not coherent to act on the basis that the unity of the market must be achieved at all costs. Europe is committed to overarching values of respect for human rights and human dignity;[70] such fundamental commitments always govern instrumental considerations, both as to the ends pursued and the means employed; and they are the essential tests of the coherence of the regulatory environment whether the spotlight is on contract law or patent law.

Finally, although the discussion of coherence in this chapter has been at some distance from technological management, it is clear that the former is far from irrelevant to the latter. To satisfy the ideal of coherence, technological management must, above all, be compatible with the preservation of the commons; and, to satisfy the Rule of Law, technological management must cohere with the background constitutive and authorising rules that form part of the compact between governments and citizens—and, in the context of fundamental politico-legal commitments in Europe, this means that technological management must cohere with both human rights and underlying respect for human dignity.[71]

---

70 Explicitly so in Article 2 of the Treaty on the European Union (the Lisbon Treaty, 2007), according to which the Union is founded on 'the values of respect for human dignity, freedom, democracy, equality, the rule of law and respect for human rights, including the rights of persons belonging to minorities.' Article 2 continues by emphasising that these 'values are common to the Member States in a society in which pluralism, non-discrimination, tolerance, justice, solidarity and equality between women and men prevail.'

71 For more on 'coherentist' thinking and its relationship with 'regulatory-instrumentalist' and 'technocratic' mind-sets, see Chapter Eight and subsequent chapters.

7

THE LIBERAL CRITIQUE OF COERCION

Law, liberty and technology

I Introduction

In the nineteenth century, John Stuart Mill famously laid down one of the cornerstones of liberal values by declaring that the use of 'coercion' (in the sense of force or compulsion using threats, and whether applied through legal, moral, or social norms and their sanctions) cannot be justified unless the target conduct is harmful to others. The prevention of harm to others is a necessary condition before command and control regulation can be legitimately employed; the prevention of harm to oneself, physical or moral, is never a sufficient reason.[1] In the twenty-first century, with the rapid emergence of new technologies, there are a number of questions that we might ask about the relationship between law and liberty;[2] and, with technological management in the mix, we need to understand how liberal values might be impacted by this non-normative form of regulatory control.

One of the headline questions is whether new technologies enhance or diminish our liberty. On the face of it, new technologies offer human agents new tools; new ways of doing old things and new things to do. With each new tool, there is a fresh option—with each 'app', so to speak, there is a further option—and, seemingly, with each option there is an enhancement of, or an extension to, human liberty. At the same time, however, with some

---

1 J.S. Mill, 'On Liberty', in J.S. Mill, *Utilitarianism* (ed Mary Warnock) (Glasgow: Collins/ Fontana, 1962, first published 1859).

2 For reflections on the balance between liberty and legal restriction (by rule rather than by technological management), see Norman Anderson, *Liberty, Law and Justice* (London: Stevens and Sons, 1978).

new technologies and their applications, we may worry that the price of a short-term gain in liberty is a longer-term loss of liberty;[3] or we may be concerned that whatever increased security comes with the technologies of the 'surveillance society' it is being traded for a diminution in our political and civil liberties.[4]

In this chapter, we respond in three steps to the general question whether new technologies impact positively or negatively on the liberty of individuals; and, at the same time, we can assess the conditions that liberals should set for the legitimate use of technological management. First, we propose a two-dimensional conception of liberty that covers both the paper options that the rules permit and the real options that are available to people. This is important for an appreciation of the options that are available to individuals both under the rules and in actual practice; and it is absolutely crucial if we are to assess the significance of technological management, where the impact is on real, not paper options. Secondly, the Millian principles are sketched—that is, the categorical exclusion of the use of coercive rules for paternalistic purposes and the stipulation of 'harm to others' as a threshold condition for the use of coercive rules—and their limitations (some more commonly rehearsed than others) are indicated. Thirdly, the use of technological management within the criminal justice system and for general health and safety purposes is assessed relative to liberal principles.

Broadly speaking my conclusions are that, even if technological management is not, strictly speaking, an example of coercion by threat—because it is an even stronger form of forcing or compelling action—liberals simply cannot afford to ignore it as a regulatory strategy; that the liberal values that set conditions for the justified use of the criminal law should be applied to the use of technological management for criminal justice purposes; and that, while liberalism is under-developed in relation to general health and safety regulation, its values need to be articulated and brought to bear on the use of technological management for such regulatory purposes. In short, if we are to re-imagine liberal values in an age of technological management, one of the priorities is to shake off the idea that brute force and coercive rules are the most dangerous expressions of regulatory power; the regulatory power to limit our practical options may be much less obvious but no less

---

3 See, e.g., Franklin Foer, *World Without Mind* (London: Jonathan Cape, 2017) 231, claiming that 'Our faith in technology is no longer fully consistent with our belief in liberty'; Andrew Keen, *The Internet is Not the Answer* (London: Atlantic Books 2015); Siva Vaidhyanathan, *The Googlization of Everything (And Why We Should Worry)* (Berkeley: University of California Press 2011); and, Jonathan Zittrain, *The Future of the Internet* (London: Penguin 2009).

4 David Lyon, *Surveillance Society* (Buckingham: Open University Press 2001); and Zygmunt Bauman and David Lyon, *Liquid Surveillance* (Cambridge: Polity Press 2013).

dangerous. Power, as Steven Lukes rightly says, 'is at its most effective when least observable'.[5]

II Liberty

From the many different and contested conceptions of liberty,[6] we can start by employing the Hohfeldian characterisation of what it is for A, in relation to B, to have a mere liberty to do x.[7] Following Hohfeld, we should treat A as having a liberty to do x (some act) if, relative to some specified normative code, the doing of x by A is neither required nor prohibited, but is simply permitted. For A, in relation to B, the liberty to do x is an option; whether A does x or does not do x, A does no wrong relative to B. This conception would allow for the possibility that relative to some legal orders, there is a liberty to do x but not so according to others—for example, whereas, relative to some legal orders, researchers have a liberty to use human embryos for state-of-the-art stem cell research, relative to others they do not. Equally, we might arrive at different judgments as to the liberty to do x depending upon whether our reference point is a code of legal, moral, religious, or social norms—for example, even where, relative to a particular national legal order, researchers may have a liberty to use human embryos for stem cell research, relative to, say, a particular religious or moral code, they may not. In this way, as we explained in our earlier remarks about the tensions within and between regulatory environments, we can find that, within some regulatory spaces, there are quite different (and contradictory) signals as to whether or not A has the liberty to do x.

Applying such a conception of liberty to emerging technologies, we might map the pattern of normative optionality. If we stick simply to legal codes, we will find that, in many cases the use of a particular technology—for example, whether or not to use a mobile phone or a tablet, or a particular app on the phone or tablet, or to watch television—is pretty much optional; but, in some communities, there may be social norms that almost require their use or, conversely, prohibit their use in certain contexts (such as the use of mobile phones at meetings, or in 'quiet' coaches on trains, or at the family dinner table). Where we find divergence, further questions concerning the regulatory environment will be invited. Is the explanation for the divergence, perhaps, that different expert judgments are being made about the safety of a technology or that different methodologies of risk assessment are being employed (as seemed to be the case with GM crops), or does the difference

---

5 Steven Lukes, *Power: A Radical View* (2nd edn) (Basingstoke: Palgrave Macmillan, 2005) 1.
6 Joseph Raz, *The Morality of Freedom* (Oxford: Clarendon Press, 1986); and Ronald Dworkin *Justice for Hedgehogs* (Cambridge, Mass.: Harvard University Press 2011).
7 Wesley N. Hohfeld, *Fundamental Legal Conceptions* (New Haven: Yale University Press 1964).

go deeper, to basic values, to considerations of human rights and human dignity (as was, again, one of the key factors that explained the patchwork of views concerning the acceptability of GM crops)?[8]

For comparatists, as well as for sociologists and philosophers of law, an analysis of this kind might have some attractions. Moreover, as the underlying reasons for different normative positions are probed, we might find that we can make connections with familiar distinctions drawn in the liberty literature, most obviously with Sir Iasaiah Berlin's famous distinction between negative and positive liberty.[9] In Berlin's terminology, where a State respects the negative liberty of its citizens, it gives them space for self-development and allows them to be judges of what is in their own best interests. By contrast, where a State operates with a notion of positive liberty, it enforces a vision of the real or higher interests of citizens even though these are interests that citizens do not identify with as being in their own self-interest. To be sure, this is a blunt distinction.[10] Nevertheless, where a State denies its citizens access to a certain technology (for example, where the State filters or blocks access to the Internet, as with the great Chinese firewall) because it judges that it is not in the interest of citizens to have such access, then this contrasts quite dramatically with the situation where a State permits citizens to access technologies and leaves it to them to judge whether it is in their self-interest.[11] For the former State to justify its denial of access in the language of liberty, it needs to draw on a positive conception of the kind that Berlin criticised; while, for the latter, if respect for negative liberty is the test, there is nothing to justify.

According to such a view of liberty, the relationship between liberty and particular technologies or their applications, will depend upon the position taken by the reference normative code. As some technology or application moves onto the normative radar, a position will be taken as to its permissibility and, in the light of that position, we can speak to how liberty is impacted. However, this analysis is somewhat limited. It does not speak at all to the

---

8 Sheila Jasanoff, *Designs on Nature: Science and Democracy in Europe and the United States* (Princeton: Princeton University Press, 2005); Maria Lee, *EU Regulation of GMOs: Law, Decision-making and New Technology* (Cheltenham: Edward Elgar, 2008); and Naveen Thayyil, *Biotechnology Regulation and GMOs: Law, Technology and Public Contestations in Europe* (Cheltenham: Edward Elgar, 2014).

9 Isaiah Berlin, 'Two Concepts of Liberty' in Isaiah Berlin, *Four Essays on Liberty* (Oxford: Oxford University Press, 1969).

10 See, Gerald C. MacCallum, Jr., 'Negative and Positive Freedom' (1967) 76 *The Philosophical Review* 312; and C.B. Macpherson, *Democratic Theory: Essays in Retrieval* (Oxford: Clarendon, 1973).

11 Brian W. Esler, 'Filtering, Blocking, and Rating: Chaperones or Censorship?' in Mathias Klang and Andrew Murray (eds), *Human Rights in the Digital Age* (London: The Glass-House Press, 2005) 995; and Jack Goldsmith and Tim Wu, *Who Controls the Internet?* (Oxford: Oxford University Press, 2006).

way in which normative (paper) liberties relate to the (real) actual options available to individuals—for example, it does not speak to concerns about the so-called 'digital divide'; and, critically for our purposes, because liberty is locked in the normative dimension of the regulatory environment, it does not help us to understand the impact of technological management on real options.

It follows that, if we are to enrich this account, we need to employ a conception of liberty that draws on not only the formal normative position but also the practical possibility of A doing x. For example, if we ask whether we have a liberty to fly to the moon or to be transported there on nano-technologically engineered wires, then relative to many normative codes we would seem to have such a liberty—or, at any rate, if we read the absence of express prohibition or requirement as implying a permission, then this is the case. However, given the current state of space technology, travelling on nanowires is not yet a technical option; and, even if travelling in a spacecraft is technically possible, it is prohibitively expensive for most persons. But, who knows, at some time in the future, human agents may be able to fly to the moon in fully automated spacecraft in much the way that they will be travelling along Californian freeways in driverless cars.[12] In other words, the significance of new technologies and their applications is that they present new technical options and, in this sense, they expand the practical liberty of humans—or, at any rate, the real liberty of some humans—subject always to normative codes reacting (in a paper liberty-restricting manner) by prohibiting or requiring the acts in question.

Equipped with this combination of the normative and the practical dimensions of liberty, we have more scope for our discussion. On this approach, for example, we can say that, before the development of modern technologies for assisted conception, couples who wanted to have their own genetically related children might have been frustrated by their unsuccessful attempts at natural reproduction. In this state of frustration, they enjoyed a 'paper' liberty to make use of assisted conception because such use was not prohibited or required; but, before reliable IVF and ICSI technologies were developed, they had no real practical liberty to make use of assisted conception. Even when assisted conception became available, the expense involved in accessing the technology might have meant that, for many human agents, its use remained only a paper liberty.

These remarks suggest that we should employ a two-dimensional concept of liberty that incorporates both the normative coding(s) and the practical possibility of accessing and using the technology. So, to take another example, if the question concerns the liberty of prospective file-sharers to share their music with one another, then normative coding will give mixed

---

12 Eric Schmidt and Jared Cohen, *The New Digital Age* (New York: Alfred A. Knopf, 2013).

responses: according to many copyright codes, file-sharing is not permitted and, thus, there is no liberty to engage in this activity; but, according to the norms of 'open-sourcers' or the social code of the file-sharers, this activity might be regarded as perfectly permissible. However, the practical possibilities will not necessarily align with a particular normative coding: when the normative coding is set for prohibition, it might still be possible in practice for some agents (indeed, millions of agents) to file-share;[13] and, when the normative coding is set for permission, some youngsters might nevertheless be in a position where it is simply not possible to access file-sharing sites. In other words, normative options do not necessarily correlate with practical options, and vice versa.[14] Finally, if technological management is used to prevent file-sharing, this may or may not align with the normative coding; but, unless file-sharers are able to find a workround for the technological fix, there will be no real practical option to file-share.

III The Millian principles

Millians seek to defend liberty, to preserve (competent) A's option to do x, by setting out two supposedly simple constitutive principles for the justified use of coercion (whether in support of legal, moral, or social rules). These principles are: first, that the paternalistic use of coercive measures to prohibit an option for the sake of the individual's own good (physical or moral) should be categorically excluded; and, secondly, that coercive rules should not be used to prevent the doing of some act unless the doing of that act is plainly harmful to others. Hence, coercion should never be applied to restrict (competent) A's option to do x simply because this is judged to be in A's best interest; and it is only where the doing of x is harmful to others that coercive restriction may be justifiably applied to A.

Before we attempt to apply these principles to the use of technological management, we need to be aware of the obvious limitations of these principles and try to express them in a way that engages with the full range of Twenty-First Century regulatory instruments.

---

13 See *MGM Studios Inc v Grokster Ltd* 125 S.Ct 2764 (2005); 545 US 913 (2005). There, the evidence collected by MGM (the copyright holders) indicated that some 90% of the files shared by using Grokster's software involved copyright-infringing material. As Justice Souter, delivering the Opinion of the Court, put it, '[Given that] well over 100 million copies of the software in question are known to have been downloaded, and billions of files are shared across the ... networks each month, the probable scope of copyright infringement is staggering' (at 923).

14 It should also be noted that where the law treats file-sharing as impermissible because it infringes the rights of IP proprietors, there is a bit more to say. Although the law does not treat file-sharing as a liberty, it allows for the permission to be bought (by paying royalties or negotiating a licence). Again, however, for some groups, the price of permission might be too high and, in practice, they do not enjoy this conditional liberty.

(i) The categorical exclusion of paternalistic reasons for the use of coercive rules

Millians draw a distinction between, on the one hand, those humans who have the capacity to judge what is in their own best interests and to choose the kind of life that they wish to live and, on the other, those humans who do not have this capacity (including children who do not yet have this capacity and the elderly who have lost this capacity). While paternalistic measures might be justifiable in relation to the latter class of humans, coercive rules that seek to control fully capacitous humans for their own good, whether physical or moral, are simply not justifiable. Accordingly, there would be no justification to criminalise activities that are potentially *physically* harmful to the individual (from extreme sports to driving without a seat belt engaged, from consuming excessive amounts of alcohol to overeating) provided that it is only that individual who is harmed; and there is no justification to criminalise activities that (even if consensual) are judged to be *morally* harmful to the individual (such as, in the textbook discussions, gambling, adultery, homosexuality, prostitution; and, nowadays, the use of online dating services, and so on).[15]

It has to be conceded that there will be cases where it is difficult to be confident in drawing the line between those who do and those who do not have the relevant capacity. When precisely is a particular child capable of making his or her own decisions? When precisely is a person suffering from dementia no longer capable of making their own decisions? However, this is not a fundamental weakness of the Millian philosophy and liberals can handle many of the difficult cases by defaulting to the principle that, if there is a doubt about whether a person has the relevant capacity, then they should be treated as though they do have that capacity.

A more fundamental question arises about the way in which Millians seem to treat *physical* paternalism as no different from *moral* paternalism. Even those who follow Millians in holding that it is always wrong to try to coerce the moral interests of an individual (for their own moral salvation) may wonder whether it is unjustifiable to coerce individuals where this is necessary to prevent them from self-harming physically. As H.L.A Hart put it, even those with liberal instincts might draw back from the 'extreme' position that Millians seem to take on this point.[16] Surely, the argument goes, where humans can become dangerously addicted to drink or drugs, it cannot be wrong to rechannel their conduct (for their own physical well-being) by criminalising the use (and supply) of the most harmful substances. Given what we now know about the susceptibility and impulsivity of humans, even those with the capacity to judge what is in their own interest, is it not irresponsible to leave individuals to their own self-destructive experiments in living?

---

15 For a well-known example in the English jurisprudence, see *R v Brown* [1993] 2 All ER 75.
16 H.L.A. Hart, *Law, Liberty and Morality* (Stanford: Stanford University Press, 1963).

In response to the charge that they take their stand against physical paternalism to implausible lengths, Millians might want to plead a number of points in mitigation—notably, that they allow that there are some situations where, in order to alert an individual to the existence of risks to their health and safety, some short-term coercive measures might exceptionally be justified. However, a possibly more attractive response is that the absence of a coercive intervention does not imply that there should not be other attempts to steer regulatees away from acts that might cause them physical harm. For example, smart regulators might try to encourage regulatees to lead more healthy life-styles by educating them, by offering various kinds of incentive, by nudging them towards healthier diets and activities,[17] or by some combination of such non-coercive measures. While Millians recognise that it is perfectly legitimate to debate the prudence of various kinds of life-style and, indeed, to try to persuade a person to change their ways (for their own physical good), liberals might wonder how far the pressure on a person or the manipulation of their preferences can go before this is either tantamount to coercion or, if not coercive, then nevertheless antithetical to deep liberal respect for a person's right to self-determination.[18]

For present purposes, we do not need to resolve these difficulties and rewrite liberalism. Our focal interest is in whether technological management, when applied for paternalistic reasons, offends Millian liberal values. By definition, technological management eliminates an option. It is not in the business of persuading, cajoling or even coercing. Technological management is about force and compulsion pure and simple: it does not threaten 'Your money or your life'; if regulators want to take your money, technological management will ensure that your money is taken; and, if regulators want to take your life, technological management will ensure that your life is taken. If liberals are concerned about the use of coercion (qua the use of threats to force or compel), then their concerns must be heightened where regulators are able to force or compel—as they can do with the use of technological management—without issuing threats.

(ii) Harm to others as a necessary condition for the use of coercive rules

Expressed as a threshold condition for the justifiable use of coercive rules, paradigmatically for the justifiable use of the criminal law, the Millian principle

---

17 Seminally, see Richard H. Thaler and Cass R. Sunstein, *Nudge: Improving Decisions about Health, Wealth, and Happiness* (New Haven: Yale University Press, 2008); and compare Cass R. Sunstein, *Simpler: The Future of Government* (New York: Simon and Schuster, 2013) 38–39 for an indicative list of possible 'nudges'.

18 See Roger Brownsword, 'Criminal Law, Regulatory Frameworks and Public Health' in A.M. Viens, John Coggon, and Anthony S. Kessel (eds) *Criminal Law, Philosophy and Public Health Practice* (Cambridge: Cambridge University Press, 2013) 19; and Karen Yeung, 'Nudge as Fudge' (2012) 75 *Modern Law Review* 122.

has it that the target act must be harmful to others. If acts, however harmful, are merely 'self-regarding', they should not be penalised; unless acts are 'other-regarding' and harmful, they should not be criminalised. Even then, there might be additional considerations that militate against criminalisation or enforcement; but, at least where the acts in question are harmful to others, liberal values are not offended by the application of coercion.

Given the propensity in many societies to respond to novel acts that shock and offend by calling for there to be a 'criminal law against it', liberals do well to demand that we pause before legislating for new crimes and penalties. However, as is all too well known, the guiding liberal standard of whether the offending act causes harm to others breaks down quite quickly. The key elements of what counts as a 'harm', who counts as an 'other', and what counts as a sufficiently proximate 'cause' are all eminently contestable. In particular, if offence to the conservative interests of society can qualify as a relevant 'harm', then the Millian principle will fail to defend its liberal manifesto.[19]

To give one example, any attempt to square criminal prohibitions on smoking in public parks (as is now the case in New York City and in some other parts of the US) will struggle to find a convincing kind of harm as well as a proximate cause. Generally, it is not convincing to argue that passive smoking is a risk in open spaces; and, even if discarded cigarette butts are a nuisance, the standard prohibition on litter surely suffices. This leaves the argument that smokers fail to meet their responsibilities as role models in public places. However, liberals will rightly ask how far this goes. When we are in public places, should we stop eating candy or high-calorie food, should we make a point of walking rather than riding on public transport, of using stairs rather than escalators or elevators, and so on? Public health enthusiasts might not like these liberal constraints; but, if we profess to be liberals, we need to respect the limits set by our commitments.[20]

Similarly, in debates about the use of technologies for the enhancement of human capacity or performance, we soon run into problems with the characterisation of competitive advantage and disadvantage (or 'positional harm'). One of the leading advocates of enhancement, John Harris, defines enhancement as necessarily bringing about some improvement in human functioning. Thus:

> Enhancements of course are good if and only if those things we call
> enhancements do good, and make us better, not perhaps simply by

---

19 Compare Roger Brownsword and Jeff Wale, 'Testing Times Ahead: Non-Invasive Prenatal Testing and the Kind of Community that We Want to Be' (2018) 81 *Modern Law Review* 646.

20 See, Roger Brownsword, 'Public Health Interventions: Liberal Limits and Stewardship Responsibilities' (2013) *Public Health Ethics* (special issue on NY City's public health programme) doi: 10/1093/phe/pht030.

curing or ameliorating our ills, but because they make us better people. Enhancements will be *enhancements properly so-called* if they make us better at doing some of the things we want to do, better at experiencing the world through all of the senses, better at assimilating and processing what we experience, better at remembering and understanding things, stronger, more competent, more of everything we want to be.[21]

On this view, post-enhancement (properly so-called), there is an improvement in A's capacity to perform some task (that is, an improvement relative to A's pre-enhancement capacity). For example, post-enhancement, A has significantly improved vision or hearing; or A is able to concentrate better or stay awake longer or train harder than was the case pre-enhancement. Given that A has these improved capacities, it follows not only that 'post-enhancement A' can outperform 'pre-enhancement A' but also that A can now perform better relative to B, C, and D. This is not to say that A will be able to outperform others (this depends on the capacities of B, C, and D, and whether or not they, too, are enhanced in the relevant respect) but it does mean that, post-enhancement, A gains some positional advantage. In the context assumed by Harris, it is not A's intention to gain such a positional advantage; it is a secondary effect.[22] Nevertheless, once we switch to, say, sporting, educational, or employment contexts, the problem is that enhancement is intended precisely to gain positional advantage. Students and employees, like sportsmen, who use enhancers or enhancing technologies do so precisely for their competitive (positional) advantage; they cannot say, other than quite disingenuously, that they are happy for others to have access to the same enhancers. How does this sit with the harm principle?

Unless we already know the answer that we want to give, we can keep on repeating the question (is it harmful?) without making any progress. So much so familiar. However, there is also an equally serious, but much less well appreciated, deficit in relation to the harm principle—namely, that Mill's principle speaks only to coercive (and normative) regulatory interventions. As we have already noted, when regulators try to guide conduct by using incentives rather than threats, or by setting defaults in a particular way, it is not clear whether or how Millian liberalism is engaged. Is a 'nudge' exempt from liberal scrutiny because it falls short of coercion? But, what if the context in which the nudge is situated is itself pressurised? Or, is it the availability of the opt-out from the nudge that squares it with liberal values?

Again, I do not need to resolve these issues. Whatever the status of a nudge, technological management is a good deal more than a nudge and

---

21 John Harris, *Enhancing Evolution* (Princeton: Princeton University Press, 2007) 2 (emphasis added).
22 *Ibid.*, 28 et seq.

even more than coercion. Accordingly, even if Millians did not anticipate technological management—when regulators force, compel, or exclude certain acts or actors by design, architecture, or coding, or the like—it is surely clear that liberals would want to appeal to the harm principle to limit the occasions when regulators may legitimately resort to measures that have such a direct and negative effect on an agent's practical options.

(iii) Measures for the general health and safety of regulatees

In the twenty-first century, we are surrounded by regulations that are designed to protect and promote our general health and safety. Indeed, the density and detail of health and safety requirements has become an object of ridicule. However, it is no laughing matter; health and safety is a serious business. For example, following a tragic accident at an air show in Sussex, the regulators (the Civil Aviation Authority) revised the safety rules by banning vintage jets from performing risky manoeuvres (such as loop-the-loops) and limiting hundreds of jets from the 1950s and 1960s to fly-pasts unless they are over the sea.[23] What do liberals make of general regulatory measures of this kind (for the safety of people on 'white knuckle rides' at theme parks, for the safety of spectators at F1 motor racing or in soccer stadiums or at air shows, and so on)? Similarly, what do they make of the long-running and controversial debate about introducing fluoride into the water supply so that children have healthier teeth? Or, what do they make of the long-standing regulation of health and safety in the workplace, or of food safety, or of the regulation of road traffic, and so on? If liberals oppose such general safety measures, would this not be confirmation of their extreme and implausible stand against regulation for reasons of physical paternalism?

For liberals, much of the regulatory criminal law (to be superseded, I predict, by technological management) can be justified as consistent with the harm principle. If there is a risk that aircraft at air shows might crash and cause death and injury to persons, then to the extent that such persons are 'others' (that is, they are agents other than the pilot, consenting passengers or spectators), there is good reason to intervene—and, if necessary, to intervene coercively.[24] Much the same might be said about health and safety meas-

---

23 See, James Dean, Graeme Paton, and Danielle Sheridan, 'Jet Crash Fireball Forces Air Show Safety Rethink' *The Times*, August 24, 2015, p 1 (reporting on the fatalities and injuries caused when a vintage Hawker Hunter T7 jet crashed on the A27 main road while performing a stunt at the Shoreham Air Show); and Graeme Paton and James Dean, 'Jets Grounded as Air Shows Face Critical Safety Checks' *The Times*, August 25, 2015, p.8 (reporting on the regulatory response).

24 Some of the victims of the Sussex air show were indeed 'others' in this sense; they were simply driving past the venue on the adjacent main road.

ures on the roads and at workplaces where the potentially dangerous acts of one person might cause the death or injury of a non-consenting other.

In the much-debated case of introducing fluoride into the water supply, this may not seem directly to engage liberal principles because no rules backed by coercion are involved. However, this is a proto-typical form of technological management and, as such, it is precisely the kind of case that should concern today's liberals. Before bottled water became widely available, and when the fluoride debate was at its height, there was no real option other than to take fluoride. Irrespective of whether fluoride was of net benefit, and irrespective of whether the health of children's teeth was the real purpose of supplying fluoridated water, liberals should object that this was a case of unjustifiable regulated compulsion. The consumption of water without fluoride did not harm others; this was a clear case of physical paternalism. Moreover, to take such a stand was not extreme; it was perfectly possible to supply fluoride in a tablet form and to do so in a way that gave individuals the option of whether or not to use fluoride.[25]

While it might be difficult always to separate cleanly those elements of general safety measures that are for the protection of individuals against themselves and those that are for the protection of individuals against the harmful acts of others, the liberal line has to be that such measures should be focused on the prevention of harm to others. So far as possible, strategies should be employed that give individuals the option of acting in ways that are 'unsafe' where they alone will be 'harmed'—for example, technological management should not be used to embed filters or blocks that prevent competent adults from accessing online content that might be harmful to them; rather, competent adults should have the option of using filters or blocks, wisely or unwisely, as they see fit.

This leaves one important rider. Recalling our earlier discussion of the special nature of the essential infrastructure for human existence, liberals cannot coherently oppose the use of coercion to restrain acts that would otherwise damage the infrastructure. Such acts might not seem to cause direct harm to another but they are, by definition, harmful to everyone.

Applying these leads to the use of technological management, the liberal guidelines are that technological management may be legitimately used where this is necessary in order to protect the essential infrastructure or to prevent harm to others; but technological management should categorically not be used for any kind of paternalistic reason; and, if there are cases where the general prevention of harm to others involves a form of technological management that also limits the self-regarding (but potentially harmful) options of individuals, attempts should be made to protect those options.

---

25 For discussion, see the Nuffield Council on Bioethics, *Public Health: Ethical Issues* (London, November 2007) Ch.7.

IV Technological management and liberty

Technological management does not impinge on the paper dimension of liberty; it does not speak the language of permissions, prohibitions or penalties; rather, it speaks the language of what is practicable or possible and, thus, its impingement is on the real dimension of liberty. In the words of Adam Greenfield, what we are dealing with here is a strategy that 'resculpts the space of possibilities we're presented with.'[26] In the light of our previous discussion of the Millian limiting principles and liberal values, we can assess this resculpting, this impingement, first, in the area of criminal justice and crime control (where we might reasonably expect liberal concerns to be alerted even if technological management is neither normative nor 'coercive') and then in relation to the promotion of general health and safety, and the like (where, although liberal concerns should be raised, they are perhaps less likely to be articulated).

(i) Technological management, liberty, and crime control

If technological management is to be used within the criminal justice system, it will be compatible with liberal limiting principles provided that its use is restricted to excluding conduct that is a threat to the essential infrastructure or that is clearly harmful to others. Focusing on the latter, we should differentiate between those uses of technological management that are employed (i) to prevent acts of wilful harm that either already are by common consent criminal offences or that would otherwise be agreed to be rightly made criminal offences (such as the use of the Warwickshire's golf carts by local joy-riders) and (ii) to prevent acts that cannot be categorised unproblematically as harmful to others (perhaps, for example, the use of the 'Mosquito' to prevent groups of youngsters gathering in certain public places).

In the first case, technological management targets conduct that is agreed straightforwardly to cause significant harm to others. This seems to be precisely the kind of case where liberals would accept that the threshold condition for criminalisation has been satisfied. Indeed, on one view, this is exactly where crime control most urgently needs to adopt technological management[27]—for example, if the hardening of targets can be significantly improved by taking advantage of today's technologies, then the case for doing so might seem to be obvious. That said, liberals will have some caveats.

First, so long as liberal thinking holds that 'criminal law and punishment should attach to actions as abstracted from the personal histories and

---

26 Adam Greenfield, *Radical Technologies* (London: Verso, 2017) 256.
27 Roger Brownsword, 'Code, Control, and Choice: Why East is East and West is West' (2005) 25 *Legal Studies* 1.

propensities of those who engage in them'[28]—that is to say, that people should be punished only 'because of what they do, not by virtue of their inherent potential to do harm'[29]—then liberals will be concerned if preventative orders are routinely employed in a way that is detached from the values of punitive justice. Expressing this concern, Andrew Ashworth and Lucia Zedner[30] argue that, even if preventative measures legitimately aim to minimise overall harm, where there is a threat of the deprivation of liberty, we should not lose sight of human rights' constraints. According to Ashworth and Zedner, while it is widely accepted that the State has a core duty to protect and secure its citizens,

> [t]he key question is how this duty ... can be squared with the state's duty of justice (that is to treat persons as responsible moral agents, to respect their human rights), and the state's duty to provide a system of criminal justice ... to deal with those who transgress the criminal law. Of course, the norms guiding the fulfilment of these duties are contested, but core among them are the liberal values of respect for the autonomy of the individual, fairness, equality, tolerance of difference, and resort to coercion only where justified and as a last resort. More amorphous values such as trust also play an important role in liberal society and find their legal articulation in requirements of reasonable suspicion, proof beyond all reasonable doubt, and the presumption of innocence, Yet, the duty to prevent wrongful harms conflicts with these requirements since it may require intervention before reasonable suspicion can be established, on the basis of less conclusive evidence, or even in respect of innocent persons. The imperative to prevent, or at least to diminish the prospect of, wrongful harms thus stands in acute tension with any idealized account of a principled and parsimonious liberal criminal law.[31]

When the practice is to resolve the tension in favour of prevention, and when the latest technologies (including AI, machine learning, and profiling) offer instruments that promise to enhance predictive accuracy and preventative effectiveness, liberal concerns are inevitably heightened. If technological management is the latest manifestation of utilitarian risk management, and if the focus of criminal justice is less on harmful acts and more on agents who might act in harmful ways, liberals will fear that due process is being subordinated to crime control.

---

28 George P. Fletcher, *Basic Concepts of Criminal Law* (Oxford: Oxford University Press, 1998) 180.
29 *Ibid*.
30 Andrew Ashworth and Lucia Zedner, *Preventive Justice* (Oxford: Oxford University Press, 2014).
31 Ashworth and Zedner (n 30) 251–252.

Secondly, following on from the first caveat, there might be questions about the accuracy of the particular technology (about the way that it maps onto offences, or in its predictive and pre-emptive identification of 'true positive' prospective offenders, and so on).[32] If technological management is characterised as an exercise in risk management, a certain number of false positives might be deemed to be 'acceptable'. However, liberals should resist any such characterisation that purports to diminish the injustice to those individual agents who are wrongly subjected to restrictive measures;[33] the values of liberty and due process are not to be by-passed in this way.[34]

Thirdly, we know that, when locked doors replace open doors, or biometrically secured zones replace open spaces, the context for human interactions is affected: security replaces trust as the default and this might have some impact on the real options that are available.

Fourthly, as we have emphasised, when technological management is applied in order to prevent or exclude intentional wrongdoing, important questions are raised about the compromising of the conditions for moral

---

32 See, further, Christina M. Mulligan, 'Perfect Enforcement of Law: When to Limit and When to Use Technology' (2008) 14 *Richmond Journal of Law and Technology* 1–49, available at law.richmond.edu/jolt/v14i4/article13.pdf (last accessed 4 November 2018); and Ian Kerr, 'Prediction, Pre-emption, Presumption' in Mireille Hildebrandt and Katja de Vries (eds), *Privacy, Due Process and the Computational Turn* (Abingdon: Routledge, 2013) 91.

33 Compare *United States v Salerno and Cafaro* (1987) 481 US 739, where one of the questions was whether detention under the Bail Reform Act 1984 was to be treated as punishment or as a 'regulatory' measure. The majority concluded that 'the detention imposed by the Act falls on the regulatory side of the dichotomy. The legislative history of the Bail Reform Act clearly indicates that Congress did not formulate the pretrial detention provisions as punishment for dangerous individuals ... Congress instead perceived pretrial detention as a potential solution to a pressing societal problem ... There is no doubt that preventing danger to the community is a legitimate regulatory goal' (at 747).

34 Compare the minority opinion of Marshall J (joined by Brennan J) in *US v Salerno and Cafaro* (n 33) 760:

   Let us apply the majority's reasoning to a similar, hypothetical case. After investigation, Congress determines (not unrealistically) that a large proportion of violent crime is perpetrated by persons who are unemployed. It also determines, equally reasonably, that much violent crime is committed at night. From amongst the panoply of 'potential solutions', Congress chooses a statute which permits, after judicial proceedings, the imposition of a dusk-to-dawn curfew on anyone who is unemployed. Since this is not a measure enacted for the purpose of punishing the unemployed, and since the majority finds that preventing danger to the community is a legitimate regulatory goal, the curfew statute would, according to the majority's analysis, be a mere 'regulatory' detention statute, entirely compatible with the substantive components of the Due Process Clause.

   The absurdity of this conclusion arises, of course, from the majority's cramped concept of substantive due process. The majority proceeds as though the only substantive right protected by the Due Process Clause is a right to be free from punishment before conviction. The majority's technique for infringing this right is simple: merely redefine any measure which is claimed to be punishment as 'regulation', and, magically, the Constitution no longer prohibits its imposition.

community. So, even if a particular use of a particular mode of technological management satisfies the threshold liberal condition for the adoption of coercive measures, there are other considerations to be taken into account before its use might be judged appropriate.

In the second case, where acts are not clearly harmful to others, criminalisation is controversial—for example, with reference to the Mosquito device, liberals might object that gatherings of youths do not necessarily cause harm to others. To the extent that liberals would oppose a criminal law to this effect, they should oppose *a fortiori* the use of technological management. Suppose, for example, that an act of 'loitering unreasonably in a public place' is controversially made a criminal offence. If property owners now employ technological management, such as the Mosquito, to keep their areas clear of groups of teenagers, this will add to the controversy. First, there is a risk that technological management will overreach by excluding acts that are beyond the scope of the offence or that should be excused—for example, the technology might prevent groups of youths congregating even when this is perfectly reasonable. Secondly, without the opportunity to reflect on cases as they move through the criminal justice system, the public might not be prompted to revisit the law (the risk of 'stasis'); and, for those teenagers who wish to protest peacefully against the law by disobeying it, they cannot actually do so—their liberty to protest has been diminished.[35] The point is that, by the time that technological management is in place, it is too late; for most citizens, non-compliance is no longer an option.[36]

If it is agreed that technological management is an appropriate measure for the prevention of crime, this is not quite the end of the matter. Suppose, for example, that it is agreed that it is appropriate to use technological management to prevent golf carts or supermarket trolleys being taken off site. Further, suppose that, even though it is known that the carts will be immobilised at the permitted limit, some might still attempt to drive the carts or push the trolleys beyond the permitted zone. What should liberals make of a proposal that attempts to commit such (technologically managed) impossible acts should be criminalised? If liberals believe that the essence of a crime is that an agent, with intent, *causes harm to another*, rather than that an agent, with intent, *attempts* to cause harm to another, then liberals should resist the criminalisation of this conduct. On what basis should such innocuous, if irrational, acts be penalised? For liberals to accept either that conduct of this kind is 'harmful' (because it evinces a lack of respect for the law) or that these are 'dangerous' agents who need to be controlled by the criminal justice system

---

35 Danny Rosenthal, 'Assessing Digital Preemption (And the Future of Law Enforcement?)' (2011) 14 *New Criminal Law Review* 576.

36 Compare Evgeny Morozov, *To Save Everything, Click Here* (London: Allen Lane, 2013), and the discussion in Chapter Three above.

is to make major concessions to anti-liberal principles. As George Fletcher explains in a normative context:

> [T]wo anti-liberal principles convince many jurists that they should punish innocuous attempts—namely, attempts that are perfectly inno-cent and nonthreatening on their face. The first is the principle of gear-ing the criminal law to an attitude of hostility toward the norms of the legal system. The second is changing the focus of the criminal law from acts to actors. These two are linked by the inference: people who display an attitude of hostility towards the norms of the system show themselves to be dangerous and therefore should be subject to impris-onment to protect the interests of others.[37]

If instead of hostility towards 'the norms of the system', we read hostility towards 'legally authorised technological management', we can see how anti-liberal thinking might engage with attempts to commit the systematically impossible—not merely the happenstance impossibility of the gun that is not loaded or the pocket that is empty—in a context of widespread technological management.

(ii) Technological management, liberty, and HSE purposes

As we have said, the world of regulation—and particularly the world of health and safety regulation—has changed a good deal since the time of Mill. With the industrialisation of societies and the development of trans-port systems, new machines and technologies presented many dangers to their operators, to their users, and to third parties which regulators tried to manage by introducing health and safety rules.[38] The principal instruments of risk management were a body of 'regulatory' criminal laws, character-istically featuring absolute or strict liability, in conjunction with a body of 'regulatory' tort law, again often featuring no-fault liability but also some-times immunising business against liability. However, in the twenty-first cen-tury, we have the technological capability to manage the relevant risks: for example, in dangerous workplaces we can replace humans with robots; we can create safer environments where humans continue to operate; and, as 'green' issues become more urgent, we can introduce smart grids and various energy-saving devices.[39]

Seemingly, Millian liberals do not have a well worked-out position in rela-tion to the regulation of general health and safety. Filling this gap, I have

---

37 George P. Fletcher (n 28) 180.
38 Susan W. Brenner, *Law in an Era of 'Smart' Technology* (Oxford: Oxford University Press, 2007).
39 Giuseppe Bellantuono, 'Comparing Smart Grid Policies in the USA and EU' (2014) 6 *Law, Innovation and Technology* 221.

suggested that modern liberals should not have a fundamental objection to the use of technological management in order to improve the conditions for human health and safety, to conserve energy, to protect the environment, and the like—provided always that the regulatory purpose is to protect the essential infrastructure or to prevent the causing of harm to others. In such cases, the use of technological management should be treated as permissible (other things being equal) and, possibly, even required. What remains categorically prohibited is the use of technological management in relation to competent humans where the purpose of the intervention is paternalistic. Management of the risk of harm to oneself is for the individual, not for the criminal law and not for technological management.

Nevertheless, as with criminal justice measures that satisfy the liberal threshold condition, health and safety measures can have extremely disruptive effects on the options that are available to regulatees (especially, where automation takes out jobs), and this is something that liberals need to evaluate. Technological management might be permitted but, before it is used, there needs to be an inclusive debate about the acceptability of the risk-management package that is to be adopted.

V Conclusion

Employing a two-dimensional conception of liberty, it has been suggested in this chapter that, in order to understand how technology relates to liberty, we need to consider both whether the paper position is set for permission as well as the real options available to persons. In practice, a paper permission might not be matched by a real option; and, conversely, a paper prohibition might not be reflected by effective practical restraints (e.g. file sharing). Accordingly, any assessment of new technologies on our liberty needs to be sensitive to both the paper and the real options.

In liberal communities, respect for negative liberty (in Berlin's sense) will be one of the relevant considerations in conducting the prudential and moral assessment of a new technology. In such communities, there will be pressure to keep options open unless the application or use of a technology is clearly harmful to others. However, it should always be borne in mind that liberalism is not the only view, or indeed the only rendition of what is required if regulators are to respect the human rights and human dignity of their regulatees. In other communities, where liberty does not enter the calculation in this way, the impact of new technologies on liberty will be determined by rather different considerations of prudence and morality—such as the considerations of human dignity and solidarity that drive Francis Fukuyama's judgment that we should be deeply concerned about the impact of new red biotechnologies[40]

---

40 Francis Fukuyama, *Our Posthuman Future: Consequences of the Biotechnology Revolution* (London: Profile Books, 2002).

or the communitarian values that characterise Michael Sandel's critique of technologies of enhancement.[41]

Most importantly, it has been argued that the impact of new technologies on our real options needs to be particularly carefully monitored where technological management is used—and that, even if technological management is a post-Millian phenomenon, it is a regulatory strategy that needs to engage the urgent attention of liberals. Technological management, it bears repetition, is more than command; it is more than coercive threats; it is complete control, operating ex ante rather than ex post, and closing the gap between prescription and practical possibility. Unless the acts or practices that are being controlled in this way are a background threat to the essential infrastructure, or involve a particular foreground harm to others, they should not be controlled by technological management; technologically managing the conduct of the competent for paternalistic reasons is categorically excluded.

Finally, we can summarise how the re-imagining of liberty sits alongside the re-imagining of the ideals of the Rule of Law and of coherence. Such re-imagining, of course, takes place against the backcloth of the re-imagined ideas that we have sketched in the first part of the book—in particular, a reworked and extended understanding of the 'regulatory environment', the idea of the complexion of that environment, and an appreciation of the three-tiered nature of regulatory responsibilities. Given that backcloth, we can envisage regulatory power being exercised through normative (rules and standards) or non-normative means. The fact that regulatory power can be exercised in these different ways has important implications for both the Rule of Law and liberty. While these ideas may have been crafted in a world where power was exercised through coercive rules, the assumption in this book is that our world is likely to move on. To carry forward the spirit of the Rule of Law and liberty, these ideals need to be reworked so that they engage fully with non-normative regulatory instruments. As for coherence, to the extent that we think that this is a distinctive concern of private lawyers who value the integrity of the body of legal rules, we need to make two adjustments. One adjustment is to relate coherence to the full range of regulatory responsibilities so that we can check that regulatory interventions are compatible with the maintenance of the essential conditions for any form of human social existence as well as with the distinctive values that constitute a particular community; and the other adjustment is to apply this check for coherence to all modes of regulation, including measures of technological management whether employed by public officials or by private corporations.

---

41 Michael J. Sandel, *The Case Against Perfection* (Cambridge, Mass.: The Belknap Press of Harvard University Press, 2007).

PART THREE
Re-imagining legal rules

8

LEGAL RULES, TECHNOLOGICAL DISRUPTION, AND LEGAL/ REGULATORY MIND-SETS

I Introduction

As originally conceived, the central question for this part of the book was this: In an age of technological management, what happens to the traditional rules of law, what happens to the rules of the criminal law, of torts, and of contract law?

Why this question? Quite simply, the prospect of technological management implies that rules of any kind have a limited future. Put bluntly, to the extent that technological management takes on the regulatory roles traditionally performed by legal rules, those rules seem to be replaced and redundant—except perhaps as benchmarks for the acceptable scope of the technological measures that are introduced in place of the rules, or possibly as continuing guides for regulatees who are habituated to following rules during a transitional period.[1] To the extent that technological management co-exists with legal rules, while some rules will be redirected, others will need to be refined and revised. Accordingly, the short answer to the question for this part of the book is that the destiny of legal rules is to be found somewhere in the range of redundancy, replacement, redirection, revision and refinement. Precisely which rules are replaced, which refined, which revised and so on, will depend on both technological development and the way in which particular communities respond to the idea that technologies, as much as rules,

---

1 In other words, we can anticipate some contexts in which, although 'rule compliance' is technologically guaranteed, agents will continue to be guided by rules that are familiar or by a rule-book. See, further, Roger Brownsword, 'Technological Management and the Rule of Law' (2016) 8 *Law, Innovation and Technology* 100.

are available as regulatory instruments—indeed, that legal rules are just one species of regulatory technology.

This short answer, however, does not do justice to the deeper disruptive effects of technological development on both legal rules and the regulatory mind-set. Accordingly, in this prelude to the particular discussions in this part of the book, I want to sketch a backstory that features two overarching ideas: one is the idea of a double technological disruption and the other is the idea of a divided legal/regulatory mind-set. The disruptions in question are, first, to the substance of traditional legal rules and then to the use—or, rather, non-use—of legal rules; and the ensuing three-way legal and regulatory mind-set is divided between (i) traditional concerns for coherence in the law, (ii) modern concerns with instrumental effectiveness (relative to specified regulatory purposes) and particularly seeking an acceptable balance of the interests in beneficial innovation and management of risk, and (iii) a continuing concern with instrumental effectiveness and risk management but now focused on the possibility of employing technocratic solutions.

If what the first disruption tells us is that the old rules are no longer fit for purpose and need to be revised and renewed, the second disruption tells us that, even if the rules have been changed, regulators might now be able to dispense with the use of rules (the rules are redundant) and rely instead on technological instruments.

II Law and technology: a double disruption

It is trite that new technologies are economically and socially disruptive, impacting positively on some but negatively on others.[2] Famously, Instagram, a small start-up in San Francisco, disrupted the photographic market in a way that benefited millions but wiped out Eastman Kodak, one of the biggest corporations in the world.[3] However, it is not just economies and social practices that are disrupted by the emergence of new technologies; the law and legal practice, too, is disrupted.[4] Currently, law firms are taking up new technologies that enable much routine documentary checking to be automated; and new technologies promise to make legal services more accessible and cheaper.[5] Without doubt, these developments will shake up employ-

---

2 Compare, e.g., Clayton M. Christensen, *The Innovator's Dilemma: When New Technologies Cause Great Firms to Fail* (Boston: Harvard Business Review Press, 1997); and Monroe E. Price, 'The Newness of New Technology' (2001) 22 *Cardozo Law Review* 1885.

3 Evidently, in its final years, Kodak closed 13 factories and 130 photolabs, and cut 47,000 jobs. See, Andrew Keen, *The Internet is not the Answer* (London: Atlantic Books, 2015) 87–88.

4 See, e.g., Richard Susskind and Daniel Susskind, *The Future of the Professions* (Oxford: Oxford University Press, 2015).

5 See, e.g., *The Future of Legal Services* (London: Law Society, 2016) 38, where technology is said to be impacting on legal services in the following ways: enabling suppliers to become more efficient at procedural and commodity work; reducing costs by replacing salaried

ment patterns in the legal sector. My focus, though, is somewhat different. The double disruption to which I am drawing attention concerns, first, the substance of legal rules and, secondly, the use of technological management rather than legal rules. While this is not an essay in legal or social history, I take it that signs of the first disruption emerge in the industrialised societies of the nineteenth century and signs of the second are with us right now.

(i) The first disruption

The first disruption is highlighted in a seminal article by Francis Sayre.[6] In this paper, Sayre remarks on the 'steadily growing stream of offenses punishable without any criminal intent whatsoever'.[7] In what was apparently a parallel, but independent, development in both England and the United States, from the middle of the nineteenth century, the courts accepted that, so far as 'public welfare' offences were concerned, it was acceptable to dispense with proof of intent or negligence.[8] If the food sold was adulterated, or if employees polluted waterways, and so on, sellers and employers were simply held to account. For the most part, this was no more than a tax on business; it relieved the prosecutors of having to invest time and resource in proving intent or negligence; and, as Sayre reads the development, it reflected 'the trend of the day away from nineteenth-century individualism towards a new sense of the importance of collective interests'.[9]

Although there was no mistaking this development, and although in a modernising world it was not clearly mistaken—as Sayre recognises, the 'invention and extensive use of high-powered automobiles require new forms of traffic regulation; ... the growth of modern factories requires new forms of labor regulation; the development of modern building construction and the growth of skyscrapers require new forms of building regulation',[10] and so on—Sayre emphatically rejects any suggestion that it would, or should, 'presage the abandonment of the classic requirement of a *mens rea* as an essential of criminality'.[11] In a key passage, Sayre says:

> The group of offenses punishable without proof of any criminal intent must be sharply limited. The sense of justice of the community will not

---

humans with machine-read or AI systems; creating ideas for new models of firm and process innovation; generating work around cybersecurity, data protection and new technology laws; and, supporting changes to consumer decision-making and purchasing behaviours.

6 F.B Sayre, 'Public Welfare Offences' (1933) 33 *Columbia Law Review* 55.

7 Sayre (n 6) 55.

8 So far as the development in English law is concerned, illustrative cases include *R v Stephens* LR 1 QB 702 (1866); *Hobbs v Winchester* [1910] 2 KB 471; and *Provincial Motor Cab Co v Dunning* [1909] 2 KB 599.

9 Sayre (n 6) 67.

10 *Ibid.*, 68–69.

11 *Ibid.*, 55.

tolerate the infliction of punishment which is substantial upon those innocent of intentional or negligent wrongdoing; and law in the last analysis must reflect the general community sense of justice.[12]

In other words, so long as there is no stigmatisation or serious punishment of those (largely business people) who act in ways that deviate from public welfare regulatory requirements, dispensing with mens rea is tolerable. However, what is not to be tolerated is any attempt to dispense with mens rea where the community sees the law as concerned with serious moral delinquency and where serious punishments follow on conviction. As Sayre puts it, 'For true crimes it is imperative that courts should not relax the classic requirement of *mens rea* or guilty intent.'[13] False analogies with public welfare offences, in order to ease the way for the prosecution to secure a conviction, should be resisted. He concludes with the warning that the courts should avoid extending the doctrines applicable to public welfare offences to 'true crimes', because this would be to 'sap the vitality of the criminal law'.[14]

Similarly, in their Preface to Miquel Martin-Casals' edited volume, *The Development of Liability in Relation to Technological Change*,[15] John Bell and David Ibbetson remark that, as new technologies developed from the mid-nineteenth century, we can see the beginnings of a movement from 'tort' to 'regulation'. Thus, they say:

> We see the way in which regulatory law, private insurance and state-run compensation schemes developed to deal with the issues the law now confronted. Regulatory law and inspections by officials and private insurers and associations dealt with many of the issues of preventing accidents. Compensation systems outside tort offered remedies to

---

12 *Ibid.*, 70.

13 *Ibid.*, 80.

14 *Ibid.*, 84. Compare, R.A. Duff, 'Perversions and Subversions of Criminal Law' in R.A. Duff, Lindsay Farmer, S.E. Marshall, Massimo Renzo, and Victor Tadros (eds), *The Boundaries of the Criminal Law* (Oxford: Oxford University Press, 2010) 88 at 104: 'We must ask about the terms in which the state should address its citizens when it seeks to regulate their conduct, and whether the tones of criminal law, speaking of wrongs that are to be condemned, are more appropriate than those of a regulatory regime that speaks only of rules and penalties for their breach.' According to Duff, where the conduct in question is a serious public wrong, it would be a 'subversion' of the criminal law if offenders were not to be held to account and condemned. For questions that might arise relative to the 'fair trial' provisions of the European Convention on Human Rights where a state decides to transfer less serious offences from the criminal courts to administrative procedures (as with minor road traffic infringements), see e.g., *Öztürk v Germany* (1984) 6 EHRR 409.

15 Miquel Martin-Casals (ed), *The Development of Liability in Relation to Technological Change* (Cambridge: Cambridge University Press, 2010).

many of the victims of accidents. In this matrix of legal interventions, we can see that the place of tort law and of fault in particular changes. We become aware of [tort law's] limitations.[16]

Thus, as Geneviève Viney and Anne Guégan-Lécuyer observe, a tort regime 'which seemed entirely normal in an agrarian, small-scale society, revealed itself rather quickly at the end of the nineteenth century to be unsuitable'.[17] So, for example, in Sweden, following a railway accident in 1864 when seven people died and eleven were seriously injured, and with no realistic claim for compensation (the train driver being in no position to satisfy a personal tort claim), a petition was presented to parliament to respond to the special needs created by the operation of the railways.[18] Again, in England, when a train derailed near Reading early in the morning on Christmas Eve 1841, with nine fatalities and 16 serious injuries, there was no adequate scheme of compensation (other than the ancient 'deodand' which treated the responsible 'things', the railway engine and trucks, as forfeited).[19] Five years later, the Deodands Act 1846 abolished the deodand and, more importantly, the Fatal Accidents Act 1846 laid the foundation for a modern compensatory approach in the event of accidents on the railways and elsewhere. As Mr Justice Frankfurter was to put it a century later, these nineteenth-century societies were learning that 'the outmoded concept of "negligence" [was an inadequate] working principle for the adjustments of injuries inevitable under technological circumstances of modern industry'.[20]

In contract law, too, there was a significant shift from a 'subjective' consensual (purely transactional) model to an 'objective' approach. In the United States, against the background of an 'increasingly national corporate economy, the goal of standardization of commercial transactions began to overwhelm the desire to conceive of contract law as expressing the subjective desires of individuals';[21] and, in English law, in addition to the general shift to an objective approach, there was a particularly significant shift to a reasonable notice model in relation to the incorporation of the terms and conditions

---

16 Martin-Casals (n 15) viii.
17 Geneviève Viney and Anne Guégan-Lécuyer, 'The Development of Traffic Liability in France' in Martin-Casals (n 15) 50.
18 See, Sandra Friberg and Bill W. Dufwa, 'The Development of Traffic Liability in Sweden', in Martin-Casals (n 15) 190.
19 See, en.wikipedia.org/wiki/Sonning_Cutting_railway_accident (on the accident) and en.wikipedia.org/wiki/Deodand (on the historic Deodand) (both last accessed 4 November 2018).
20 *Wilkerson v McCarthy* 336 US 53 at 65 (1949).
21 Morton J. Horwitz, *The Transformation of American Law 1870–1960* (Oxford: Oxford University Press, 1992) 37. At 48–49, Horwitz notes a parallel transformation in relation to both corporate forms and agency.

on which carriers (of both goods and persons) purported to contract. In the jurisprudence, this latter shift is symbolised by Mellish LJ's direction to the jury in *Parker v South Eastern Railway Co*,[22] where the legal test is said to be not so much whether a customer actually was aware of the terms and had agreed to them but whether the company had given reasonable notice. In effect, this introduced an objective test of consent and agreement; but, as Stephen Waddams has pointed out, there was an even more radical view, this being expressed in Bramwell LJ's judgment, the emphasis of which is 'entirely on the reasonableness of the railway's conduct of its business and on the unreasonableness of the customers' claims; there is no concession whatever to the notion that they could only be bound by their actual consent.'[23] So, while intentionality and fault were set aside in the regulatory parts of criminal law and torts, classical transactionalist ideas of consent and agreement were marginalised in the *mainstream* of contract law being replaced by 'objective' tests and standards set by reasonable business practice. In short, as Morton Horwitz puts it, there was a dawning sense that 'all law was a reflection of collective determination, and thus inherently regulatory and coercive'.[24]

What we see across these developments is a pattern of disruption to legal doctrines that were organically expressed in smaller-scale non-industrialised communities. Here, the legal rules presuppose very straightforward ideas about holding those who engage intentionally in injurious or dishonest acts to account, about expecting others to act with reasonable care and attention, and about holding others to their word. Once new technologies disrupt these ideas, we see the move to strict or absolute criminal liability without proof of intent, to tortious liability without proof of fault (whether formally as strict liability or in the form of negligence without fault),[25] and to contractual liability (or limitation of liability) without proof of actual intent, agreement or consent. Even if the development in contract is less clear at this stage, in both criminal law and torts we can see the early signs of a risk management approach to liability. Moreover, we also see the early signs of doctrinal bifurcation,[26] with some parts of criminal law, tort law and contract law

---

22 (1877) 2 CPD 416.

23 Stephen Waddams, *Principle and Policy in Contract Law* (Cambridge: Cambridge University Press, 2011) 39.

24 Horwitz (n 21) 50.

25 See Albert A. Ehrenzweig, 'Negligence without Fault' (1966) 54 *California Law Review* 1422.

26 As recognised, for example, in the Canadian Supreme Court case of *R. v Sault Ste Marie* [1978] 2 SCR 1299, where (at 1302–1303) Dickson J remarks:

> In the present appeal the Court is concerned with offences variously referred to as 'statutory', 'public welfare', 'regulatory', 'absolute liability', or 'strict responsibility', which are not criminal in any real sense, but are prohibited in the public interest ... Although enforced as penal laws through the utilization of the machinery of the criminal law, the

resting on traditional principles (and representing, so to speak, 'real' crime, tort and contract) while others deviate from these principles—often holding enterprises to account more readily but also sometimes easing the burden on business for the sake of beneficial innovation[27]—in order to strike a more acceptable balance of the benefits and risks that technological development brings with it.

(ii) The second disruption

Arguably, the second disruption is as old as the (defensive) architecture of the pyramids and the target-hardening use of locks. However, I want to say that the significant moment of disruption is right now, with the dense, sophisticated and various instruments of technological management that distinguish our circumstances, quantitatively and qualitatively, from those of both pre-industrial and early industrial societies. Whether or not this amounts to a difference of kind or degree scarcely seems important; we live in different times, with significantly different regulatory technologies. In particular, there is much more to technological management than traditional target-hardening: the management involved might—by designing products and places, or by coding products and people—disable or exclude potential wrongdoers as much as harden targets or immunise potential victims; and, crucially, there is now the prospect of widespread automation that takes humans altogether out of the regulatory equation. Crucially, with a risk management approach well established, regulators now find that they have the option of responding by employing various technological instruments rather than rules. This is the moment when, so to speak, we see a very clear contrast between the legal and regulatory style of the East coast (whether traditional or progressive) and the style of the West coast.[28]

Two things are characteristic of technological management. First, as we have emphasised in earlier chapters, unlike rules, the focus of the regulatory intervention is on the practical (not the paper) options of regulatees. Secondly, whereas legal rules back their prescriptions with ex post penal,

---

offences are in substance of a civil nature and might well be regarded as a branch of administrative law to which traditional principles of criminal law have but limited application. They relate to such everyday matters as traffic infractions, sales of impure food, violations of liquor laws, and the like. In this appeal we are concerned with pollution.

27 For example, in the United States, the interests of the farming community were subordinated to the greater good promised by the development of the railroad network: see Morton J. Horwitz, *The Transformation of American Law 1780–1860* (Cambridge, Mass.: Harvard University Press, 1977).

28 Seminally, see Lawrence Lessig, *Code and Other Laws of Cyberspace* (New York: Basic Books, 1999). See, too, Roger Brownsword, 'Code, Control, and Choice: Why East is East and West is West' (2005) 25 *Legal Studies* 1.

compensatory, or restorative measures, the focus of technological management is entirely ex ante, aiming to anticipate and prevent wrongdoing rather than punish or compensate after the event. As Lee Bygrave puts it in the context of the design of information systems and the protection of both IPRs and privacy, the assumption is that, by embedding norms in the architecture, there is 'the promise of a significantly increased *ex ante* application of the norms and a corresponding reduction in relying on their application *ex post facto*'.[29]

This evolution in regulatory thinking is not surprising. If, instead of accidents on the railways, or sparks from steam engines setting fire to land adjacent to the railroad, or boilers exploding, we think about the risks arising from the adoption of driverless cars and drones, from devices that communicate with one another, from 3D printers, and so on, then we might expect there to be a continuation of whatever movement there already has been from 'tort' to 'regulation'. Moreover, as we become increasingly aware of the limitations of legal rules, and as technological management offers itself as a more effective means of securing the conditions for human health and safety, we can expect the movement to be towards a risk-management approach that takes advantage of the technological instruments that increasingly become available to regulators.

For example, with the development of computers and then the Internet and World Wide Web, supporting a myriad of applications, it is clear that, when individuals operate in online environments, they are at risk in relation to both their 'privacy' and the fair processing of their personal data. Initially, regulators assumed that 'transactionalism' would suffice to protect individuals: in other words, unless the relevant individuals agreed to, or consented to, the processing of their details, it would not be lawful. However, once it was evident that consumers in online environments routinely signalled their agreement or consent in a mechanical way, without doing so on a free and informed basis, a more robust risk-management approach invited consideration. As Eliza Mik, writing about the privacy policies of internet companies, puts the alternative:

> What could be done … is to cease treating privacy policies *as if* they were contracts and evaluate consent and disclosure requirements from a purely regulatory perspective. Enhanced, or express, consent requirements may constitute a good first step. It could, however, also be claimed that the only solution lies in an outright prohibition of certain

---

29 Lee A. Bygrave, 'Hardwiring Privacy' in Roger Brownsword, Eloise Scotford, and Karen Yeung (eds), *The Oxford Handbook of Law, Regulation and Technology* (Oxford: Oxford University Press, 2017) 754 at 755.

technologies or practices. In this context, the difficulty lies in regulatory target setting. The first overriding question is what is it that we are trying to protect? It can hardly be assumed that the 'protection of autonomy' is sufficiently precise to provide regulatory guidance.[30]

We might, however, take this a step further. Once we are thinking about the protection of the autonomy of Internet-users or about the protection of their privacy, why not also consider the use of technological instruments in service of the regulatory objectives (provided that they can be specified in a sufficiently precise way)? Indeed, this is just what we find in the General Data Protection Regulation (GDPR).[31]

Following Recital 75 of the Regulation, which lays out a catalogue of risks and harms that might impact on individuals as a result of the processing of their data,[32] we have in Recital 78 an enjoinder to data controllers to

---

30 Eliza Mik, 'Persuasive Technologies—From Loss of Privacy to Loss of Autonomy' in Kit Barker, Karen Fairweather, and Ross Grantham (eds), *Private Law in the 21st Century* (Oxford: Hart, 2017) 363 at 386.

31 Regulation (EU) 2016/679. Similarly, in the UK, .s.57(1) of the Data Protection Act 2018 provides that each data controller 'must implement appropriate technical and organisational measures which are designed—(a) to implement the data protection principles in an effective manner, and (b) to integrate into the processing itself the safeguards necessary for that purpose.' This duty applies 'both at the time of the determination of the means of processing the data and at the time of the processing itself' (s.57(2)). Then, according to s.57(3), each data controller 'must implement appropriate technical and organisational measures for ensuring that, by default, only personal data which is necessary for each specific purpose of the processing is processed'—this applying to the amount of personal data that is collected, the extent of its processing, the period of its storage, and its accessibility (s.57(4)). To this, s.57(5) adds: 'In particular, the measures implemented to comply with the duty under subsection (3) must ensure that, by default, personal data is not made accessible to an indefinite number of people without an individual's intervention.'

32 According to Recital 75:

> The risk to the rights and freedoms of natural persons, of varying likelihood and severity, may result from personal data processing which could lead to physical, material or non-material damage, in particular: where the processing may give rise to discrimination, identity theft or fraud, financial loss, damage to the reputation, loss of confidentiality of personal data protected by professional secrecy, unauthorised reversal of pseudonymisation, or any other significant economic or social disadvantage; where data subjects might be deprived of their rights and freedoms or prevented from exercising control over their personal data; where personal data are processed which reveal racial or ethnic origin, political opinions, religion or philosophical beliefs, trade union membership, and the processing of genetic data, data concerning health or data concerning sex life or criminal convictions and offences or related security measures; where personal aspects are evaluated, in particular analysing or predicting aspects concerning performance at work, economic situation, health, personal preferences or interests, reliability or behaviour, location or movements, in order to create or use personal profiles; where personal data of vulnerable natural persons, in particular of children, are processed; or where processing involves a large amount of personal data and affects a large number of data subjects.

take 'appropriate technical and organisational measures' to ensure that the requirements of the Regulation are met. In the body of the Regulation, this is expressed as follows in Article 25 (Data protection by design and by default):

> 1. Taking into account the state of the art, the cost of implementation and the nature, scope, context and purposes of processing as well as the risks of varying likelihood and severity for rights and freedoms of natural persons posed by the processing, the controller shall, both at the time of the determination of the means for processing and at the time of the processing itself, implement appropriate technical and organisational measures, such as pseudonymisation, which are designed to implement data-protection principles, such as data minimisation, in an effective manner and to integrate the necessary safeguards into the processing in order to meet the requirements of this Regulation and protect the rights of data subjects.
>
> 2. The controller shall implement appropriate technical and organisational measures for ensuring that, by default, only personal data which are necessary for each specific purpose of the processing are processed. That obligation applies to the amount of personal data collected, the extent of their processing, the period of their storage and their accessibility. In particular, such measures shall ensure that by default personal data are not made accessible without the individual's intervention to an indefinite number of natural persons.

While talk of 'privacy enhancing technologies' and 'privacy by design' has been around for some time,[33] in the GDPR we see that this is more than talk; it is not just that the regulatory discourse is more technocratic, there are signs that the second disruption is beginning to impact on regulatory practice—although how far this particular impact will penetrate remains to be seen.[34]

---

33 See, Bygrave (n 29); and Ann Cavoukian, *Privacy by Design: The Seven Foundational Principles* (Information and Privacy Commissioner of Ontario, 2009, rev edn 2011) available at www.ipc.on.ca/images/Resources/7foundationalprinciples.pdf (last accessed 4 November 2018)..

34 Bygrave (n 29) argues, at 756, that, despite explicit legal backing, 'the privacy-hardwiring enterprise will continue to struggle to gain broad traction.' Most importantly, this is because this enterprise 'is at odds with powerful business and state interests, and simultaneously remains peripheral to the concerns of most consumers and engineers' (*ibid*). So far as the engineering community is concerned, see Adamantia Rachovitsa, 'Engineering and Lawyering Privacy by Design: Understanding Online Privacy both as a Technical and an International Human Rights Issue' (2016) 24 *International Journal of Law and Information Technology* 374.

III Three legal/regulatory mind-sets: coherentist, regulatory-instrumentalist, and technocratic

According to Edward Rubin, we live in the age of modern administrative states where the law is used 'as a means of implementing the policies that [each particular state] adopts. The rules that are declared, and the statutes that enact them, have no necessary relationship with one another; they are all individual and separate acts of will.'[35] In other words,

> Regulations enacted by administrative agencies that the legislature or elected chief executive has authorized are related to the authorizing statute, but have no necessary connection with each other or to regulations promulgated under a different exercise of legislative or executive authority.[36]

In the modern administrative state, the 'standard for judging the value of law is not whether it is coherent but rather whether it is effective, that is, effective in establishing and implementing the policy goals of the modern state'.[37] By contrast, the distinctive feature of 'coherentism' is the idea that law forms 'a coherent system, a set of rules that are connected by some sort of logical relationship to each other'[38]—or 'a system of rules that fit together in a consistent logically elaborated pattern'.[39] Moreover, within the modern administrative state, the value of coherence itself is transformed: coherence, like the law, is viewed as 'an instrumental device that is deployed only when it can be effective'.[40] In a concluding call to arms, Rubin insists that legal scholarship needs to 'wake from its coherentist reveries';[41] and that legal scholars 'need to relinquish their commitment to coherence and concern themselves with the effectiveness of law and its ability to achieve our democratically determined purposes'.[42]

For my purposes, we can draw on Rubin to construct two ideal-typical mind-sets in thinking about the way that the law should engage with new

---

35 Edward L. Rubin, 'From Coherence to Effectiveness' in Rob van Gestel, Hans-W Micklitz, and Edward L. Rubin (eds), *Rethinking Legal Scholarship* (New York: Cambridge University Press, 2017) 310 at 311.

36 *Ibid.*, 311.

37 *Ibid.*, 328.

38 *Ibid.*, 312.

39 *Ibid.*, 313.

40 *Ibid.*, 328.

41 *Ibid.*, 349. For scholarly concerns that include but also go beyond coherentism, see Roger Brownsword, 'Maps, Critiques, and Methodologies: Confessions of a Contract Lawyer' in Mark van Hoecke (ed), *Methodologies of Legal Research* (Oxford: Hart, 2011) 133.

42 Rubin (n 35) 350; and, compare the seminal ideas in Hugh Collins, *Regulating Contracts* (Oxford: Oxford University Press, 1999).

technologies and, more generally, about the reform and renewal of the law. One ideal-type, 'regulatory-instrumentalism', views legal rules as a means to implement whatever policy goals have been adopted by the State; the adequacy and utility of the law is to be assessed by its effectiveness in delivering these goals. The other ideal-type is 'coherentism', according to which the adequacy of the law is to be assessed by reference to the doctrinal consistency and integrity of its rules. Where 'regulatory-instrumentalism' informs a proposal for reform, the argument will be that some part of the law 'does not work' relative to desired policy goals. By contrast, where 'coherentism' informs a proposal for reform, the argument will be that there is a lack of clarity in the law or that there are internal inconsistencies or tensions within the law that need to be resolved.

Although Rubin does not suggest that the shift from a coherentist to a regulatory-instrumentalist mind-set is associated with the emergence of technologies, it is of course precisely this shift that I am suggesting reflects the first technological disruption of the law. We can say a bit more about both coherentist and regulatory-instrumentalist views before focusing on the technocratic mind-set that is provoked by the second disruption.

(i) The coherentist mind-set

It is axiomatic within coherentism that the law should be formally consistent; and, while there may be some confusion, uncertainty and inefficiency if legal rules are contradictory or in tension, the coherence of legal doctrine is typically viewed as desirable in and of itself.[43] However, coherentism also has a substantive dimension. Thus, in Rubin's account of coherentism, the law not only displays an internal consistency and integrity, it also expresses and concretises higher 'natural law' principles, all this being distilled by an intellectual elite applying their rational wisdom.[44] Although, even now, we might detect traces of such top-down 'pre-modern' thinking (as Rubin puts it), this is not a necessary characteristic. Rather, coherentists draw on simple traditional principles that are generally judged to be both reasonable and workable. The law, on this view, is about responding to 'wrongs', whether

---

43 See Chapter Six. Alongside coherentist thinking, we might also place what Neil Walker terms 'traditionality': see Neil Walker, *Intimations of Global Law* (Cambridge: Cambridge University Press, 2015). At 152, Walker describes the standard triad of traditionality as follows:

> [F]irst, default orientation towards and contemplation of past legal processes and events as a basis for any new legal claim; secondly, selective specification of that past as an authoritative foundation for the present; and, thirdly, the conduct of that selective specification in accordance with a recognised transmission procedure or formula which consciously ensures continuity between past and present.

44 Rubin (n 35).

by punishing wrongdoers or by compensating victims; it is about correction and rectification, and holding wrongdoers to account. In the field of transactions, there are echoes of this idea in the notion that the law of contract should be guided, as Lord Steyn has put it, by the simple ideal of fulfilling the expectations of honest and reasonable people;[45] and, in the field of interactions, it almost goes without saying that the law of tort should be guided by the standards and expectations of these same honest and reasonable people.

Anticipating the contrast between this coherentist mind-set and mindsets that are more instrumental and/or technocratic, we should emphasise that the formal and substantive dimensions of coherentism betray little or no sense of the direction in which the law should be trying to move things. Coherentism looks up and down, backwards, and even sideways, but not forward. It is not instrumental; it is not about engineering change. Moreover, insofar as coherentists are focused on righting wrongs, their gaze is not on prevention and certainly not on the elimination of practical options.

There is one further important aspect of coherentist thinking, a feature that manifests itself quite regularly now that new technologies and their applications present themselves for classification and characterisation relative to established legal concepts and categories. Here, we find a coherentist reluctance to abandon existing classifications and categories and contemplate bespoke responses. For example, rather than recognise new types of intellectual property, coherentists will prefer to tweak existing laws of patents and copyright.[46] Similarly, we will recall Lord Wilberforce's much-cited remarks on the heroic efforts made by the courts—confronted by modern forms of transport, various kinds of automation, and novel business practices—to force 'the facts to fit uneasily into the marked slots of offer, acceptance and consideration'[47] or whatever other traditional categories of the law of contract might be applicable. And, in transactions, this story continues; coherentism persists. So, for example, coherentists will want to classify emails as either instantaneous or non-instantaneous forms of communication (or transmission),[48] they will want to apply the standard formation template to online shopping sites, they will want to draw on traditional notions of agency in order to engage

---

45 Seminally, see Johan Steyn, 'Contract Law: Fulfilling the Reasonable Expectations of Honest Men' (1997) 113 *Law Quarterly Review* 433.

46 Compare the analysis of multi-media devices in Tanya Aplin, *Copyright Law in the Digital Society: the Challenges of Multimedia* (Oxford: Hart, 2005).

47 As Lord Wilberforce put it in *New Zealand Shipping Co Ltd v A.M. Satterthwaite and Co Ltd: The Eurymedon* [1975] AC 154, 167.

48 See, e.g., Andrew Murray, 'Entering into Contracts Electronically: the Real WWW' in Lilian Edwards and Charlotte Waelde (eds), *Law and the Internet: A Framework for Electronic Commerce* (Oxford: Hart, 2000) 17; and Eliza Mik, 'The Effectiveness of Acceptances Communicated by Electronic Means, Or—Does the Postal Acceptance Rule Apply to Email?' (2009) 26 *Journal of Contract Law* 68 (concluding that such classificatory attempts should be abandoned).

electronic agents and smart machines,[49] they will want to classify individual 'prosumers' and 'hobbyists' who buy and sell on new platforms (such as platforms that support trade in 3D printed goods) as either business sellers or consumers,[50] and they will be challenged by agreements for the supply of digital content being unsure whether to classify such contracts as being for the supply of goods or the supply of services.[51] As the infrastructure for transactions becomes ever more technological the tension between this strand of common law coherentism and regulatory-instrumentalism becomes all the more apparent.

(ii) The regulatory-instrumentalist mind-set

'Regulation' is generally understood as a process of directing regulatees, monitoring and detecting deviation, and correcting for non-compliance, all of this relative to specified regulatory purposes. The regulatory mind-set is, at all stages, instrumental. The question is: what works? When a regulatory intervention does not work, it is not enough to restore the status quo; rather, further regulatory measures should be taken, learning from previous experience, with a view to realising the regulatory purposes more effectively. Hence, the purpose of the criminal law is not simply to respond to wrongdoing (as corrective justice demands) but to reduce crime by adopting whatever measures of deterrence promise to work.[52] Similarly, in a safety-conscious community, the purpose of tort law is not simply to respond to wrongdoing but to deter practices and acts where agents could easily avoid creating risks of injury and damage. For regulatory-instrumentalists, the path of the law

---

49 Compare, e.g., Emily Weitzenboeck, 'Electronic Agents and the Formation of Contracts' (2001) 9 *International Journal of Law and Information Technology* 204; and, see, Gunther Teubner, 'Digital Personhood: The Status of Autonomous Software Agents in Private Law' *Ancilla Juris* (2018) 107, 130 (contrasting the somewhat conservative German view with the view of the European Parliament that the traditional rules of contract law are inadequate)..

50 Compare, e.g., Christian Twigg-Flesner, 'Conformity of 3D Prints—Can Current Sales Law Cope?' in R. Schulze and D. Staudenmayer (eds), *Digital Revolution: Challenges for Contract Law in Practice* (Baden-Baden: Nomos, 2016) 35.

51 As Christian Twigg-Flesner points out in 'Disruptive Technology—Disrupted Law?' in Alberto De Franceschi (ed), *European Contract Law and the Digital Single Market* (Cambridge: Intersentia, 2016) 21, 32, '[n]either [classification] seems to make particularly good sense'.

52 Compare David Garland, *The Culture of Control: Crime and Social Order in Contemporary Society* (Oxford: Oxford University Press, 2001); and Amber Marks, Benjamin Bowling, and Colman Keenan, 'Automatic Justice? Technology, Crime, and Social Control' in Roger Brownsword, Eloise Scotford, and Karen Yeung (eds), *The Oxford Handbook of Law, Regulation and Technology* (Oxford: Oxford University Press, 2017) 705.

should be progressive: we should be getting better at regulating crime and improving levels of safety.[53]

One of the striking features of the European Union has been the single market project, a project that the Commission has pursued in a spirit of conspicuous regulatory-instrumentalism. Here, the regulatory objectives are: (i) to remove obstacles to consumers shopping across historic borders; (ii) to remove obstacles to businesses (especially small businesses) trading across historic borders; and (iii) to achieve a high level of consumer protection. In order to realise this project, it has been essential to channel the increasing number of Member States towards convergent legal positions. Initially, minimum harmonisation Directives were employed, leaving it to Member States to express the spirit and intent of Directives in their own doctrinal way. To this extent, a degree of divergence was tolerated in the way that the regional inputs were translated into national outputs that, in turn, might become the relevant legal material for interpretation and application. However, where the Commission needed a stronger steer, it could (and did) resort to the use of maximum harmonisation measures (restricting the scope for local glosses on the law); and, where Directives did not work, then Regulations could be used (a case in point being the recent General Data Protection Regulation[54]), leaving Member States with even less room, formally speaking, for local divergence.

As the single market project evolves into the digital Europe project, the Commission's regulatory-instrumentalist mind-set is perfectly clear. As the Commission puts it:

> The pace of commercial and technological change due to digitalisation is very fast, not only in the EU, but worldwide. The EU needs to act

---

53 The parallel development of a risk-management ideology in both criminal law and tort is noted by Malcolm Feeley and Jonathan Simon, 'Actuarial Justice: The Emerging New Criminal Law' in David Nelken (ed), *The Futures of Criminology* (London: Sage, 1994) 173. At 186, Feeley and Simon say:

> Although social utility analysis or actuarial thinking is commonplace enough in modern life…in recent years this mode of thinking has gained ascendancy in legal discourse, a system of reasoning that traditionally has employed the language of morality and focused on individuals …
>
> Thus, for instance, it is by now the conventional mode of reasoning in tort law. Traditional concerns with fault and negligence standards—which require a focus on the individual and concern with closely contextual causality—have given way to strict liability and no-fault. One sees this in both doctrines, and even more clearly in the social vision that constitutes the discourse about modern torts. The new doctrines ask, how do we 'manage' accidents and public safety. They employ the language of social utility and management, not individual responsibility.

54 Regulation (EU) 2016/679.

now to ensure that business standards and consumer rights will be set according to common EU rules respecting a high-level of consumer protection and providing for a modern business friendly environment. It is of utmost necessity to create the framework allowing the benefits of digitalisation to materialise, so that EU businesses can become more competitive and consumers can have trust in high-level EU consumer protection standards. By acting now, the EU will set the policy trend and the standards according to which this important part of digitalisation will happen.[55]

In this context, coherentist thoughts about tidying up and standardising the lexicon of the consumer acquis, or pushing ahead with a proposed Common European Sales Law,[56] or codifying European contract law drop down the list of priorities. For regulatory-instrumentalists, when we question the fitness of the law, we are not asking whether legal doctrine is consistent, we are asking whether it is fit for delivering the regulatory purposes.[57]

Last but not least, I take it to be characteristic of the regulatory-instrumentalist mind-set that the thinking becomes much more risk-focused. In the criminal law and in torts, the risks that need to be assessed and managed relate primarily to physical and psychological injury and to damage to property and reputation; in contract law, it is economic risks that are relevant. So, for example, we see in the development of product liability a scheme of acceptable risk management that responds to the circulation of products (such as cars or new drugs) that are beneficial but also potentially dangerous. However, this response is still in the form of a revised *rule* (it is not yet technocratic); and it is still in the nature of an ex post correction (it is not yet ex ante preventative). Nevertheless, it is only a short step from here to a greater investment in ex ante regulatory checks (for food and drugs, chemicals, and so on) and to the use of new technologies as preventative regulatory instruments.

---

55 European Commission, Communication from the Commission to the European Parliament, the Council and the European Economic and Social Committee, 'Digital contracts for Europe—Unleashing the potential of e-commerce' COM(2015) 633 final (Brussels, 9.12.2015), p.7.

56 Despite a considerable investment of legislative time, the proposal was quietly dropped at the end of 2014. This also, seemingly, signalled the end of the project on the Common Frame of Reference in which, for about a decade, there had been a huge investment of time and resource.

57 Compare, too, the approach of the European Parliament in its Report on AI and robotics together with recommendations to the Commission on Civil Law Rules on Robotics (January 27, 2017); available at www.europarl.europa.eu/sides/getDoc.do?pubRef=-//EP//TEXT+REPORT+A8-2017-0005+0+DOC+XML+V0//EN (last accessed 4 November 2018).

(iii) The technocratic mind-set

Perhaps the technocratic mind-set is best captured by Joshua Fairfield when, writing in the context of non-negotiable terms and conditions in online consumer contracts, he remarks that 'if courts [or, we might say, the rules of contract law] will not protect consumers, robots will'.[58] However, the seeds of technocratic thinking long pre-date the latest robotics.

As is well known, there was a major debate in the United Kingdom at the time that seat belts were fitted in cars and it became a criminal offence to drive without engaging the belt. Tort law responded, too, by treating claimant drivers or passengers who failed to engage their seat belts as, in part, contributing to their injuries.[59] Critics saw this as a serious infringement of their liberty—namely, their option to drive with or without the seat belt engaged. Over time, though, motorists became encultured into compliance. So far, we might say, so regulatory-instrumentalist.

Suppose, though, that motorists had not become encultured into compliance. Given the difficulty of enforcing a rule requiring seat belts to be engaged, regulatory-instrumentalism might have taken a more technocratic turn. For example, there might have been a proposal to design vehicles so that cars were simply immobilised if seat belts were not worn. As we described in Chapter One, regulators in the USA sought (unsuccessfully as it proved at that time) to introduce precisely such a measure. Despite the optimistic (safety-improving) forecasts of its proponents, not only does technological management of this kind aspire to limit the practical options of motorists, including removing the real possibility of non-compliance with the law, there is a sense in which it supersedes the rules of law themselves. Some might object to this technocratic approach in itself; and many more might object if the particular regulatory purposes that guide the use of technological measures are unpopular.

Nowadays, similar debates might be had about the use of mobile phones by motorists. There are clear and dramatic safety implications but many drivers persist in using their phones while they are in their cars. If we are to be technocratic in our approach, perhaps we might seek a design solution that disables phones within cars, or while the user is driving. However, once automated vehicles relieve 'drivers' of their safety responsibilities, it seems that the problem will drop away—rules that penalise humans who use their mobile phones while driving will become redundant; humans will simply be transported in vehicles and the one-time problem of driving while phoning will no longer be an issue.

---

58 Joshua Fairfield, 'Smart Contracts, Bitcoin Bots, and Consumer Protection' (2014) 71 *Washington and Lee Law Review Online* 36, 39.

59 *Froom v Butcher* [1976] QB 286.

So, unlike coherentists, technocrats are not concerned with doctrinal integrity and their focus is not on restoring the status quo prior to wrongdoing; and, unlike regulatory-instrumentalists who do view the law in a purposive way (and who might employ various technologies in support of the rules), technocrats are wholly concerned with preventing or precluding wrongdoing and employing technological measures or solutions, rather than rules or standards, to achieve their objectives.

(iv) Which mind-set to engage

In the opening chapter of the book, I introduced a hypothetical scenario in which a child slipped the attention of a robot carer at a shopping mall, leading to a collision between the child and an elderly lady. At that point, I indicated that, should the child's parents be sued, there might be two kinds of response. One response—which we now see is a coherentist response—would be to be guided by principles of corrective justice: liability would be assessed by reference to what communities judge to be fair, just and reasonable—and different communities might have different ideas about whether it would be fair, just and reasonable to hold the parents liable in the hypothetical circumstances.[60] The other response would be more regulatory-instrumentalist. Here, the thinking would be that before retailers, such as the shop at the mall, are to be licensed to introduce robot babysitters, and parents permitted to make use of robocarers, there needs to be a collectively agreed scheme of compensation should something 'go wrong'. On this view, the responsibilities and liabilities of the parents would be determined by the agreed terms of the risk management package. However, we might also imagine a third response, a response of a technocratic nature. Quite what measures of technological management might be suggested is anyone's guess—perhaps an invisible 'fence' at the edge of the care zone so that children simply could not stray beyond the limits. However, thinking about the puzzle in this way, the question would be entirely about designing the machines and the space in such a way that collisions between children and mall-goers could not happen.

Which of these responses is appropriate? On the face of it, coherentism belongs to relatively static and stable communities, not to the turbulent times of the twenty-first century. To assume that traditional legal frameworks enable regulators to ask the right questions and answer them in a rational way seems over-optimistic. If we reject coherentism, we will see regulatory-instrumentalism as a plausible default with the option of a technocratic

---

60 E.g., Ugo Pagallo, *The Laws of Robots* (Dordrecht: Springer, 2013) 124–130.

resolution always to be considered.[61] However, the use of technological management introduces new concerns.

In Chapter Four, we said that the paramount responsibility of regulators has to be to maintain and protect the 'commons', the essential pre-conditions for human social existence. While technological management might protect some aspects of the commons, it might also compromise others. When regulators think in a coherentist or regulatory-instrumentalist way, they might fail to take the necessary protective steps; but, it is with a technocratic approach that the mode of intervention presents this distinctive new concern. As we have already suggested, this invites the articulation of a 'new coherentism', reminding those regulators who think in technocratic ways that, whatever their interventions, they must always be compatible with the preservation of the commons—or, at any rate, with the relevant principles for the use of technological measures that have been agreed in a particular community (and which, as we said in Chapter Five, represent a key dimension of the local articulation of the Rule of Law).

IV Institutional roles and responsibilities

In the late 1970s, when techniques for assisted conception were being developed and applied, but also being seriously questioned, the response of the UK Government was to set up a Committee of Inquiry chaired by Mary Warnock. In 1984, the committee's report (the Warnock Report) was published.[62] However, it was not until 1990, and after much debate in Parliament, that the framework legislation, the Human Fertilisation and Embryology Act 1990, was enacted. This process, taking the best part of a decade, is regularly held up as an example of best practice when dealing with emerging technologies. Nevertheless, this methodology is not in any sense the standard operating procedure for engaging with new technologies—indeed, there is no such procedure.

The fact of the matter is that legal and regulatory responses to emerging technologies vary from one technology to another, from one legal system to another, and from one time to another. Sometimes, there is extensive public engagement, sometimes not. On occasion, special Commissions (such as the Human Genetics Commission in the UK) have been set up with a dedicated oversight remit; and there have been examples of standing technology

---

61 For a discussion in point, see David S. Wall, *Cybercrime* (Cambridge: Polity Press, 2007) where a number of strategies for dealing with 'spamming' are considered. As Wall says, if the choice is between ineffective legal rules and a technological fix (filters and the like), then most would go for the latter (at 201).

62 *Report of the Committee of Inquiry into Human Fertilisation and Embryology* (London: HMSO, Cm. 9314, 1984).

foresight commissions (such as the US Office of Technology Assessment);[63] but, often, there is nothing of this kind. Most importantly, questions about new technologies sometimes surface, first, in litigation (leaving it to the courts to determine how to respond) and, at other times, they are presented to the legislature (as was the case with assisted conception).

With regard to the question of which regulatory body engages with new technologies and how, there can of course be some local agency features that shape the answers. Where, as in the United States, there is a particular regulatory array with each agency having its own remit, a new technology might be considered in just one lead agency or it might be assessed in several agencies.[64] Once again, there is a degree of happenstance about this. Nevertheless, in a preliminary way, and with reference to the analysis of legal/regulatory mind-sets in this chapter, we can make five points.

First, if the question is put to the courts, their responsibility for the integrity of the law will push them towards a coherentist (whether of a traditional or new kind) assessment. Typically, courts are neither sufficiently resourced nor mandated to undertake a risk assessment, let alone adopt a risk management strategy (unless the legislature has already put in place a scheme that delegates such a responsibility to the courts).[65]

Secondly, if the question finds its way into the legislative arena, it is much more likely that politicians will engage with it in a regulatory-instrumentalist way; and, once the possibility of technological measures gets onto the radar, it is much more likely that (as with the institutions in the EU) we will see a more technocratic mind-set.

Thirdly, in some regulatory agencies there may already be a mind-set that is focused on making use of new technologies to improve performance relative to the agency's particular objectives (such those relating to consumer protection or competition). For example, the UK Financial Conduct Authority is conspicuously sensitised to 'reg tech' (in the sense of utilising information technologies to enhance regulatory processes with a particular emphasis

---

63 On which, see Bruce Bimber, *The Politics of Expertise in Congress* (Albany: State University of New York Press, 1996) charting the rise and fall of the Office and drawing out some important tensions between 'neutrality' and 'politicisation' in the work of such agencies.

64 Compare, Albert C. Lin, 'Size Matters: Regulating Nanotechnology' (2007) 31 *Harvard Environmental Law Review* 349.

65 Perhaps we should view Patent Offices in this light. In the 1980s, there were major decisions to be made about the patentability of biotechnological products and processes, models of which could not be brought into the Office to demonstrate how they worked and which also raised complex moral issues. For extended discussion, see Alain Pottage and Brad Sherman, *Figures of Invention: A History of Modern Patent Law* (Oxford: Oxford University Press, 2010); and, on the moral dimension of these debates, see Deryck Beyleveld and Roger Brownsword, *Mice, Morality and Patents* (London: Common Law Institute of Intellectual Property, 1993).

on monitoring, reporting and compliance).[66] Thus, the Authority recently announced that it was seeking views on the use of technology to make it easier for firms to meet their regulatory reporting requirements and, with that, to improve the quality of the information available to the agency.[67] Elaborating on this, the Authority said:

> The FCA regularly explores how technology can make our regulations more efficient and reduce the regulatory burden on firms. One of the ways we do this is through 'TechSprints' that bring together financial services providers, technology companies and subject matter experts to develop solutions to regulatory challenges.
>
> In November 2017, the FCA and the Bank of England, held a two-week TechSprint to examine how technology can make the current system of regulatory reporting more accurate, efficient and consistent. All regulated firms submit data to the FCA based on their financial activities. The data received from these regulatory reports are critical to our ability to deliver effective supervision, monitor markets and detect financial crime.
>
> At the TechSprint, participants successfully developed a 'proof of concept' which could make regulatory reporting requirements machine-readable and executable. This means that firms could map the reporting requirements directly to the data that they hold, creating the potential for automated, straight-through processing of regulatory returns.
>
> This could benefit both firms and regulators. For example, the accuracy of data submissions could be improved and their costs reduced, changes to regulatory requirements could be implemented more quickly, and a reduction in compliance costs could lower barriers to entry and promote competition.[68]

As this elaboration highlights, when a technocratic approach is applied to the regulatory enterprise, it can impact on thinking in all dimensions of that enterprise—on rule-making and standard-setting (including whether a technological measure will perform better than a rule), on monitoring compliance, and on detecting and correcting non-compliance.

Fourthly, the technological disruptions of the last couple of centuries have provoked a fractured legal and regulatory discourse which we can expect to be directed at emerging technologies. There are historic conventions about which institutions engage in which kinds of conversation (at any rate,

---

66 See en.wikipedia.org/wiki/Regulatory_technology (last accessed 4 November 2018).

67 FCA Press Release 20/02/2018. Available at www.fca.org.uk/news/press-releases/fca-launches-call-input-use-technology-achieve-smarter-regulatory-reporting (last accessed 4 November 2018).

68 *Ibid.*

explicitly)[69] but whether or not these institutional arrangements are now fit for purpose is a question that is open for debate.

Fifthly, if leaving so much to chance seems unsatisfactory, then it is arguable that there needs to be a body that is charged with undertaking the preliminary engagement with new technologies. The task of such a body would be to ensure that such technologies are channelled to our most urgent needs (relative to the commons); and, for each community, the challenge is to address the basic question of the kind of society that it distinctively wants to be—and, to do that, moreover, in a context of rapid social and technological change. As Wendell Wallach rightly insists:

> Bowing to political and economic imperatives is not sufficient. Nor is it acceptable to defer to the mechanistic unfolding of technological possibilities. In a democratic society, we—the public—should give approval to the futures being created. At this critical juncture in history, an informed conversation must take place before we can properly give our assent or dissent.[70]

Granted, the notion that we can build agencies that are fit for such purposes might be an impossible dream. Nevertheless, I join those who argue that this is the right time to set up a suitably constituted body[71]—possibly along the lines of the Centre for Data Ethics and Innovation (to set standards for the ethical use of AI and data) as announced by the UK Government in late 2017[72]—that would underline our responsibilities for the commons as well as facilitating the development of each community's regulatory and social licence for these technologies.[73]

---

69 The extent to which the courts do, or should, depart from questions of coherence and principle in order to purse questions of policy (in the fashion of legislators and regulators) is a long-standing bone of contention. Seminally, see Ronald Dworkin, *Taking Rights Seriously* (rev ed) (London: Duckworth, 1978); and, compare, Albert A. Ehrenzweig, 'Negligence without Fault' (1966) 54 *California Law Review* 1422 at 1476–1477:

> The negligence rule, though phrased in terms of fault, has with regard to tort liabilities for dangerous enterprise, come to exercise a function of loss distribution previously developed mainly within rules of strict liability. This new function of 'fault' liability has transformed its central concept of reprehensible conduct and 'foreseeability' of harm in a way foreign to its language and original rationale and has thus produced in our present 'negligence' language a series of misleading equivocations.

70 See, Wendell Wallach, *A Dangerous Master* (New York: Basic Books, 2015) 10.
71 Amongst many matters that invite further discussion, the composition of such a Commission invites debate. See, too, Wallach (n 70) Chs 14–15.
72 See 'Autumn Budget 2017: 25 things you need to know' (H.M. Treasury, November 22, 2017) point 16: available at www.gov.uk/government/news/autumn-budget-2017-25-things-you-need-to-know (last accessed 4 November 2018).
73 Compare Geoff Mulgan's proposal for the establishment of a Machine Intelligence Commission: available at www.nesta.org.uk/blog/machine-intelligence-commission-uk (blog 'A machine intelligence commission for the UK', February 22, 2016: last accessed 4

V Conclusion

In this chapter, I have sketched two modes of technological disruption, impacting on both the substance of legal rules and the form of regulation, and generating, in turn, three mind-sets—coherentist, regulatory-instrumentalist, and technocratic—that may manifest themselves in legal and regulatory discourse and debates. The bearing of these mind-sets on the replacement, refinement, and revision of legal rules is significant, possibly critical. For example, to the extent that the technocratic mind-set dominates, we can expect rules to be replaced and rendered redundant; to the extent that regulatory-instrumentalism dominates coherentism, we can expect new rules to be adopted in place of older traditional rules; and, to the extent that coherentism persists, we can expect there to be some tweaking of traditional rules and concepts to accommodate new technologies as well as resistance to both regulatory-instrumentalism and technocracy.

That said, the reception of new technologies is likely to differ from one place to another. The interaction between global and local politics is hard to predict. Technological management may not be the only game in town; and, we should not assume that the technocratic approach will be universally acceptable.

Nevertheless, unless we follow the example of Samuel Butler's eponymous Erewhonians,[74] who thought it appropriate to punish those who fall ill while sympathising with those who commit crimes, and who destroyed their machines, human agents will co-exist and evolve with their technologies. In the regulatory sphere, the direction of travel, I have suggested, is towards technological management; but, so long as the regulatory mind-set is divided in the way that I have sketched, the future of legal rules is unpredictable. Some will be replaced; others will be revised; and others will be renewed; but when, where, and how precisely this will happen is impossible to predict.[75]

November 2018); Olly Bustom et al, *An Intelligent Future? Maximising the Opportunities and Minimising the Risks of Artificial Intelligence in the UK* (Future Advocacy, London, October 2016) (proposing a Standing Commission on AI to examine the social, ethical, and legal implications of recent and potential developments in AI); HC Science and Technology Committee, *Robotics and Artificial Intelligence* HC 145 2016–17.

74 Samuel Butler, *Erewhon*, first published 1872. Available at www.planetebook.com (last accessed 4 November 2018).

75 There is also the possibility that rules and standards will be replaced by what Anthony Casey and Anthony Niblett term 'micro-directives': see 'Self-Driving Laws' (2016) 66 *University of Toronto Law Journal* 429 and 'The Death of Rules and Standards' (2017) 92 *Indiana Law Journal* 1401. Combining the 'certainty' of rules with the 'flexibility' of standards, micro-directives are able to give precise personalised guidance as agents move from one situation to another. For the most part, Casey and Niblett foresee such micro-directives as normative and assistive elements of the regulatory environment. However, they also anticipate the possibility that the death of rules and standards will eventually lead to the death of norms, making the point with which we started this book, namely that 'There will be no

Coda

BookWorld (a fictitious bookshop) has operated for a long time with a particular classificatory scheme, one that has served it well. The scheme starts
with fiction and non-fiction but then it employs various sub-classes. When
new books arrive at the shop they are shelved in accordance with this scheme.
Occasionally, there will need to be some discussion about the right place to
shelve a particular book, but this is pretty exceptional. In general, booksellers
and customers alike know where to find the titles in which they are interested. However, with an explosion of books about new technologies—about
genetics and the Internet, about neurotechnologies and nanotechnologies,
about AI and machine learning, and about blockchain and 3D printing—the
bookshop finds that it has no ready-made class for these titles. Some of these
books might be shelved, albeit imperfectly, under 'science fiction' or 'popu-
lar science', some under 'current affairs', some under 'smart thinking', and
so on. For a time, BookWorld persists with its historic classification scheme,
but these new books are shelved in ways that stretch and distort their classificatory indicators. Although the owners of BookWorld realise that their
classificatory scheme is no longer adequate, the bookshop soldiers on, determined to find a place for these new titles in its old and evidently outdated
scheme.

Eventually, BookWorld seeks the advice of an external consultant who
confirms that its shelving arrangements are no longer fit for purpose and
recommends that it should adopt a new classificatory scheme that recognises
the existence of, and divisions within, the technological literature. However,
the consultant also advises that the bookshop proprietors should think hard
about both their business objectives and whether even the new classification
scheme best serves those purposes.

Historically, BookWorld has invested heavily in building a long-term relationship with its customers. The owners have prided themselves that there
is more to their business than the selling of books. However, in a difficult
market, the bookshop proprietors agree that their priority is to sell books and
to maximise their profits. Given these purposes, they take further advice from
their consultant. The consultant advises that the bookshop should close its
bricks and mortar store and move its business online. Reluctantly, the proprietors heed this advice and close their store. The thriving street in which
BookWorld once opened its doors is now largely deserted. And, the moral
is that, when traditional coherentism gives way to more focused regulatory-
instrumentalism which then yields to efficient technocracy, community life is
not necessarily improved.

"norms" of driving when all of the vehicles are self-driving. All current norms will either
vanish or be entrenched in the vehicles' algorithms' (2016) 66 UTLJ 429, 442.

9

REGULATING CRIME

The future of the criminal law

I Introduction

The criminal law of any particular community tells us quite a lot about the particular nature of that community—about the particular kind of community that the community in question wants to be. The wrongs that each community decides to criminalise tell us not only which matters are taken most seriously—which wrongs cannot be left to the private law or informal regulation, which wrongs are to be met with penal sanctions, rather than being left to orders for compensation or injunction, which wrongdoers are to be stigmatised as 'criminals', and so on—but also which acts are viewed as being violations of the 'commons' for any community of human agents as well as which acts are treated as violations of the defining values of this particular community of human agents.

As each community articulates its understanding of public wrongdoing, its perspective on crime and punishment, we might find more than one view. For example, in some communities, the guiding perspective will be that the criminal law underwrites important moral values; it is engaged where the basic rights of agents are infringed or the basic duties are breached; and the function of the criminal justice system is to hold culpable individuals to account for their wrongs, to rebalance the moral ledger, to right moral wrongs, to correct an injustice. By contrast, in other communities, the criminal law will be viewed as being less concerned with moral accountability than with social protection and the effective management of risk; according to this view, the criminal law should be geared to protect the interests of all members of the community and it should channel behaviour in a way that minimises harm to these interests. George Fletcher detects in modern societies a movement from the former outlook to the latter:

> Modern societies may be losing their sense for criminal punishment as an imperative of justice. The tendency at the close of the twentieth century

is to focus not on the necessity that the guilty atone but on the pragmatic utility of using criminal sanctions to influence social behavior.[1]

The contrast between these two approaches—between, on the one hand, traditional coherentist and, on the other, regulatory-instrumentalist approaches to the criminal law—which reflects the shock of what I have called the 'first technological disruption', is one that threads through this chapter. However, the main business for this chapter is the 'second technological disruption' and the possibility of employing technologies that assess who is likely to commit crime, where it is likely to be committed, who are likely to be victims, and so on, this data then being used to anticipate and prevent crime. The big question is: how far is it acceptable to go with the technological management of crime? What kind of conversations should communities have before they agree a social licence for the use of technological measures to reduce the level of crime?

The chapter is in five principal parts. It starts (in Part II) with some further remarks about the direction of regulatory travel in the criminal justice system before rehearsing (in Part III) Bernard Harcourt's important caveats concerning both the effectiveness and the legitimacy of a predictive approach to policing and criminal justice.[2] We then return (in Part IV) to the question whether it is acceptable to use measures of technological management in relation to the preservation of the commons; and then (in Part V) we identify some of the headline questions that communities should be asking themselves before agreeing to the use of various kinds of technological measure intended to eliminate the practical options and opportunities for crime. Finally, in Part VI, some brief reflections are presented on the use of technological management as a way of responding to the vexed question of corporate criminal responsibility where we search in vain for an effective and acceptable accommodation of the interests of those who are victims of corporate wrongdoing and the imposition of vicarious liability on those who are the 'directing minds' of the company.

II The direction of travel

In the previous chapter, we noted Francis Sayre's seminal contribution to our understanding of the creation of absolute and strict liability offences in the nineteenth century;[3] and, we also know a good deal about the practical

---

1 George P. Fletcher, *Basic Concepts of Criminal Law* (Oxford: Oxford University Press, 1998) 203.

2 Bernard E. Harcourt, *Against Prediction* (Chicago: University of Chicago Press, 2007).

3 F.B. Sayre, 'Public Welfare Offences' (1933) 33 *Columbia Law Review* 55; and, see, too, Peter Ramsay, 'The Responsible Subject as Citizen: Criminal Law, Democracy, and the Welfare State' (2006) 69 *Modern Law Review* 29.

attitudes of the inspectorates in enforcing these regulatory laws.[4] So much for regulatory crime, but what is the direction of travel in the domain of real crime and how does the second technological disruption impact there?

Broadly speaking, the recent history of real crime reflects a tension between the politics of crime control ('law and order') and the civil libertarian demand for due process.[5] While due process requirements are seen by advocates of crime control as an unwelcome 'obstacle course' standing between law enforcement officers and the conviction of the guilty,[6] they are viewed by civil libertarians as essential protections of both innocent persons and the rights of citizens. With the rise of a regulatory-instrumentalist mind-set, and with the reduction of crime as the primary regulatory objective, we should expect there to be increasing pressure on the due process features of criminal justice systems; and, at any rate in the United Kingdom, the conventional wisdom is that these features have been weakened and watered down.[7]

With technology now in the mix,[8] it is important to appreciate that, even if the core criminal offences might remain much the same, many of the auxiliary rules of the criminal justice system—relating to police powers, the admissibility of evidence, the right to silence, jury trials, and so on—have been modified to reflect the increasing influence of a risk management mind-set. In this context, as Malcolm Feeley and Jonathan Simon summarise it in a much-cited essay, the emergent criminal law (and, concomitantly, criminal justice system) concerns itself

> with techniques for identifying, classifying and managing groups assorted by levels of dangerousness. It takes crime for granted. It accepts deviance as normal. It is sceptical that liberal interventionist crime control strategies do or can make a difference. Thus its aim is not to intervene in individuals' lives for the purpose of ascertaining responsibility, making the guilty 'pay for their crime' or changing them.

---

4 See, e.g., W. Carson, 'White Collar Crime and the Enforcement of Factory Legislation' (1970) 10 *British Journal of Criminology* 383.

5 Seminally, see Herbert L. Packer, *The Limits of the Criminal Sanction* (Stanford: Stanford University Press, 1969).

6 For example, see Sir Robert Mark's Dimbleby Lecture, *Minority Verdict*, in 1973, where hard questions were raised inter alia about juries, the role of lawyers, and technical defences.

7 For a civil libertarian assessment and critique, see Harriet Harman and J.A.G. Griffith, *Justice Deserted: The Subversion of the Jury* (London: NCCL, 1979); and, for an overview, see John N. Adams and Roger Brownsword, *Understanding Law* 4th edn (London: Sweet and Maxwell, 2006) Ch.6.

8 For the big picture, see Benjamin Bowling, Amber Marks, and Cian Murphy, 'Crime Control Technologies: Towards an Analytical Framework and Research Agenda' in Roger Brownsword and Karen Yeung (eds), *Regulating Technologies* (Oxford: Hart, 2008) 51.

Rather it seeks to regulate groups as part of a strategy of managing danger.[9]

And, sure enough, in the United Kingdom, we are no strangers to fierce debates about sentencing and penal policy in relation to 'dangerous' people where we find utilitarian and risk-management views being opposed by civil libertarian and rights-based views.[10]

What should we make of all this? According to Amber Marks, Benjamin Bowling and Colman Keenan:

> We are witnessing a gradual movement away from the traditional retrospective, individualised model of criminal justice, which prioritizes a deliberated and personalized approach to pursuing justice and truth, towards a prospective, aggregated model, which involves a more ostensibly efficient, yet impersonal and distanced, approach.[11]

Once the impact of the second technological disruption is felt, this movement becomes more technocratic: the movement that we now witness is towards 'actuarial justice' (powered by smart machines) which 'is based on a "risk management" or "actuarial" approach to the regulation of crime and the administration of justice'.[12] Looking ahead, Marks, Bowling and Keenan suggest the direction of travel is

> Towards an increasingly automated justice system that undercuts the safeguards of the traditional criminal justice model. This system favours efficiency and effectiveness over traditional due process safeguards and

---

9 Malcolm Feeley and Jonathan Simon, 'Actuarial Justice: the Emerging New Criminal Law' in David Nelken (ed), *The Futures of Criminology* (London: Sage, 1994) 173, 173.

10 See, e.g., the protagonists in the dangerousness debate of the 1980s: Jean E. Floud and Warren A. Young, *Dangerousness and Criminal Justice* (London: Heinemann, 1981); Nigel Walker, *Punishment, Danger and Stigma* (Lanham, Md: Rowman and Littlefield, 1980); and Anthony E. Bottoms and Roger Brownsword, 'Dangerousness and Rights' in J. Hinton (ed), *Dangerousness: Problems of Assessment and Prediction* (London: George Allen and Unwin, 1983) 9, and 'The Dangerousness Debate after the Floud Report' 22 *British Journal of Criminology* (1982) 229. For more recent debates about protective sentencing, in particular, for analysis of the indeterminate sentence of imprisonment for public protection, see Harry Annison, *Dangerous Politics: Risk, Political Vulnerability, and Penal Policy* (Oxford: Oxford University Press, 2015).

11 Amber Marks, Benjamin Bowling, and Colman Keenan, 'Automatic Justice? Technology, Crime, and Social Control' in Roger Brownsword, Eloise Scotford, and Karen Yeung (eds), *The Oxford Handbook of Law, Regulation and Technology* (Oxford: Oxford University Press, 2017) 705, 708.

12 Marks et al (n 11) 708.

is taking on a life of its own as it becomes increasingly mediated by certain types of technology that minimize human agency.[13]

To a considerable extent, this vision of automated justice anticipates the rapid development and deployment of smart machines. However, Marks, Bowling and Keenan specifically highlight the significance of big data, surveillance, and the 'new forensics'.

The new forensics, particularly the making, retention, and use of DNA profiles has been with us for some time. In the United Kingdom, advocates of crime control saw this biotechnological breakthrough as an important tool for the police and prosecutors; and the legislative framework was duly amended to authorise very extensive taking and retention of profiles. Even when legal proceedings were dropped or suspects were acquitted, the law authorised the retention of the profiles that had been taken. As a result, a DNA database with several million profiles soon was in place and, where DNA samples were retrieved from crime scenes, the database could be interrogated as an investigative tool (so that 'reasonable suspicion' could be cast on an individual, not by independent evidence, but by a 'match'). Precisely how much contribution to crime control was (or is) made by the profiles is hard to know. However, it was clear that the traditional rights of individuals were being subordinated to the promise of the new technology; and it was just a matter of time before the compatibility of the legislative provisions with human rights was raised in the courts. Famously, in the *Case of S. and Marper v The United Kingdom*,[14] the leading case in Europe on the taking (and retention) of DNA samples and the banking of DNA profiles for criminal justice purposes, the Grand Chamber in Strasbourg held that the legal provisions were far too wide and disproportionate in their impact on privacy. To this extent at least, individual human rights prevailed over the latest technology of crime control.

Although DNA profiling and the 'new forensics' (including digital fingerprinting) offer important investigative resources, these technologies are still operating *after* the event, after a crime has been committed. However, the key to the second technological disruption is to offer technologies that are capable of being used in ways that promise to operate *before* the event, anticipating and preventing the commission of crime. This is where Big Data, machine learning and artificial intelligence, operating in conjunction with

---

13 *Ibid.*, 705.
14 (2009) 48 EHRR 50. For the domestic UK proceedings, see [2002] EWCA Civ 1275 (Court of Appeal), and [2004] UKHL 39 (House of Lords). See, further, Roger Brownsword and Morag Goodwin, *Law and the Technologies of the Twenty-First Century* (Cambridge: Cambridge University Press, 2012) Ch.4.

human preventive agents and the required technological resources pave the way for the automation of criminal justice.

III Harcourt, prediction and prevention

As generally understood, 'actuarial justice' is characterised by four features: the normalisation of deviance, a focus on risk and the risk profiles of individuals, a managerial approach (offenders, both likely and actual, being classified in terms of a risk profile), and a forward-looking approach to the identification and incapacitation of (likely) offenders and re-offenders. However, once technological management is in play, although the emphasis is very much on prediction and prevention, the first feature changes: we might still accept that deviance is a normal human trait, but it is now compliance, not deviance, that regulators seek to normalise.

Responding to the emergence of actuarial justice, Bernard Harcourt declares himself to be against prediction. Notwithstanding that it 'has become second nature to think about just punishment through the lens of actuarial prediction',[15] Harcourt claims that we should think again; and, presenting three reasons in support of his position, he argues that we should eschew such a predictive approach to criminal justice.

First, Harcourt argues that a policy of focusing resources on high-risk groups might actually be 'counterproductive with regard to the central aim of law enforcement—to minimize crime'.[16] In some contexts, high-risk groups might be unresponsive to an increased risk of detection; in others, lower risk groups might prove to be responsive to a decreased risk of detection; and, in others, both of the unintended negative effects might be experienced.

Secondly, the effect of targeting persons with a record of criminal justice contacts (or some other proxy for the risk of criminality) is likely to be to ratchet up historic discrimination against minority groups and to lead to their over-representation in the prison population. This, Harcourt emphasises, is not simply a statistical phenomenon; in practice, there is a significant social cost because '[d]isproportionate criminal supervision and incarceration reduces work opportunities, breaks down families and communities, and disrupts education'.[17]

Thirdly, Harcourt argues that our espousal of an actuarial approach subverts our conception of just punishment. As Harcourt puts this critical argument:

> The perceived success of predictive instruments has made theories of punishment that function more smoothly with prediction seem more

---

15 Harcourt (n 2) 2.
16 *Ibid.*, 3.
17 *Ibid.*, 29.

natural. It favors theories of selective incapacitation and sentencing enhancements for offenders who are more likely to be dangerous in the future. Yet these actuarial instruments represent nothing more than fortuitous advances in technical knowledge from disciplines, such as sociology and psychology, that have no normative stake in the criminal law. These technological advances are, in effect, exogenous shocks to our legal system, and this raises very troubling questions about what theory of just punishment we would independently embrace and how it is, exactly, that we have allowed technical knowledge, somewhat arbitrarily, to dictate the path of justice.[18]

When Harcourt refers to theories of just punishment that have a normative stake in the criminal law, we can, as he rightly says, draw on a 'long history of Anglo-Saxon jurisprudence—[on] centuries of debate over the penal sanction, utilitarianism, or philosophical theories of retribution'.[19] However, I take the general point to be that, whichever of the candidate theories we 'independently embrace', we will have a view of the merits of an actuarial approach that turns on considerations other than whether the technical knowledge that we have 'works' in preventing and controlling crime. Beyond this, Harcourt has a more particular claim in mind. This is that the use of predictive approaches violates 'a core intuition of just punishment—the idea that anyone who is committing the same crime should face the same likelihood of being punished regardless of their race, sex, class, wealth, social status, or other irrelevant categories'.[20]

Since the publication of Harcourt's book, there has been a further fortuitous advance in technical knowledge, this time from data science, now offering new predictive instruments—new options for assessing and managing the risk of crime—enabled by machine-learning and AI. To the extent that these new instruments encourage the belief that we can predict more accurately who is likely to commit criminal offences, when, how, and where, this might lead to a change in the way that prospective offenders are monitored and, with that, to an increased likelihood of their being detected and punished. However, the logic of these predictive instruments does not stop at detecting and punishing crime; the logic is that prospective offenders should be prevented from committing offences in the first place.

If such a preventative strategy of risk assessment combined with risk management (let us call it 'smart prevention') means that agent A, who is assessed as being in a high risk category has less practical chance of committing, say, the offence of theft than agent B, who is assessed as being in a low risk

---

18 *Ibid.*, 3.
19 *Ibid.*, 188. For a succinct overview, see Ted Honderich. *Punishment: The Supposed Justifications* (London: Hutchinson, 1969).
20 Harcourt (n 2), 237.

category, then we might think that this violates the spirit, if not quite the letter, of the core intuition of just punishment that Harcourt identifies. In other
words, we might think that smart prevention violates, not the core principle
that there should be an equal chance of offending agents being caught and
punished, but the extended principle that agents should have an equal opportunity to commit crime. Whatever we might make of the core intuition and
its extension, Harcourt is surely right in insisting that we need an independent theory of just punishment (and, concomitantly, just prevention) if we are
to assess, not simply the effectiveness, but the legitimacy of an algorithmic
approach to criminal wrongdoing.

IV Crime, technological management and the commons

One afternoon in March 2017, an Islamist extremist intentionally drove a car
into pedestrians on Westminster Bridge; and, a few weeks later, on a busy Sat-
urday night, three men drove a white van into pedestrians on London Bridge
before jumping out carrying knives and heading for Borough Market where
they killed and injured many people. Actions of this kind clearly involve a
violation of the commons' conditions—quite simply, the primary impact was
on the rights of the agents who were killed or injured but there was also a
secondary impact insofar as the acts created an environment of fear that is
not conducive to agency. In the absence of respect for the life, for the physical integrity, and for the psychological security of agents, while some agents
might act 'defiantly', it is likely that others will not act freely but rather in a
way that is defensive, non-trusting, and inhibited. How should we respond
to such serious wrongdoing?

Already, the emphasis of the global response to 'terrorism' is on prevention. The intelligence services are expected to monitor 'high risk' persons
and intervene before they are able to translate their preparatory acts (which
themselves might be treated as serious criminal offences) into the death of
innocent persons. Concrete barriers can be installed on bridges to prevent
vehicles mounting the pavements. GPS-enabled protective fences might be
used to disable entry into high-risk areas;[21] and, in future, autonomous vehicles might be designed so that they can no longer be used as lethal weapons.[22]
So, prevention is already the name of this game and, according to the public
narrative, the smarter the prevention the better.

---

21 See, Graeme Paton, 'Digital force fields to stop terrorist vehicles', *The Times,* July 1, 2017,
p.4.
22 Apparently, when a truck was driven into a crowd at a Christmas market in Berlin in 2016,
the impact was mitigated by the vehicle's automatic braking system which was activated as
soon as a collision was registered: see Matthew Parris, 'It's wrong to say we can't stop this
terror tactic' *The Times,* August 19, 2017, p.25.

However, while smart prevention might share precisely the same regulatory purposes as counterpart rules of the criminal law, it differs from the traditional criminal justice response in three important respects. First, whereas punishment operates after the offence has been committed, smart prevention operates before offences are committed; smart prevention is an ex ante, not an ex post, strategy. Secondly, whereas traditional criminal justice might invest in more effective deterrence of crime, smart prevention invests in more effective anticipation, interception, and deflection of crime. Thirdly, and crucially, smart prevention does not give agents either moral or prudential reasons for compliance; rather, it focuses on reducing the practical options that are available to agents. The question (which is essentially a restatement of Harcourt's question) is whether, even if smart prevention is more effective in protecting the commons than the traditional rules of the criminal law, it is compatible with our general moral theory and, in particular, with our independent principles of just punishment and punitive justice.[23]

This leads to the following cluster of questions: (i) what are the costs for an aspirant community of rights if measures of smart prevention are adopted; (ii) how do the benefits and costs of smart prevention compare to other strategies for crime reduction; and (iii) even if capable of general justification, should smart prevention be treated as a strategy of last resort?

However, before we respond to these questions, an important caveat is in order. As Harcourt cautions, one of the objections to predictive criminal justice is that it tends to amplify historic discrimination against minority and ethnic groups. Indeed, questions have already been raised in the US about the hidden racial bias of apparently colour-blind algorithms used for bail and sentencing decisions.[24] If smart prevention exacerbates the unfairness that is otherwise present in criminal justice practice, it cannot be the strategy of choice. Accordingly, when comparing and contrasting smart prevention with traditional ex post criminal justice strategies, we will do so on a *ceteris paribus* basis.

(i) The costs of smart prevention

We can start with a general consideration—by now familiar from earlier chapters—that might be thought to militate against smart prevention. This

---

23 See, further, Deryck Beyleveld and Roger Brownsword. 'Punitive and Preventive Justice in an Era of Profiling, Smart Prevention and Practical Preclusion: Three Key Questions' (2019) IJLIC (forthcoming).

24 Sam Corbett-Davies, Emma Pierson, Avi Feller, and Sharad Goel, 'A computer program used for bail and sentencing decisions was labelled biased against blacks. It's actually not that clear', *The Washington Post* (October 17, 2016). NB the cautionary remarks about judicial reliance on algorithmic tools in *State of Wisconsin v Loomis* 881 N.W.2d 749 (Wis. 2016). Compare, too, Cathy O'Neil, *Weapons of Math Destruction* (London: Allen Lane, 2016).

is that, where a community has moral aspirations, it might be thought to be important—for reasons of both moral development and moral opportunity—to maintain rule-guided zones of conduct. The thinking is that it is in such zones that there is a public accounting for our conduct, that such accounting is one of the ways in which moral agents come to appreciate the nature of their most important rights and responsibilities, and that this is how in interpersonal dealings agents develop their sense of what it is to do the right thing.[25]

Although, as we have said, the potential significance of the complexion of the regulatory environment is a somewhat under-researched topic, it is clear that even effective measures of smart prevention might come at a price in an aspirant community of rights.[26] Distinctively, a State that employs smart prevention does not (to that extent) attempt to reason, either prudentially or morally, with agents. It channels the conduct of agents by restricting their practical options. In the extreme case, certain wrongdoing is not just immoral or imprudent, it is simply not a practical option. If we assume that there is a background moral awareness and understanding of the importance of the commons, we can trace how the use of smart prevention in relation to the protection of the essential commons' conditions might impact on four classes of agent in an aspirant moral community.

First, there are *moral compliers*: these are agents who are disposed to comply with commons-protective rules, and to do so for moral reasons. With the introduction of smart prevention, these agents (a) lose the opportunity to show that their compliance is morally inspired and (b) possibly no longer view compliance as a moral requirement. Arguably, in an aspirant community of rights, the latter is a more serious concern than the former. However, provided that there is an awareness of this risk, it should be possible to maintain the sense of moral obligation even in a context of reduced practical possibility for non-compliance.

Secondly, there are *prudential compliers*: these are agents who are disposed to comply with commons-protective rules, but who are so disposed only for longer-term prudential reasons. While the loss of opportunity to show that one's compliance is morally inspired does not seem to apply to such agents, and might be unimportant to them, should we be concerned that these

---

25 Compare, too, Anthony Duff's caution against changing the (rule-based) regulatory signals so that they speak less of crime and punishment and more of rules and penalties: R.A. Duff, 'Perversions and Subversions of Criminal Law' in R.A. Duff, Lindsay Farmer, S.E. Marshall, Massimo Renzo, and Victor Tadros (eds) *The Boundaries of the Criminal Law* (Oxford: Oxford University Press, 2010) 88, esp at 104. According to Duff, where the conduct in question is a serious public wrong, it would be a 'subversion' of the criminal law if offenders were not to be held to account and condemned. See, too, the argument in Alon Harel, 'The Duty to Criminalize' (2015) 34 *Law and Philosophy* 1.

26 See, Roger Brownsword, 'Lost in Translation: Legality, Regulatory Margins, and Technological Management' (2011) 26 *Berkeley Technology Law Journal* 1321.

agents are not fully morally committed to the protection of the commons? The short answer is that we should. However, these agents present no threat to the commons' conditions and it is not obvious that employing measures of smart prevention to protect these conditions (rather than relying on the longer-term disposition of prudential compliers) involves any additional loss of opportunity to engage morally with these agents.

Thirdly, there are *prudential non-compliers*: these are agents who are not disposed to comply with commons-protective rules and who will not comply if, opportunistically, they see short-term prudential gains by breach. Prima facie, the fact that smart prevention forestalls such opportunistic breach is a good thing. However, if effective prevention means that agents who are prudential non-compliers might fly below the radar, this might be a more general concern. For example, if such agents would, smart prevention aside, be convicted of offences and then engaged morally in relation to the importance of respect for the commons' conditions (perhaps in the way that offending motorists are sent to speed awareness courses or by undertaking acts of community service where the relevance of the violation to the *community* is underlined), this might be viewed as a loss.

Fourthly, there are *moral objectors*: these are agents who, even though generally disposed to comply with commons-protective rules, now resist in some respect because they believe that they have a moral objection. If what the State treats as an ostensibly commons-protective intervention is crazy, or if there is a reasonable disagreement about whether an ostensibly commons-protective measure is actually so protective, then the loss of opportunity for conscientious objection is an issue. In an aspirant community of rights, it will be a matter of concern that agents—particularly agents who are otherwise moral compliers—are forced to act against their conscience. Accordingly, in this kind of case, it is arguable that smart prevention should be avoided; regulators should stick with rules.

In practice, it has to be recognised that agents will not always fit neatly into these categories and their dispositions might not be constant.[27] Hence, assessing the impact of smart prevention is far from straightforward, the costs are uncertain, and human agents are not all alike. Nevertheless, there is no suggestion that smart prevention is such a costly strategy that it simply should not be contemplated.

(ii) Comparing the costs and benefits of smart prevention with other strategies

On the assumption that there is a background of moral education and awareness, how does smart prevention compare with moral and prudential strategies

---

27 Compare Mireille Hildebrandt, 'Proactive Forensic Profiling: Proactive Criminalization?', in Duff et al (n 25) 113.

for discouraging and dealing with serious crime? And, how do those latter strategies compare with one another?

First, how does smart prevention compare with ex post moral reason applied in a penal setting? Famously, Anthony Duff argues that the State should respect citizens as autonomous agents and should treat offending agents as ends in themselves. Whether, like Duff, we take a Kantian perspective or a rights-based view, we see that, even within traditional institutions of criminal justice, there is a further opportunity for moral education, both at the trial and, post-conviction, in rehabilitative penal institutions.[28] As Duff summarises the nature of the trial:

> The criminal law and the criminal trial are, like moral criticism, communicative enterprises which seek the assent and participation of those whom they address: the law seeks the allegiance of the citizen as a rational moral agent, by appealing to the relevant moral reasons which justify its demands; the trial seeks to engage the defendant in a rational dialogue about the justice of the charge which she faces, and to persuade her—if that charge is proved against her—to accept and make her own the condemnation which her conviction expresses.[29]

As for punishment, which follows conviction at the trial, this should be seen as a continuation of the communicative process. Thus,

> Punishment, like moral blame, respects and addresses the criminal as a rational moral agent: it seeks his understanding and his assent; it aims to bring him to repent his crime, and to reform himself, by communicating to him the reasons which justify our condemnation of his conduct.[30]

In this way, we can see the criminal law and punishment as 'a unitary enterprise of dialogue and judgment in which law-abiding citizens, defendants and convicted offenders are all called to participate'.[31] However, even when practised in an exemplary fashion, this ongoing dialogue with the criminal does not guarantee the safety of innocent agents; and, where the decisive educational intervention comes only after the commons' conditions and innocent agents have already been harmed, this may seem to be too little too late. The thought persists that smart prevention might be a better option—at any rate, it might be a better option if its preventative measures could be suitably integrated into the community's moral narrative.

---

28 See, R.A. Duff, *Trials and Punishments* (Cambridge: Cambridge University Press, 1986).
29 Duff (n 28) 233.
30 *Ibid.*, 238.
31 *Ibid.*

Secondly, a strategy that relies on adjusting the prudential disincentives against offending—for example, by intensifying surveillance or by making the penal sanctions themselves even more costly for offenders—invites a two-way comparison: first, with a strategy that relies on moral reason; and, secondly, with smart prevention. How might a prudential strategy fare if compared in this way? Although a criminal justice system that relies on prudential reasons remains a communicative enterprise, the register is no longer moral. Such a deviation from the ideal-type of a communicative process that focuses on *moral* reasons might be judged to be a cost in and of itself; and, if the practical effect of prudentialism, for both compliers and offenders, is to crowd out moral considerations,[32] the consequences involve a cost. Nevertheless, the selling point for such a prudential strategy is that agents who are capable of making reasonable judgments about what is in their own interest will respond in the desired (compliant) way and that this will protect innocent agents against avoidable harm. Of course, this sales pitch may be overstated. There is no guarantee that regulatees will respond in the desired way to a switch from moral exhortation to prudential sanctions;[33] and, as Harcourt warns, neither is there a guarantee that they will make the overall prudential calculation that regulators expect. The problem is that agents might continue to opt for non-compliance and prudentialism is powerless to stop them from so doing. At this point, smart prevention becomes the relevant comparator. If we want to reduce the possibilities for regulatees to respond in their own way to the State's prudential signals, then smart prevention looks like a serious option. To be sure, smart prevention gives up on any idea of a communicative process, moral or prudential. Practical options are simply eliminated and agents are presented with architectures, products, and processes that limit their possibilities for non-compliance. If smart prevention can outperform prudentialism, and if its restrictions can be integrated into the community's moral narrative, this looks like a serious candidate.

Thirdly, as has just been said, smart prevention may offer more effective protection of the commons than any other strategy. However, if it cannot be integrated into the community's moral narrative, this is a major cost; and if it means that we lose what is otherwise an opportunity to reinforce the moral message or, as we have already suggested, to re-educate those who have not internalised the moral principles, this is again a cost.[34] Stated shortly, this is the dilemma: if we act ex post, for some innocent agents, the State's moral

---

32 For relevant insights about the use of CCTV, see, Beatrice von Silva-Tarouca Larsen, *Setting the Watch: Privacy and the Ethics of CCTV Surveillance* (Oxford: Hart, 2011).

33 Being exhorted to do the right thing is one thing; being fined for doing something might be viewed, not as a response to moral wrong, but simply as a tax on conduct: see U. Gneezy and A. Rustichini, 'A Fine is a Price' (2009) 29 *Journal of Legal Studies* 1.

34 Compare Michael L. Rich, 'Should We Make Crime Impossible?' (2013) 36 *Harvard Journal of Law and Public Policy* 795.

reinforcement may be too late; but, if the State employs smart prevention ex ante, we may weaken the community's moral narrative and we may not realise that, for some agents, there is a need for moral reinforcement.

Provisionally, we can conclude that it is not obvious, a priori, which strategy should be prioritised or which combination of strategies will work best in protecting the commons while also assisting the community to realise its moral aspirations. Whatever strategy is adopted, its impact will need to be monitored; and, so far as smart prevention is concerned, a key challenge is to find ways of its being fully integrated into the community's moral narrative.

(iii) Even if it is capable of general justification, should smart prevention be treated as a strategy of last resort?

While there seems to be no reason, a priori, for treating smart prevention as a strategy of last resort, we say this subject to two provisos.

First, the case of moral controversy and conscientious objection is troubling.[35] To repeat a point that has been made before, for a community with moral aspirations, if a strategy that compels an agent to do x (or that prevents an agent from doing y) is morally problematic even where the agent judges that doing x (or not doing y) is the right thing to do, then it is (at least) equally problematic where the agent judges that doing x (or not doing y) is either straightforwardly morally wrong or an option that should not be taken. Accordingly, where there is any reasonable doubt about the measures ostensibly employed to protect the commons, smart prevention probably should be a last resort.

Secondly, it is important that smart prevention is employed in a way that maintains a clear and intelligible connection with the community's moral narrative. What this means is that its preventative measures are clearly designed to protect the commons, and that the members of the community retain the sense of why it is necessary to restrict agents' options in this way. Reliance on technological management might be fine, but there is a downside if agents lose their previous skills or know-how, or if they forget the moral (as well as prudential) rationale for what is now the way that things are.[36]

---

35 See, e.g., Evgeny Morozov, *To Save Everything, Click Here* (London: Allen Lane, 2013); and Roger Brownsword, '*Law as a Moral Judgment*, the Domain of Jurisprudence, and Technological Management' in Patrick Capps and Shaun D. Pattinson (eds), *Ethical Rationalism and the Law* (Oxford: Hart, 2016) 109.

36 See, Nicholas G. Carr, *The Glass Cage: Automation and Us* (London: W.W. Norton and Company, 2014). The standard example is that of pilots who sit up front monitoring the controls of automated aircraft but who forget how to actually fly the plane when it is necessary for them to take over the controls (as in the event of an emergency). We might have similar concerns about humans and driverless cars.

(iv) One final reservation

Smart preventive measures that are designed to protect the commons are not 'paternalistic' in the obviously objectionable sense of one agent (A) imposing A's view of what is in the interest of another agent (B) on B, where B is perfectly capable of making a self-interested judgment and where A has no independent justification for his view. Nevertheless, the use of smart prevention by A to protect conditions that are in the interest of all agents is paternalistic to the extent that B is prevented from acting in ways that, by compromising these conditions, are contrary to his best interests (as well as the best interests of all agents). But, of course, agents should see no problem with this: indeed, this might be the best argument for showing that paternalism is not always problematic.

What, though, if the reservation is expressed in terms of a concern that smart prevention signals that A no longer trusts B to act in ways that are consistent with the self-interest of both A and B? As we have said before, technological management affects the complexion of the regulatory environment and here, with smart prevention, we see it corroding the context in which agents trust one another. In some communities, this might be a fatal reservation about the use of smart prevention, whether in order to protect the commons' conditions or to prevent acts that violate the distinctive values of the community. In others, this might not be a fatal reservation; rather, there might be a willingness to make a trade-off between the corrosion of a culture of trust and the protection of the commons or the distinctive values of the community. Either way, these are matters for the community to debate. If the content of the criminal code says something about the particular kind of community that a group of human agents wants to be, so too does the group's willingness or reluctance to use smart prevention.

V Crime, technological management and questions for the community's social licence

The development of intelligent machines presents opportunities for new efficiencies and real benefits. In some sectors, perhaps in health research and health care, these new technologies might dramatically improve our ability to diagnose and treat serious diseases. Moreover, technological management promises transport systems (including autonomous road vehicles) that are safer, homes that are greener, and (by removing humans from dangerous production processes) workplaces that are less hazardous. If all this is achieved by a combination of automated processes, intelligent machines, and technological management, all functioning reliably and efficiently 24/7, why should we be concerned?

For example, if, instead of resorting to the legislation of strict (even absolute) liability regulatory offences (which, at the very least, calls for some explanation and justification), we have the option of relying on technological risk assessment and management to secure acceptable levels of human health and safety and environmental protection, why should we hesitate?

That said, there might be some applications of technological management where we should hesitate, where we should be concerned—and, arguably, the proposed adoption of intelligent machines in the core area of the criminal justice system is just such a case. Raising in an acute form the age-old question of the kind of society that we want to be, how far are we prepared to go with these new technological options? Are we ready to abandon rule-based proscription (subject to penalty) in favour of technological regulation of risk? How far are we prepared to accept the use of intelligent machines in at least an *advisory* capacity (for example, indicating crime hot spots to which resources should be deployed)?[37] Even though machine 'intelligence' is not directly comparable to human 'intelligence', is there any reason why humans should not make smarter decisions by taking advice from machines? What should we make of 'automated suspicion' generated by software that surveys the landscape of big data?[38] Over and above smart machines tendering advice or making provisional risk-assessments, how far are we prepared to *delegate* decision-making to intelligent machines? In the criminal justice system (if not in all safety systems), should there always be the option for a human operator to override a smart machine? If so, in what circumstances should that override be available? Last, but not least, recalling our discussion in Chapter Three, how important is the complexion of the regulatory environment; and, in particular, how important is it to the community that compliance is a matter of choice?

Without attempting to be comprehensive, in what follows we can highlight a number of debating points for a community that is discussing the terms and conditions on which it might grant a social licence for this (criminal justice) application of technological management. These points are: (i) the disruptive impact on the conventional understanding of criminal responsibility; (ii) the reinforcement of already unacceptable discrimination and bias (as Harcourt warns); (iii) the prevalence of false positives; (iv) whether decisions are transparent, explainable, and open to review; (v) the (ir)rationality of algorithmic decisions; (vi) the erosion or elimination of discretion and margins of tolerance if human regulators are no longer in the loop; and (vii) the concern that Harcourt flags up about the integrity of the community's thinking across punitive ex post and preventive ex ante criminal justice.

---

37 But NB J. Saunders, P. Hunt and J.S. Hollywood, 'Predictions put into practice: a quasi-experimental evaluation of Chicago's predictive policing pilot' (2016) 12 *Journal of Experimental Criminology* 347.

38 Compare Elizabeth E. Joh, 'The New Surveillance Discretion: Automated Suspicion, Big Data, and Policing' (Research Paper No 473, UC Davis Legal Studies Research Paper Series, December 2015); and Michael L. Rich, 'Machine Learning, Automated Suspicion Algorithms, and the Fourth Amendment' (2016) 164 *University of Pennsylvania Law Review* 871.

(i) The criminal responsibility of an agent

Already, the criminal justice system has been something of a test-bed for new technologies—for example, for new surveillance, locating, recognition, identifying, and tracking technologies—and we now have the prospect of intelligent machines insinuating themselves into this domain. While the former cluster of technologies (concerned with surveillance and identification, for example) does not disrupt either the simple idea that the rules of the criminal law presuppose that their addressees are agents who are 'responsible' in the sense that they have the capacity to follow and be guided by the rules, or the fundamental idea that justice demands that agents should be convicted and punished only for the criminal wrongs for which they are 'responsible' (in the sense that they committed these wrongs), the same may not be true of smart machines.[39] Actuarial justice might disrupt ideas of criminal responsibility in ways that trouble a community that is committed to liberal values.

In this vein, Nicola Lacey has cautioned against one direction of travel that might be indicated by the development of intelligent machines:

> More speculatively, and potentially more nightmarishly, new technologies in fields such as neuroscience and genetics, and computer programs that identify crime 'hot spots' that might be taken to indicate 'postcode presumptive criminality', have potential implications for criminal responsibility. They will offer, or perhaps threaten, yet more sophisticated mechanisms of responsibility-attribution based on notions of character essentialism combined with assessments of character-based risk, just as the emerging sciences of the mind, the brain, and statistics did in the late nineteenth century. Moreover, several of these new scientific classifications exhibit more extreme forms of character essentialism than did their nineteenth century forbears.[40]

This takes some unpacking. Neuroscience and new imaging technologies certainly challenge some of the details of criminal responsibility—for example, they raise questions about the age at which the brains of young persons have developed to the point at which the evidence supports treating them as having the capacity to follow the rules;[41] and, they push for an extension of

---

39 The language of 'responsibility' is notoriously ambiguous. See the classic discussions in HLA Hart, *Punishment and Responsibility* (2nd edn) (Oxford: Oxford University Press, 2008).

40 Nicola Lacey, *In Search of Criminal Responsibility* (Oxford: Oxford University Press, 2016) 170–171 (and for 'capacity', 'character', 'outcome', and 'risk'-based ideas of responsibility, see Ch.2).

41 See Lisa Claydon, 'Criminal Law and the Evolving Technological Understanding of Behaviour' in Roger Brownsword, Eloise Scotford, and Karen Yeung (eds), *The Oxford Handbook of Law, Regulation and Technology* (Oxford: Oxford University Press, 2017) 338, 348–350.

defences such as diminished responsibility[42] as well as a broader appreciation of mitigating factors. These neuro-technologies and their research results can also be invoked in a much more radical way to challenge the whole project of holding agents criminally responsible for their actions. If human conduct is entirely mechanistic, if humans cannot act otherwise than they do, then (so the argument goes) it cannot be right to punish them; they simply do not deserve to be punished.[43] Similarly, developments in human genetics can be cited in support of the claim that human decision-making is mechanistic and that the criminal justice system operates on entirely the wrong premises.[44] By contrast, intelligent machines put to actuarial use do not challenge this aspect of responsibility. Indeed, they promise to avoid the hard questions raised by neuro-imaging and genetic technologies. We do not need to agonise about the age of responsibility, about whether teenagers who do terrible things should face the death penalty,[45] because we can simply prevent these people (whatever their age) from doing wrong in the first place. Similarly, we do not need to worry about the range of defences because, once again, we can prevent the crimes being committed. So, if we are able through technological measures to prevent crime, we do not need to worry about whether human agents have the capacity to follow rules. Now, this might be a 'nightmare' for some communities because it disrupts their self-understanding as beings whose dignity demands (i) that they be treated not just as risks to be managed but as responsible agents who (ii) are guided by rules; but, elsewhere, there might be other, quite different, nightmares.

---

42 For some redrafting of the notion of diminished responsibility in English law, see s.52 of the Coroners and Justice Act, 2009. See, further, Lisa Claydon, 'Law, Neuroscience, and Criminal Culpability' in Michael Freeman (ed), *Law and Neuroscience* (Oxford: Oxford University Press, 2011) 141, 168–169.

43 Famously, see Joshua Greene and Jonathan Cohen, 'For the Law, Neuroscience Changes Nothing and Everything' (2004) 359 *Philosophical Transactions of the Royal Society B: Biological Sciences* 1775, 1784:

   Neuroscience is unlikely to tell us anything that will challenge the law's stated assumptions. However, we maintain that advances in neuroscience are likely to change the way people think about human action and criminal responsibility by vividly illustrating lessons that some people appreciated long ago. Free will as we ordinarily understand it is an illusion generated by our cognitive architecture. Retributivist notions of criminal responsibility ultimately depend on this illusion, and, if we are lucky, they will give way to consequentialist ones, thus radically transforming our approach to criminal justice. At this time, the law deals firmly but mercifully with individuals whose behaviour is obviously the product of forces that are ultimately beyond their control. Some day, the law may treat all convicted criminals this way. That is, humanely.

44 For a compelling assessment of these disruptive claims, see Stephen J. Morse, 'Law, Responsibility, and the Sciences of the Brain/Mind' in Roger Brownsword, Eloise Scotford, and Karen Yeung (eds), *The Oxford Handbook of Law, Regulation and Technology* (Oxford: Oxford University Press, 2017) 153.

45 The highest profile example is *Roper v Simmons* 543 US 551 (2005); see, for an impressive survey and analysis, O. Carter Snead, 'Neuroimaging and the "Complexity" of Capital Punishment' (2007) 82 *New York University Law Review* 1265.

One distinct possibility is that the disruption of criminal responsibility raises the concern that agents would be held to be responsible for wrong-doing, and punished, without there being any direct engagement with the question whether they actually did the wrong. For example, if a defendant's DNA profile matches DNA recovered from a crime scene, a jury might jump to the conclusion that the defendant must have committed the crime—and, particularly so, if prosecutors misrepresent the significance of the match. Rather than asking whether the defendant actually did the crime, the DNA match is taken uncritically (without thinking about the number of other agents who might match, without thinking that the defendant might have been present at the scene of the crime without actually committing the crime, and without thinking about the possibility that the defendant's samples were placed at the scene by a third party) as a proxy for guilt. Similarly, if a defendant, having been neuro-imaged, is found to have a neuro-profile that indicates guilt, this would suffice for conviction. So, if the classifications and characterisations produced by smart machines were sufficient to secure a conviction, this really would be a nightmare. It would mean that defendants were being held criminally responsible and punished for wrongs that they not only might not have committed, but simply on the basis that a smart machine places them in a 'high risk' category. Again, though, if the application of intelligent machines and technological management means that crimes are prevented, there will be no need to ask whether agents actually committed the crime or simply have a profile that matches that of whoever committed the crime, because the commission of crime will be a thing of the past.

Can we sleep easy? Perhaps not, because the real concern surely is that the smart machines will place agents in categories that, in some cases, then trigger various technological measures that restrict their practical options. Here, agents find themselves restricted on the basis of a risk profile that indicates that they are likely to commit a crime. If not punishment as we know it, there is a sense, as Mireille Hildebrandt has noted, in which this is, so to speak, 'proactive criminalization'.[46] If there is no possibility for an individual agent to challenge their classification (or, even a reversal of the usual burden of proof); or if there are large numbers of false positives; or if there is an unacceptable bias in the classifications, then a community might view this as a nightmare and withhold its social licence.

(ii) Discrimination

One of Harcourt's concerns is that the algorithms that operate in smart machines might amplify the pre-existing criminal justice bias against certain sectors of the community (for example, against young males from the inner

---

46 Hildebrandt (n 27).

cities). Even if the variables that are built into the algorithms are not directly discriminatory in a way that clearly puts them on the wrong side of the law, there might be less obvious biases that the community views as unacceptable.

Such cautions and concerns are not merely academic. As we have already noted, there are questions being raised in the US about the hidden racial bias of apparently colour-blind algorithms used for bail and sentencing decisions.[47] The COMPAS tool that is at the centre of one particular storm uses more than one hundred factors (including age, sex and criminal history) to score defendants on a 1–10 scale: defendants scored 1–4 are treated as low risk; defendants with scores of 5–10 are treated as medium or high risk. Although the factors do not include race, it is alleged that the algorithms implicitly discriminate against black defendants by assigning them higher risk scores (largely because, as a class, they have significant criminal histories and higher rates of recidivism). This means that there are significantly more black than white persons amongst those defendants who are classified as higher risk and who are then risk managed accordingly.

While there may be tricky questions about whether a particular algorithm is discriminatory, the general principle is clear. Quite simply, if the technological management of crime means that discrimination is perpetuated, whether overtly or covertly, which the community has judged to be unfair, this is not acceptable and no covering social licence should be given.

(iii) False positives

In communities that subscribe to due process values, the worst thing that the criminal justice system can do is to convict an innocent person of a crime. In the processes leading to conviction, there will be safeguards against such a false positive error being made; and, post-conviction there will be procedures for review and correction. In practice, everyone accepts that errors will be made but every effort should be made to avoid the punishment of the innocent.

If technological management can prevent crime, the risk of convicting and punishing the innocent is eliminated. However, the risk of false positives is still very much part of the actuarial approach.[48] In those communities that are committed to due process values and that believe, following Harcourt, that these values should also apply to preventative measures, any social licence for the technological management of crime will incorporate stringent safeguards against the risk of innocent agents having their practical options restricted. Such agents might not be incarcerated but their freedom of movement and other practical opportunities might be restricted.

---

47 See, Corbett-Davies et al (n 24), *State of Wisconsin v Loomis* (n 24), and O'Neil (n 24).
48 See Beyleveld and Brownsword (n 23) for discussion of the intelligibility of the notion of false positives in the context of preventive measures.

By contrast, in communities that are less committed to due process values or, indeed, are strongly committed to crime control, the concern about false positives will be less acute. Indeed, in these communities, where the public are only interested in crime reduction, and where there is a tendency for politicians and criminal justice professionals to be more concerned about false negatives (about the guilty who escape prosecution, conviction, or punishment) than false positives, we can expect there to be an uneven approach to the adoption of new technologies. As Andrea Roth pointedly argues:

> [A]lthough the motivation of law enforcement, lawmakers, and interest groups who promote 'truth machines', mechanical proxies, and mechanical sentencing regimes, is often a desire for objectivity and accuracy, it is typically a desire for a particular type of accuracy: the reduction of false negatives.[49]

What price, then, so-called 'safer societies' if their profiling, predictive, and preemptive technologies of crime control unfairly restrict or exclude swathes of agents who have not yet committed a crime and who would not have done so?

Once again, we find the recurrent tension in the design of criminal justice systems between a utilitarian and deontological ethics (particularly ethics that privilege human rights and due process). To the extent that communities approach penal justice in a Benthamite spirit, they will probably approach preventative justice in the same way and be willing to sign up to a social licence for actuarial justice provided that the distress caused by the restriction of false positives is broadly acceptable.[50] To the extent that due process values persist in the community, it is less likely that a social licence for actuarial justice will be granted.

(iv) Transparency and review

If the technologies (such as 'deep learning' or 'neural networks') deployed by the criminal justice system make decisions that cannot be understood or explained, this could be a cause for concern even in a community that is relatively untroubled by false positives. As Max Tegmark, noting the recent success of neural networks which outperform traditional easy-to-understand algorithms, asks:

> If defendants wish to know *why* they were convicted, shouldn't they have the right to a better answer than 'we trained the system on lots of data, and this is what it decided'?[51]

---

49 Andrea Roth, 'Trial by Machine' (2016) 104 *Georgetown Law Journal* 1245, 1252.
50 Compare the discussion in Bottoms and Brownsword (n 10), particularly with regard to the confinement of non-offenders, such as those suspected of carrying a dangerous disease but also internment during 'the troubles' in Northern Ireland.
51 Max Tegmark, *Life 3.0: Being Human in the Age of Artificial Intelligence* (London: Allen Lane, 2017) 108.

Suppose, for example, while driven cars are still on the roads, road traffic offences are detected by surveillance and recognition technologies and that smart machines automatically enforce financial penalties. If an agent is noti-fied that he has been detected committing an offence and that the penalty payment has been duly debited to his account, imagine the frustration of that agent if no information is given as to the time and place at which the offence was supposedly committed. Imagine the even greater frustration if the agent does not drive and that the only grounds for an appeal that are recognised by the system are of a medical kind—the system simply does not recognise the possibility of an error being made.[52] Finally, imagine the ultimate frustration: the agent is notified that the system has identified him as a high risk and that, given this profile, the agent's licence to drive is withdrawn with immediate effect. The agent's car is remotely immobilised. There is no appeal, no expla-nation, no review. Would any community, even a community of Benthamites, be willing to license such a system?

To be sure, once the community decides that there must be transparency (and the possibility of review), there is then much devil in the detail of the social licence.[53] In some applications, it makes no sense to disclose how the technology works lest regulatees are able to game the system (for example, if the tax authorities were to disclose to taxpayers how they identify those tax returns that raise suspicion and merit further investigation). In short, to say that there must be transparency is just the beginning; there need to be debates about how much transparency and in which contexts.[54]

(v) The rationality of algorithmic decisions

It might seem obvious that, when a smart machine is judging the character of John Doe, it should be referring to the profile of John Doe and not that of Richard Roe. Suppose, though, that John Doe, wishing to upgrade his smart

---

52 For this example, I am indebted to Daniel Seng in his response to my closing keynote at the 'Future of Law' conference held at Singapore Management University, October 26–27, 2107.

53 Compare Elizabeth Zima, 'Could New York City's Transparency Bill Be a Model for the Country?' *Government Technology* (January 4, 2018): available at www.govtech.com/pol-icy/Could-New-York-Citys-AI-Transparency-Bill-Be-a-Model-for-the-Country.html (last accessed 5 November 2018).

54 Compare, e.g.,Tal Zarsky, 'Transparent Predictions' (2013) *University of Illinois Law Review* 1503; and, *State of Wisconsin v Loomis* (2016) WI 68 for guidance on the transparency of the process: importantly, the accuracy of predictions was necessary but not sufficient, see [87–92]. For a rather different take on the transparency issue, see Joshua A. Kroll, Joanna Huey, Solon Barocas, Edward W. Felten, Joel R. Reidenberg, David G. Robinson, and Har-lan Yu, 'Accountable Algorithms' (2017) 165 *University of Pennsylvania Law Review* 633, proposing the use of a technological tool kit to enable verification that an algorithm oper-ates in ways that are compliant with legal standards of procedural fairness and due process (consistent and congruent application).

car, applies for a credit facility but that he is turned down by a smart machine that classifies him as a bad risk. When John Doe challenges the decision, he learns that one of the previous occupiers of his house, one Richard Roe, had a record of non-payment of loans. But, why, Doe asks, should the credit record of an unrelated third party, Roe, count against my application? Is that not unfair and irrational? To which the response is that the machine makes more accurate decisions when it uses third-party data in this way; and that, if such data were to be excluded from the calculation, the cost of credit would increase.

In fact, this is not a novel issue. In the English case of *CCN Systems Ltd v Data Protection Registrar*,[55] on facts of this kind, the tribunal held that, while it accepted that such third-party information might have general predictive value and utility, its use was unfair to the individual and could not be permitted. Similarly, Doe might argue that he has been treated unfairly if his application for credit is successful but the terms and conditions of the facility reflect the fact that (because of unrelated third-party data) he is classified as a higher-than-average risk; and, once again, the response will be that the costs of credit will be increased if such data are excluded. How is the choice to be made between the general utility of the credit algorithms and the unfairness of particular decisions?

Now, while it is one thing for a smart machine to deny an agent access to credit, it is another matter for intelligent machines to make risk assessments in the criminal justice system where exclusionary or pre-emptive decisions are likely to have more serious consequences for agents. For example, smart machines might be deployed to initiate pre-emptive action against agents who are judged to be high risk, to deny bail to arrestees who are assessed as high risk, and to extend custodial terms for offenders who, at the point of release, are still judged to be 'dangerous'. If, in making these decisions, unrelated third-party data are used, this seems to be contrary to due process. Yet, in all these cases, smart machines churn out decisions that are in line with Benthamite principles and that are generated by the logic of big data but that depart from the ideal of a 'justice' system.

Once again, we find a conflict between, on the one hand, deontological ethics and due process and, on the other, utilitarian principles that will endorse a system of crime control so long as it performs better than any rivals in maximising overall utility. The distress caused to Doe and others is not ignored; but it is treated as simply reducing the net utility of a system entrusted to smart machines (or of a system that has such machines operating alongside human decision-makers).[56] For Benthamite communities, there is

---

55 Case DA/90 25/4/9, judgment delivered 25 February 1991. I am grateful to Chris Reed for drawing my attention to this case.

56 For a relatively favourable report on a bail tool developed by the Arnold Foundation, see Shaila Dewan, 'Judges Replacing Conjecture with Formula for Bail' *New York Times*, 26 June 2015: available at http://www.nytimes.com/2015/06/27/us/

no categorical reason to reject the use of apparently irrelevant variables; provided that the algorithms 'perform' in utility-maximising ways, they can be given a social licence. For communities that subscribe to other moralities, the view is likely to be very different.

(vi) The erosion of discretion and margins of tolerance: taking humans out of the regulatory loop

Technological management might take humans out of the loop on both sides of the regulatory relationship. Indeed, with humans out of the loop on both sides, it might not be appropriate to characterise this as a relationship. If this means that appeals against preventative decisions are fully automated, we might end up with a Kafkaesque dystopia that no community would license. So, for some time, European regulators have accepted that data subjects have a right not to be subject to decisions based solely on automated processing (including profiling) where, in the language of Article 22 of the General Data Protection Regulation (GDPR), such decisions produce 'legal effects or similarly significantly affects him or her.' While this provision is extremely vague and open to all manner of interpretations, it puts down a marker: Europeans are not willing to grant a social licence for automated decision-making without some heavy qualification about bringing human agents back into the loop.

It is not just the possibility of review that is a concern. Human regulatees do have a relationship with human enforcement agents, in which the former might have a reasonable expectation of some margin of tolerance and the latter a discretion to deal with the case in a way that does not trigger criminal proceedings. It might seem paradoxical that communities want some latitude to commit crime; but communities debating the use of technological management and who declare that they would like to see crime-free societies might need to be careful that they do not get what they wish for.[57] It is not just a matter of having some workable wiggle-room, there are major questions here that we have already aired about matters of conscience.[58]

---

turning-the-granting-of-bail-into-a-science.html?_r=0 (last accessed 5 November 2018). According to Dewan, although the tool does not take into account some of the factors that human judges and prosecutors tend to treat as material (such as the defendant's employment status, community ties, and a history of drug and alcohol abuse) it improves accuracy by focusing on fewer than ten factors (principally, age, criminal record, previous failures to appear in court) and by giving recent offences greater weight.
57 Compare Michael L. Rich (n 34).
58 See Chapter Three.

(vii) Harcourt's concern for coherence

Finally, we can revisit Harcourt's concern for coherence. As we have already indicated, in some communities it might be seen as important that there is an integrity across penal and preventative criminal policies and practices. At the heart of this is the principle that the innocent should not be convicted.

In relation to penal policy, if legal officials impose penalties on an agent when they know that no crime has been committed or when they know the agent to be innocent, there are two things that we can say: one is that this is not a case of punishment at all; and the other is that it is obviously unjust. Where legal officials believe in good faith that they are imposing penalties on an agent who has committed a crime, but where the agent is actually innocent, what should we say? Strictly speaking, this is not a case of punishment and again it is unjust; but it is different from the first case of sham punishment. Whereas, in the former case, there is no belief on the part of legal officials that the facts correspond to our understanding of punishment, in the latter, they believe in good faith (albeit mistakenly) that they are imposing a punishment on an offender for the crime committed. If the former is a clear abuse of authority, the latter is a good faith error.

Whatever we make of these puzzles, we need to elaborate on some principles that guide our thinking about just punishment. Building on earlier discussions, a plausible starting point for any community is that the general justification for punishment is to protect the commons conditions; and, if punishment is to protect rather than compromise the commons conditions it should be applied in a way that is neither random nor unfair. Taking this approach, it is suggested that, in order to be just, the application of punishment should be guided by the following principles:

- the principle of *generic relevance* (the wrongdoing at which penal sanctions are directed should be such as touches and concerns the commons' conditions);
- the principle of *accuracy* (penal sanctions should be applied only to those agents who have committed the relevant offence);
- the principle of *proportionality* (the particular penal sanction that is applied—that is to say, the particular disposition chosen from the range of penal options—should be proportionate to the seriousness of the particular wrongdoing);
- the principle of *least restrictiveness* (a penal restriction should be imposed only to the extent that it is necessary); and,
- the principle of *precaution* (appropriate safeguards should be adopted lest penal sanctions are applied to agents who have not committed the relevant offence).

Provided that these principles are observed, we should treat a particular penal response as just.

Now, if these principles are to be translated across to preventative measures, then technological measures need to be very carefully and precisely targeted—relating to the protection of the commons, and either reducing only illegitimate (commons-compromising) options or reducing legitimate options for true positives, and only true positives, where this is necessary in order to prevent violations of the commons' conditions; or, reducing legitimate options for false positives where there are compelling reasons to believe that they are true positives and where this is a necessary and proportionate measure for the protection of the commons' conditions; or, in an emergency, where the measures are a necessary and proportionate response to a clear and present danger to the commons.

Similarly, in communities that are willing to license the use of technological management to prevent crimes that do not touch and concern the commons' conditions, they might wish to maintain an integrity between the principles that guide their ex post penal approach to such wrongs and wrongdoers and the principles that guide their preventative strategy. If the guiding principles of penal policy are Benthamite, these same principles will generally support the use of effective preventive measures; but, if the guiding principles are deontological, we can expect them significantly to constrain whatever licence is agreed.

There is one final consideration. If the due process constraints on the use of smart prevention seem unduly restrictive, imagine that we operate with unconstrained smart prevention as setting the benchmark for ex post criminal convictions. Granted, we can never be absolutely certain that agent A committed crime x. Nevertheless, is there not a fundamental difference between asking directly whether A did x and asking whether A has a 'profile' that fits with the doing of x? Or, do we think that it is simply a matter of which approach 'performs' better?

(viii) Taking stock

In almost all these debating points, a familiar pattern—indeed, an old story—emerges. In those quarters where liberal values and due process are prioritised, there will be resistance to proposals that the criminal justice system should be run as an exercise in risk management, underpinned by utilitarian principles, and now able to take advantage of new technological measures. In other words, we can expect the particular social licences that are agreed for the adoption of actuarial justice to be narrow where coherentist due process values hold and more expansive where the mind-set is utilitarian, regulatory-instrumental and technocratic.

VI Corporate criminal responsibility and technological management

Following the lethal fire at Grenfell Tower in London, it was reported that the police were considering possible charges of corporate manslaughter.[59] The difficulty of obtaining convictions for this offence, as indeed for a whole raft of corporate crime, is notorious. We need only recall the failed prosecutions following the sinking of the Herald of Free Enterprise as it left Zeebrugge ferry terminal (resulting in the loss of 193 lives) and the Hatfield rail disaster. Since that time, though, the common law has been superseded by a bespoke statutory offence. According to section 1(1) of the Corporate Manslaughter and Corporate Homicide Act 2007, which was designed to improve the chances of securing a conviction for crimes of this kind, the offence is committed where the way in which the corporation's activities are managed or organised '(a) causes a person's death, and (b) amounts to a gross breach of a relevant duty of care owed by the organisation to the deceased'. For the purposes of the Act, section 1(4)(b) defines a breach as gross if the conduct that is alleged to amount to a breach of a relevant duty 'falls far below what can reasonably be expected of the organisation in the circumstances'. Since the enactment of the new law, there have been a few successful prosecutions, but there is a question mark about whether the rule changes that have been made will make much difference.[60] If this risk management legislative package does not prove to make a significant difference, should we be thinking about a technocratic approach?

What kind of technological measures might be contemplated? Following the capsize of the Herald of Free Enterprise, P&O fitted the bridges on its ferries with warning lights so that, in future, ships' masters would be alerted if the bow doors were not closed. Similarly, after the Southall rail crash, Great Western installed Automated Train Protection systems so that any train passing through a red light would be brought to a halt.[61] Following these examples, if one of the causes of the fire at Grenfell Tower was that safety standards were not observed, might it be possible to design materials so that they cannot be fitted unless they are compliant? Might it be possible, for example, to learn from design practices in other sectors?[62]

---

59 See www.bbc.co.uk/news/uk-40747241 (last accessed 5 November 2018).
60 See, James Gobert, 'The Corporate Manslaughter and Corporate Homicide Act 2007— Thirteen Years in the Making but was it Worth the Wait?' (2008) 71 *Modern Law Review* 413; and Sarah Field and Lucy Jones, 'Five years on: the impact of the Corporate Manslaughter and Corporate Homicide Act 2007: plus ça change?' (2013) 24 *International Company and Commercial Law Review* 239.
61 For discussion, see Gobert (n 60) 418 and 424–425.
62 On design in health care, compare Karen Yeung and Mary Dixon-Woods, 'Design-Based Regulation and Patient Safety: A Regulatory Studies Perspective' (2010) 71 *Social Science and Medicine* 502.

If it were possible to prevent fires such as that at Grenfell Tower by embedding safety in the only available materials (or by taking humans out of the loop in construction) how would communities view the proposed technological measures? Arguably, they would see some parallels here between the replacement of absolute and strict liability health and safety rules with measures of technological management and the replacement of vicarious corporate liability with measures of technological management that are designed to maintain the health and safety of both the occupiers of tower blocks and the employees of large corporations.

On this analysis, if communities tend towards a positive view of technological measures that are used in this way, then the challenge is to develop effective technological tools rather than making the case for the legitimacy of such use.

VII Conclusion

There are four questions that each community of human agents must ask itself. First, what acts does it treat as infringing upon the conditions of the commons; secondly, to what extent is it prepared to use measures of technological management, rather than rules, to prevent such acts and how does this square with its view of penal sanctions; thirdly, what acts, not touching and concerning the commons' conditions does it view as going to the distinctive identity of the community such as to warrant the use of the criminal law; and, fourthly, to what extent is it prepared to use measures of technological management, rather than rules, to prevent such acts?

At first blush, one might wonder who could reasonably gainsay the proposal that, by using technological management, crime could be prevented. Such a strategy, however, even if effective, restricts the autonomy of agents and it might interfere with their moral development. If the purpose of employing technological management is to protect the commons' conditions, a community might accept that a reasonable trade-off is being made. In particular, it might accept that technological management of the existence conditions and of the general conditions for human health and safety is acceptable. Where, however, the purpose is less fundamental (in other words, where it is not related to the preservation of the commons' conditions) communities should develop their own social licence for resort to technological measures. If the criminal justice system is to speak for liberal values and for the community, it is imperative that there is an inclusive conversation before the terms of any such social licence are agreed.

10

REGULATING INTERACTIONS

The future of tort law

I Introduction

In many ways, the development of tort law runs parallel with that of criminal law. The 'first disruption' signals the emergence of a more regulatory approach. Capturing this development in an historical review of the civil liability regimes in Germany, the USA, and Russia, Gert Brüggemeier observes that, as we leave the 'classical world of individual responsibility' and envisage a 'different world of industrial modernity', the traditional model of liability loses its appeal.[1] In its place, we find that:

> [a] variety of depersonalised 'stricter' forms of enterprise liability has been developed, both openly and calmly, by special legislation and by case-law. In part liability has been substituted by insurance … Balancing of probabilities and risk absorption by the best loss spreader have replaced the fault rationale. As far as physical injuries by the materialisation of technical risks are concerned the law of delict has in continental Europe been replaced by insurance schemes or liability law operates in the meantime *like* an insurance system. Finally compensation of damages ends as a loss distribution by collectivities (*via* damage division agreements between insurers).[2]

---

1   Gert Brüggemeier, 'Risk and Strict Liability: The Distinct Examples of Germany, the US and Russia' EU1 Working Paper LAW 2012/29, p.26. I am grateful to Ronan Condon for drawing this paper to my attention as well as for making other helpful suggestions in relation to this chapter.

2   *Ibid*.

Although, as Brüggemeier points out, the general public pay twice (once as taxpayers and then again as consumers of goods and services) for this modern liability regime, its virtue is that 'neither the business risk-taker nor the fortuitous risk-exposed victim is left with the full burden of the damages or losses'.[3]

As regulatory law emerges, tort law is not just supplemented by health and safety inspectorates and by insurance schemes, it is reworked in support of background policy objectives. This generates some dissonance within the law—for example, as strict liability and enterprise liability in negligence replaces traditional fault-based liability,[4] and as individuals who act unreasonably are able to 'depersonalise' or transfer their liability. It is in the nature of this transformation that the private law of torts is co-opted for the purposes of general safety and public risk management. Thus, we read that, in Sweden, tort law is viewed as the 'cement' that operates within those regulatory risk-management regimes that now provide for and channel compensatory claims in relation to road traffic, workplace injuries, medical accidents and dangerous products.[5]

When, following the 'second disruption', technological management arrives on the scene, there need to be decisions about who is responsible for what, if, and when something 'goes wrong'—for example, if an autonomous vehicle is involved in a collision with another vehicle or with pedestrians. Unless these questions have been comprehensively addressed in a risk management protocol, claimants are likely to turn to tort law in order to seek redress and the dissonance that predates technological management is likely to persist, resulting in some incoherence in the law.

These themes are developed in the chapter in four principal sections. First, the contrast between the two ideal-typical liability paradigms—the simple tort model and the model of regulatory risk management—is developed.[6]

---

3 *Ibid.*
4 Compare Albert A. Ehrenzweig, 'Negligence without Fault' (1966) 54 *California Law Review* 1422, 1476–77:

> The negligence rule, though phrased in terms of fault, has with regard to tort liabilities for dangerous enterprise, come to exercise a function of loss distribution previously developed mainly within rules of strict liability. This new function of 'fault' liability has transformed its central concept of reprehensible conduct and 'foreseeability' of harm in a way foreign to its language and original rationale and has thus produced in our present 'negligence' language a series of misleading equivocations.

5 Sandra Friberg and Bill W. Dufwa, 'The Development of Traffic Liability in Sweden' in Miquel Martin-Casals (ed), *The Development of Liability in Relation to Technological Change* (Cambridge: Cambridge University Press, 2010) 190, 215.
6 Compare F. Patrick Hubbard, ' "Sophisticated Robots": Balancing Liability, Regulation, and Innovation' (2014) 66 *Florida Law Review* 1803, 1812:

> [Currently, there are] two distinct approaches to safety. The first uses a judicially imposed liability system of corrective justice that requires wrongdoers to compensate victims for

Secondly, in the light of that contrast, we sketch the general direction of regulatory travel but emphasise that there is no smooth transition from the former model to the latter: not only can there be turbulence within (that is, internal to) each model that leads to rule revision but also resistance to the subordination of the former model to the latter. Thirdly, we consider the ways in which the three regulatory mind-sets that form the backcloth to this part of the book bear on debates about the regulation of autonomous vehicles. Finally, by focusing on the challenge of getting the regulatory framework right for patient safety in hospitals, we see how the simple tort model, the model of regulatory risk management, and the pressure for a more technocratic approach come into tension, illustrating again the potential significance of the complexion of the regulatory environment and the quest for coherence (now in the sense of having a coherent view about when regulators should engage a traditional mind-set, when a regulatory-instrumentalist mind-set, and when a technocratic mind-set).

II Two liability models

In this part of the chapter, we can elaborate somewhat the contrast between the simple coherentist fault-based model of liability for unintentional harms and the regulatory-instrumentalist model that seeks to support innovation while managing and maintaining risks to human health and safety at acceptable levels.

(i) The simple tort model

In *Donoghue v Stevenson*,[7] in what is perhaps the best-known judgment in the common law jurisprudence of the last century, Lord Atkin opened up the range of potential tortious liability for negligence. His Lordship, together with the other majority Law Lords, held that it should no longer be a sufficient answer for a defendant to plead that, even if acting carelessly and causing harm to the claimant, it was acting pursuant to a contract with a third party. Accordingly, in *Donoghue* itself, it was of no assistance to the defendant—a local manufacturer of ginger beer and lemonade—to plead, in response to the claimant consumer of one of its drinks, that there was no case to answer simply because the product was manufactured and supplied in performance of a contract to which the claimant was not a party. Having removed this contractual restriction, on what principle was

---

injuries caused by a wrong ... The second approach involves collective determinations of the best way to address safety and imposes this determination with regulatory commands backed by the threat of sanctions for violations.

7 [1932] AC 562.

the now extended liability in negligence to be regulated? Famously, Lord Atkin said:

> [In] English law there must be, and is, some general conception of relations giving rise to a duty of care, of which the particular cases found in the books are but instances. The liability for negligence, whether you style it such or treat it as in other systems as a species of 'culpa', is no doubt based upon a general public sentiment of moral wrongdoing for which the offender must pay. But acts or omissions which any moral code would censure cannot, in a practical world, be treated so as to give a right to every person injured by them to demand relief. In this way rules of law arise which limit the range of complainants and the extent of their remedy. The rule that you are to love your neighbour becomes in law, you must not injure your neighbour; and the lawyer's question, Who is my neighbour? receives a restricted reply. You must take reasonable care to avoid acts or omissions which you can reasonably foresee would be likely to injure your neighbour. Who, then, in law is my neighbour? The answer seems to be—persons who are so closely and directly affected by my act that I ought reasonably to have them in contemplation as being so affected when I am directing my mind to the acts or omissions which are called in question.[8]

For lawyers who are trying to advise clients on their potential exposure to negligence liability, the principal task is to differentiate between those whom one might reasonably foresee being injured by one's activities and those who seem to lie beyond the zone of reasonably foreseeable risk. For practitioners, whether or not this responsibility for one's careless acts or omissions can be anchored in some primitive sense of moral wrongdoing is of marginal interest—the immediate interest will be in how the test of reasonable foreseeability (or the implicit test of proximity) is applied in the cases. However, for our purposes, the idea that tort law is a practical expression of simple moral sentiment is of the first importance. For, it suggests that communities develop in an organic way their sense of what it is to respect one another and this is then translated, in a practical world, into actionable tort laws. Of course, what one's neighbours in a literal sense might reasonably expect of one another might not be the same as what one's neighbours in Lord Atkin's figurative use might reasonably expect: quite possibly, for example, one's immediate friends and neighbours will expect more in the way of positive

---

8 [1932] AC 562, 580. Compare, however, Ehrenzweig (n 4). While Lord Atkin talks the talk of traditional fault liability, the actual outcome in *Donoghue* is to achieve a fairer balance between the interests of manufacturers and the ultimate consumers of their products (whether ginger beer or automobiles).

assistance and cooperation than might be expected by strangers. However, even if there is some variation in the intensity of one's obligations depending upon one's relationship to a particular other, Lord Atkin is interpreting the general moral sentiment as being that, if you can reasonably foresee that your acts might injure a particular person or class of persons, then the right thing to do is to take reasonable care to avoid that injury eventuating. While injuring a neighbour through lack of due care and attention is not as culpable as injuring a neighbour intentionally or recklessly (in the sense that one does not care whether or not one's neighbour is injured) it is a wrongdoing for which one is expected to answer. In other words, in the language of the modern law, it is prima facie 'fair, just and reasonable' to hold those who are careless liable to compensate those who are injured as a reasonably foreseeable result of their carelessness.

Summing up, the salient features of this traditional coherentist model of tort law are that:

- The wrong done is relative to the community's standards of reasonable conduct but it is essentially a matter for private settlement (whether by apology, restoration, or compensation) as between the wrongdoer and the victim.
- Liability is personal (not vicarious or collective, and not simply a call on insurance).
- Liability is based on fault or unreasonable acts/omissions (not no-fault or strict liability).
- Liability implies a negative (i.e., critical) judgment relative to the community's standards of reasonable conduct (to 'neighbourliness' as Lord Atkin put it in *Donoghue*).

By contrast, as we shall now explain, the ideal-typical regulatory-instrumental approach operates on very different assumptions.

(ii) The regulatory-instrumentalist approach

Regulatory-instrumentalists seek to balance the sometimes competing demands of the community (i) that there should be support for beneficial innovation (or, at any rate, that innovative businesses should not be over-exposed to liability) and (ii) that the risks to human health and safety, property and so on that are presented by innovative technologies should be managed at an acceptable level. As Albert Ehrenzweig notes, the development of stricter liability in the nineteenth century can be explained by

a certain sentiment of hostility against innovations caused by the increase of industrial risks and financial failures; ...the humanitarian demand for broader protection in a more social minded era; and finally ...the fact that the growing industrial wealth and stability, coupled with a spreading system of liability insurance, made it easier to dispense with the injurer's protection afforded by a liability law primarily based on fault.[9]

In this way, a traditional coherentist notion of moral fault and personal liability gave way to a regulatory-instrumentalist espousal of a stricter liability for enterprises and a reworking of the law of negligence.

Product liability regimes are a nice example of this kind of approach. For example, in the Recitals to Directive 85/374/EEC (on liability for defective products) we read:

> Whereas liability without fault on the part of the producer is the sole means of adequately solving the problem, peculiar to our age of increasing technicality, of a fair apportionment of the risks inherent in modern technological production.

However, although the pathway to compensation is eased by removing the requirement that fault be proved, it is no part of this approach to the apportionment of risks to name, shame and blame defendants where accidents happen or unintentional harm occurs. So, for example, in one of the leading US product liability cases, *Beshada v Johns-Manville Products Corp.*,[10] the defendant asbestos manufacturers argued that they could not reasonably know that their finished products supplied for use in the shipyards were still hazardous and could cause dust-diseases. However, as the Supreme Court of New Jersey emphasised, in a no-fault product liability regime, arguments of this kind were not relevant. In his concluding remarks, Pashman J put the point thus:

> Defendants have argued that it is unreasonable to impose a duty on them to warn of the unknowable. Failure to warn of a risk which one could not have known existed is not unreasonable conduct. But this argument is based on negligence principles. We are not saying what defendants should have done. That is negligence. We are saying that defendants' products were not reasonably safe because they did not have a warning. Without a warning, users of the product were unaware of its hazards and could not protect themselves from injury. We impose strict liability because it is unfair for the distributors of a defective product not to compensate its victims. As between those innocent victims and the distributors, it is the distributors—and the public which consumes their products—which should bear the unforeseen costs of the product.[11]

---

9 Ehrenzweig (n 4), 1431.
10 90 N.J. 191, 447 A.2d 539 (1982). For discussion (and defence of *Beshada*), see Christopher M. Placitella and Alan M. Darnell, '*Beshada v Johns-Manville Products Corp.*: Evolution or Revolution in Strict Products Liability?' (1983) 51 *Fordham Law Review* 801.
11 90 N.J. 191, 209.

In other words, product liability is not about individual fault but about collective risk management; as the Court expressed it, 'Strict liability focuses on the product, not the fault of the manufacturer'.[12] A finding that the defendants have been assigned the compensatory responsibility in the circumstances does not imply a negative judgment about the way in which they have conducted their business.

We might also derive some insight into a regulatory-instrumentalist mind-set in relation to torts and road traffic accidents. Paradigmatically, of course, regulatory-instrumentalists will see little sense in dealing with road traffic accidents on a fault basis, particularly if this means that too much of the resource dedicated to responding to such accidents finds its way into the pockets of lawyers and too little into the compensation payments to victims. However, where fault is still the name of the game, what standard of care should we expect of learner drivers? Within the community, there might be some support for the view that we should be slow to blame learner drivers who fail to drive with the skill and care of an experienced driver. However, what will concern a regulatory-instrumentalist is that learner drivers can present a level of risk that is otherwise unacceptable; that risk needs to be managed; and so there has to be compensation for victims. If this means that, in a traditional tort regime, learner drivers have to be held to the same standard of care as an experienced driver, then so be it. As Lord Denning expressed it in the well-known case of *Nettleship v Weston*:[13]

> Thus we are, in this branch of the law, moving away from the concept: 'No liability without fault'. We are beginning to apply the test: 'On whom should the risk fall?' Morally the learner driver is not at fault; but legally she is liable to be because she is insured and the risk should fall on her.[14]

In other words, while it would be wrong, morally, to treat the learner driver as being 'at fault', that is not the question. The question is about legal liability; and, once you know that the claim will be met by an insurance company, reservations about holding the defendant to be 'at fault' are reduced. Putting this squarely in a regulatory-instrumentalist way, we would say that, without compensation, the risk presented by learner drivers is unacceptable; so there has to be liability backed by insurance; and, no doubt, all motorists (learners who will become experienced drivers and experienced drivers who once were learners) will contribute to the funding of this scheme. This is an acceptable balance of interests.[15]

---

12 *Ibid.* 204.
13 [1971] 3 All ER 581.
14 [1971] 3 All ER 581, 586.
15 Mrs Weston, the learner-driver in *Nettleship v Weston*, was not only held liable in the tort action, she was also convicted of driving without due care and attention and fined £10.

Summing up, we can identify the following as salient features of a regulatory-instrumentalist approach:

- The general aim of the law is to achieve an acceptable balance between the community's interest in the support of beneficial innovation (and innovating businesses) and the proportionate management of risk (particularly risks to person, property and personality).
- 'Wrongs' are unacceptable risks that eventuate, the victims of which need to be compensated; that is to say, what calls for a compensatory response is not an individual act of carelessness (fault) but the eventuation of an unacceptable risk; liability does not imply a negative (critical) judgment.
- There is no necessary reason to hold the agent who did the relevant act associated with the wrong to be personally liable to compensate the victim; who did the wrongful act and who compensates are distinct questions.

While we can say that, following the first technological disruption, we have a significant movement towards a regulatory-instrumentalist mind-set, the way in which this affects the articulation, interpretation, and application of particular legal rules needs further examination. This is our next task in the chapter.

III The direction of travel: resistance and turbulence

In the context of the rules of tort law, the transition from a traditional coherentist to a regulatory-instrumentalist mind-set is not entirely smooth. There is some turbulence and some resistance. We can start by considering turbulence that is internal to, first, the coherentist and then the regulatory-instrumentalist models, before considering the resistance that fault-based thinking can present to a fully regulatory-instrumentalist (let alone a technocratic) approach.

(i) Turbulence within the traditional tort model

In the twentieth century, the common law systems of the world moved towards a patient-centred view of the relationship between clinicians and patients.[16] Physicians are, of course, the medical experts but it does not follow, so the thinking goes, that the law should endorse an entirely paternalistic view of their relationship with patients, particularly with regard to how much or how little they tell patients about their treatment options and the risks and benefits associated with various procedures. The modern legal view is that the

---

16 See, e.g., *Canterbury v Spence* 464 F 2d 772 (DC Cir 1972); *Reibl v Hughes* (1980) 114 DLR (3d) 1; and *Rogers v Whitaker* (1992) 67 ALJR 47.

relationship should be more in the nature of a partnership; patients should be in a position to make informed choices; and patients should be treated as having rights to reinforce these aspects of the relationship. None of this turbulence, however, owes anything to a regulatory-instrumental mind-set; it simply reflects a shift in the community's culture, as it becomes less deferential to 'experts', less anchored to the idea of duties (rather than rights), more 'consumerist' and less tolerant of paternalism.

In the United Kingdom, the latest chapter in this doctrinal evolution is the Supreme Court's decision in *Montgomery v Lanarkshire Health Board*[17]—a decision that is seen as a potential 'landmark'.[18] However, it bears repetition that, to the extent that the 'new approach' in *Montgomery* is such a significant development, it is so because it reflects a shift internal to the coherentist view. At all events, the principal question in *Montgomery* was whether a pregnant woman who was a diabetic, and whose pregnancy was regarded as high-risk requiring intensive monitoring, should have been informed that there was a risk of shoulder dystocia and given the option of delivery by caesarean section. Instead, she was not made aware of this particular risk; the risk eventuated during an attempted vaginal delivery; and, as a result, the baby was born with severe disabilities. The lower courts, following the traditional *Bolam*[19] principle, held that the acts of the consultant obstetrician and gynaecologist who did not disclose the risk, and who was by her own admission reluctant to steer women towards a caesarean section, was sufficiently supported by medical practice. However, the UK Supreme Court, resoundingly rejecting the applicability of the *Bolam* test to such matters of patient information and physician disclosure, held that the relationship between clinicians and patients must be rights-respecting rather than paternalistic and that patients have a right to be informed about their options (together with their relative benefits and risks).

In a few paragraphs, the Supreme Court rewrote the legal framework governing the relationship between physicians and patients. First, the Court recognised that 'patients are now widely regarded as persons holding rights, rather than as the passive recipients of the care of the medical profession'.[20] Secondly, the Court noted that patients, while not medical experts, are not wholly uninformed. Accordingly, it would be 'a mistake to view patients as uninformed, incapable of understanding medical matters, or wholly dependent upon a flow of information from doctors', from which it followed that it would now be 'manifestly untenable' to make this 'the default assumption

---

17 [2015] UKSC 11.
18 Compare R. Heywood, 'R.I.P. *Sidaway*: Patient-Orientated Disclosure—A Standard Worth Waiting For?' (2015) 23 *Medical Law Review* 455.
19 *Bolam v Friern Hospital Management Committee* [1957] 2 All ER 118.
20 [2015] UKSC 11, [75].

on which the law is to be based'.[21] Thirdly, professional guidance to doctors already reflects these changes by encouraging 'an approach based upon the informed involvement of patients in their treatment'.[22] Signalling a distinct movement away from medical paternalism and patient-dependence, the new approach is built on mutual rights and responsibilities, treating patients 'so far as possible as adults who are capable of understanding that medical treatment is uncertain of success and may involve risks, accepting responsibility for the taking of risks affecting their own lives, and living with the consequences of their choices'.[23] That said, *Montgomery* recognises that, in exceptional circumstances, doctors may legitimately withhold information under cover of the so-called 'therapeutic privilege'. However, the Court emphasises that this exception 'is not intended to subvert [the general principle] by enabling the doctor to prevent the patient from making an informed choice where she is liable to make a choice which the doctor considers to be contrary to her best interests'.[24] In short, patients have a right to make their own judgments, prudential and moral, of what is in their best interests;[25] and it is the responsibility of doctors not to override these judgments but to assist patients by ensuring that their choices are suitably informed.

(ii) Turbulence within the regulatory-instrumentalist model

A broad brush contrast is sometimes drawn between the European and the US models of access to new drugs and required safety. The communities on both sides of the Atlantic have experienced the harmful effects of some drugs. However, while the European response has been to invest heavily in ex ante safety checking prior to approval, the US has relied on a mix of ex ante precaution coupled with a very aggressive and sometimes punitive tort regime to compensate victims ex post. In both jurisdictions, patients can experience frustration as they are denied access to last-hope innovative drugs or procedures (prompting claims such as the 'right to try' as well as health-care tourism) but the contrast suggests that the interests of particular patients in Europe are more likely to be subordinated to the maintenance of the general safety thresholds.

For regulators, taking a regulatory-instrumentalist view, there is a difficult balance to be struck between, on the one hand, the need to protect patients against unsafe medicines, devices, and procedures and, on the other, to support the development of (and access to) innovative health care processes and products. The liability regime should not inhibit innovation but neither can regulators ignore the exposure of patients to highly risky and uncertain drugs

---

21 *Ibid.*, [76].
22 *Ibid.*, [78].
23 *Ibid.*, [81].
24 *Ibid.*, [91].
25 NB Lady Hale in *Montgomery* (n 17), [115].

and procedures, not to mention the headline-grabbing experimental treatments offered by some physicians.[26] As regulators try to adjust the balance between these competing demands, there is inevitably some turbulence—again, though, this is entirely turbulence within a particular model of liability. The legal rules might be tweaked or transformed but the underlying framing of the issues is given by a regulatory-instrumentalist mind-set.

In the United Kingdom, the proponents of what was to become the Access to Medical Treatments (Innovation) Act 2016 were anxious to ensure that the liability rules did not inhibit innovative, potentially life-saving, medical treatment. According to some critics of the draft law, such anxiety was unnecessary because the common law already immunised doctors against claims for negligence where they departed from accepted medical standards in a 'responsible' manner.[27] In the event, the legislation, as enacted, does not even purport to achieve that particular objective. Rather, section 1 of the Act declares that the purpose of the legislation is:

> to promote access to innovative medical treatments (including treatments consisting in the off-label use of medicines or the use of unlicensed medicines) by providing for—
>
> (a) the establishment of a database of innovative medical treatments, and
> (b) access to information contained in the database.

While this might be of some assistance in helping to inform both patients and physicians about the latest innovative developments, it does not explicitly address the balance of interests. On the other hand, it does not expose patients to some of the risks that critics of the Bill feared might be licensed by 'responsible' departures. Given the fanfares that accompanied the promotion of the Bill, this might have turned out to be very mild rather than more serious turbulence.

By contrast, in the United States, many States (and, most recently, Congress—with strong support from the President's office[28]) have enacted 'right to try' laws which, broadly speaking, are designed to give terminally ill patients access to investigational drugs (or, at any rate, drugs that have completed Phase I testing on humans) without the authorisation of the regulatory agency

---

26 See, e.g, William Kremer, 'Paolo Macchiarini: A surgeon's downfall' (*BBC News*, September 10, 2016): available at www.bbc.co.uk/news/magazine-37311038 (last accessed 5 November 2018).

27 See Margaret Brazier and Emma Cave, *Medicine, Patients and the Law* (6th edn) (Manchester: Manchester University Press, 2016) 205.

28 See the Statement of Administration Policy (May 21, 2018) supporting the passage of s.204 of the Right to Try Act 2017, which would amend the Federal Food, Drug, and Cosmetic

(the FDA).[29] According to these laws, manufacturers are permitted to supply such drugs to patients where various conditions relating to the recommendations of the consulting physician and informed consent are satisfied; and there are various immunities (for physicians) against disciplinary action and (for physicians and manufacturers) against civil liability. While access advocates can point to some compelling individual cases there are also some cautionary individual tales. However, even if an individual were sufficiently informed to make a sound assessment of the potential risks and benefits of resorting to a drug that has not gone through standard clinical trials, there is still the question whether individual access of this kind would undermine collective efforts to establish the safety and effectiveness of drugs. As Rebecca Dresser observes:

> The right-to-try campaign may be a small policy development, but it raises fundamental questions about our nation's attitudes toward death and dying. Right-to-try laws portray unproven interventions as desirable, even praiseworthy, responses to life-threatening illness. A more informed debate could reveal the human costs of this approach, drawing attention to alternative policies offering more meaningful help to people near the end of their lives.[30]

Unless a State takes a decisive step towards a libertarian culture or, in the other direction, to a culture of solidarity, announcing to the world that this is the particular kind of community that it wants to be, the 'right to try'

Act to create a new, alternative pathway for a broad range of patients diagnosed with life-threatening diseases or conditions. According to the Statement:

> Far too many patients in our country are faced with terminal illnesses for which there are no treatments approved by the Food and Drug Administration (FDA), or for which they have exhausted all approved treatment options. Biomedical research into treatments for debilitating and deadly diseases, including clinical trials, while proceeding faster than ever, may nonetheless take too long to help patients who are currently sick and may soon die. The Administration believes that these patients and their families should be able to seek access to potentially lifesaving therapies while those treatments are still under review by the FDA.
> Since the late 1980s, FDA has facilitated access to investigational drugs, devices, and biological products for the treatment of seriously ill patients. Families in these situations have sometimes found this process challenging, and FDA is constantly striving to make improvements to its expanded access program. Some patients and their families, however, still have challenges accessing investigational treatments. The Administration believes that treatment decisions for those facing terminal illnesses are best made by the patients with the support and guidance of their treating physicians. This legislation will advance these principles

This Statement from the White House is available at https://www.whitehouse.gov/wp-content/uploads/2018/05/saps204r_20180521.pdf (last accessed 5 November 2018).

29 See Rebecca Dresser, 'The "Right to Try" Investigational Drugs: Science and the Stories in the Access Debates' (2015) 93 *Texas Law Review* 1630.

30 Dresser (n 29), 1657.

will not be either constitutionally privileged or prohibited. Instead, it will be one element in an ongoing negotiation of the various interests pressed by stakeholders in health care and research, with the balance of interests and the acceptability of risk being subject to routine regulatory-instrumentalist assessment and adjustment.

(iii) Dissonance and resistance

Where ideas of fault-based liability persist, there can be some tension with an approach that is guided by regulatory-instrumentalist thinking. Even though agents who have been harmed might be compensated for their loss or injury, this is not enough—or, it is perceived to miss the point. The point is that someone was at fault and should be held responsible. Similarly, defendants who are not at fault (in the standard tort sense) might feel aggrieved by, and resist, those regulatory regimes that impose liability on them on a no-fault basis.

An example of the first kind of dissonance can be seen in the ex post response to cases of medical negligence. For patients who view the hospital as part of their local community (such as one of the older cottage hospitals), medical negligence is an occasion for identifying the party at fault, holding them to account and exacting an apology. However, where hospitals are part of publicly-funded health care regimes, as in the NHS, heavy with bureaucracy, and with a culture that depersonalises the incident, neither the victims nor those who are careless will feel that they are central to the process. Even if the tilt of the regime is towards quick settlement and compensation, this will not satisfy those patients who feel that this is no way to engage with a moral wrong. Ironically, of course, the co-existence of fault-based tortious liability with an administrative scheme for dealing with complaints and compensation militates against those who are potential defendants in the former offering admissions and apologies in the latter.[31]

As we have seen already, in the *Beshada* case, the development risks defence in product liability regimes is another example of potential dissonance. In Europe, Article 7(e) of Directive 85/374/EEC provides that the producer shall not be liable if 'the state of scientific and technical knowledge at the time when he put the product into circulation was not such as to enable the existence of the defect to be discovered'. However, Article 15(1)(b) permits a Member State to derogate from this defence by providing that 'the producer shall be liable even if he proves that the state of scientific and technical knowledge at the time when he put the product into circulation was not such as to enable the existence of the defect to be discovered'. The question whether or not to derogate might be viewed as simply a fine-tuning of the

---

31 See Brazier and Cave (n 27) Ch.9 (complaints and redress).

balance of interests between producers and consumers within a regulatory-instrumentalist approach. However, if a Member State does not derogate, and if its implementation of the development risks defence, or the jurisprudence that builds around it, invites a more 'fault-sensitive' approach, we may find that the negligence of the producer is once again central to liability. In other words, we may find that defendants are again pleading a lack of (moral) fault as well as rediscovering the language of traditional coherentist tort law.

IV Regulating autonomous vehicles

At Williston, Florida, in May 2016, a 40-year-old man, Joshua Brown, was killed while driving his Tesla S car in autopilot mode.[32] In bright sunlight, the car's sensors apparently failed to detect a white-sided truck ahead and the car ploughed at full speed into the side of the truck. Even though, as Tesla pointed out, this was the first such incident in some 130 million miles driven by its customers (and bearing in mind that, in the US, there is a fatality every 94 million miles), it only takes one such incident to heighten concerns about the safety of autonomous vehicles and to raise questions about their regulation.

Given the case of Joshua Brown, and other similar cases,[33] the first questions are about liability and responsibility when accidents occur. When accident liability on the roads largely hinges on whether human drivers have taken reasonable care, who is to be held liable, and on what basis, when human drivers are taken out of the equation? After that, we can speak to the much-debated moral dilemmas that have been highlighted by the prospect of human drivers being taken out of the loop.

(i) Liability issues

Perhaps the first question for the community is to determine its minimum safety standard. On the face of it, there is no reason for the community to settle for less than the safety of driven vehicles. Generally, this will be tested relative to big statistical numbers—how many cars on the road, how many miles driven, how many lives lost, how many persons injured, how much property damage, and so on. Some might argue for a different kind of test. For example, Jeremy Clarkson has said that the test should be whether a driverless vehicle could safely navigate the Death Road in Bolivia, squeezing past

---

32 Reported at www.theguardian.com/technology/2016/jun/30/tesla-autopilot-death-self-driving-car-elon-musk (last accessed 5 November 2018).

33 For example, the case involving the death of Elaine Herzberg in Tempe, Arizona: see www.theguardian.com/technology/2018/mar/27/arizona-suspends-ubers-self-driving-car-testing-after-fatality (last accessed 5 November 2018).

a lorry 'with half the tyre hanging over a 1,000ft drop'.[34] Whether a community adopts a Clarkson-type standard or something a bit less demanding, we can assume that there will be no social licence for autonomous vehicles unless they are judged to be at least as safe as driven cars. So, it was no surprise that Uber immediately suspended road-testing of its fleet of self-driving cars when one of its vehicles (being tested in autonomous mode in Tempe, Arizona) hit a pedestrian who was wheeling her bicycle across the road and killed her.[35]

Even if autonomous vehicles, are licensed, we cannot assume that this will be the end of accidents on the roads. For example, there might be unexpected problems with the on board technologies. Where this happens, who is to be liable? According to Jonathan Morgan:

> From the industry perspective, liability is arguably the innovative manufacturer's greatest concern. The clashing interests raise in acute form the classic dilemma for tort and technology: how to reconcile reduction in the number of accidents (deterrence) and compensation of the injured with the encouragement of socially beneficial innovation? Not surprisingly there have been calls for stricter liability (to serve the former goals), and for immunities (to foster innovation). But in the absence of any radical legislative reform, the existing principles of tort will apply—if only faute de mieux.[36]

This is surely right. As we said in Chapter Eight, if the regulation of an emerging technology is presented in a legislative setting, a risk-management approach is likely to prevail, with regulators trying to accommodate inter alia the interest in beneficial innovation together with the interest in risks being managed at an acceptable level. On the other hand, if the question arises in a court setting, it is more likely that both litigants and judges will talk the coherentist talk of negligence and fault even though, as Morgan observes, the common law technique of reasoning by analogy via existing categories, far from being 'common sense', is 'obfuscatory'.[37] In this light, we can say just four things.

First, it is no wonder that we find the law relating to autonomous vehicles puzzling when we try, in a coherentist way, to apply the principles for judging the negligence of human drivers to questions of liability concerning vehicles

---

34 Nick Rufford and Tim Shipman, 'Clarkson's driverless close shave' *The Sunday Times*, November 19, 2017, p.1.

35 See Ben Hoyle, 'Uber suspends self-driving tests after car kills pedestrian' *The Times*, March 20, 2018, p.1.

36 Jonathan Morgan, 'Torts and Technology', in Roger Brownsword, Eloise Scotford, and Karen Yeung (eds), *The Oxford Handbook of Law, Regulation and Technology* (Oxford: Oxford University Press, 2017) 522, 537.

37 Morgan (n 36) 539. Compare, too, Ehrenzweig (n 4).

in which there is no human in control. Perhaps we are asking ourselves the wrong questions.

Secondly, if liability questions are taken up in the courts, judges (reasoning like coherentists) will try to apply notions of a reasonable standard of care to responsibility for very complex technological failures.[38] In the Joshua Brown case, Tesla (presumably anticipating litigation or a discourse of fault and responsibility) were quick to suggest that drivers of their cars needed to remain alert and that they themselves were not careless in any way (hence their remarks about the safety record of their cars). This is worse than puzzling: it threatens to turn a human tragedy into a farce.[39]

Thirdly, if regulators in a legislative setting approach the question of liability and compensation with a risk-management mind-set, they will not need to chase after questions of fault. Rather, the challenge will be to articulate the most acceptable (and financially workable) compensatory arrangements that accommodate the interests in transport innovation and the safety of passengers and pedestrians. For example, one proposal is that there should be an autonomous vehicle crash victim compensation fund to be financed by a sales tax on such vehicles.[40] Of course, as with any such no-fault compensation scheme, much of the devil is in the detail—for example, there are important questions to be settled about the level of compensation, whether the option of pursuing a tort claim remains open to victims, and what kinds of injury or loss are covered by the scheme.[41]

Fourthly, as Morgan says, the better way of determining the liability arrangements for autonomous vehicles is not by litigation but 'for regulators to make the relevant choices of public policy openly after suitable democratic discussion of which robotics applications to allow and which to stimulate, which applications to discourage and which to prohibit'.[42] Even better, in my view, regulators should make these choices after an independent emerging technologies body (of the kind that we do not, but should, have) has informed and stimulated public debate.

Putting all this more generally, we can say that it is important that communities ask themselves the right questions, understanding, first, that coherentist framings are very different from regulatory-instrumentalist ones and,

---

38 I take it that, if autonomous vehicles have to be at least as safe as driven vehicles, there would be a difficulty in presenting them as 'dangerous' in a way that would get a product liability claim to first base.

39 See, too Adam Greenfield, *Radical Technologies* (London: Verso, 2017) 222–226.

40 See Tracy Pearl, 'Compensation at the Crossroads: Autonomous Vehicles and Alternative Victim Compensation Schemes' (2018) 60 *William and Mary Law Review* —: available at papers.ssrn.com/sol3/papers.cfm?abstract_id=3148162 (last accessed 5 November 2018).

41 One of the interesting features of Pearl's proposal (n 39) is that manufacturers of autonomous vehicles, in return for receiving protection from the fund, would be required to participate in a data-sharing and design improvement programme.

42 Morgan (n 36) 539.

secondly, that the availability of technological solutions invites reflection not only on adjustments that might be made to the liability rules but whether such rules are really necessary at all.[43]

(ii) Moral dilemmas

In the early debates about the social licensing and regulation of autonomous vehicles, a common question has been: how would such a vehicle deal with a moral dilemma[44]—for example, the kind of dilemma presented by the trolley problem (where one option is to kill or injure one innocent human and the only other option is to kill or injure more than one innocent human)[45] or by the tunnel problem (where the choice is between killing a passenger in the vehicle or a child outside the vehicle)[46]? For example, no sooner had it been reported that Uber were to pilot driverless taxis in Pittsburg than these very questions were raised.[47] Let me suggest a number of ways of responding to such questions, leading to the conclusion that, while the trolley problem is not itself a serious difficulty, there is a significant question—more effectively raised by the tunnel problem—to be asked about the way in which moral responses are programmed into autonomous vehicles or other smart technologies.

A first, and short, response is that the particular moral dilemma presented by the trolley problem (at any rate, as I have described it) is open to only two plausible answers. A moralist will either say that killing just the one person

---

43 For helpful discussions, see Ronald Leenes and Federica Lucivero, 'Law on Robots, Laws by Robots, Laws in Robots: Regulating Robot Behaviour by Design' (2014) 6 *Law, Innovation and Technology* 193; and Ronald Leenes, Erica Palmerini, Bert-Jaap Koops, Andrea Bertolini, Pericle Salvini, and Federica Lucivero, 'Regulatory Challenges of Robotics: Some Guidelines for Addressing Legal and Ethical Issues' (2017) 9 *Law, Innovation and Technology* 1.

44 See, e.g., Patrick Lin, 'The Ethics of Saving Lives with Autonomous Cars are Far Murkier than You Think' *WIRED*, 30 July, 2013: available at www.wired.com/2013/07/the-surprising-ethics-of-robot-cars/ (last accessed 5 November 2018); and, 'The Robot Car of Tomorrow May be Just Programmed to Hit You' *WIRED*, 6 May 2014: available at www.wired.com/2014/05/the-robot-car-of-tomorrow-might-just-be-programmed-to-hit-you/ (last accessed 5 November 2018).

45 For the original, see Judith Jarvis Thomson, 'The Trolley Problem' (1985) 94 *Yale Law Journal* 1395.

46 For the tunnel problem in relation to autonomous vehicles, see Jason Millar, 'You should have a say in your robot car's code of ethics' *Wired* 09.02.2014 (available at: www.wired.com/2014/09/set-the-ethics-robot-car/) (last accessed 5 November 2018). See, further, Meg Leta Jones and Jason Millar, 'Hacking Metaphors in the Anticipatory Governance of Emerging Technology: The Case of Regulating Robots' in Roger Brownsword, Eloise Scotford, and Karen Yeung (n 36) 597.

47 See, e.g., Will Pavia, 'Driverless Ubers take to road' *The Times*, September 13, 2016, p.36; and, Raphael Hogarth, 'Driverless cars will take us into a moral maze' *The Times*, September 17, 2016, p.28.

is clearly the lesser of two evils and is morally required; or it will be argued that, because the loss of one innocent life weighs as heavily as the loss of many innocent lives, neither option is better than the other—from which it follows that killing just the one person is neither better nor (crucially) worse, morally speaking, than killing many. Accordingly, if autonomous vehicles are programmed to minimise the number of humans who are killed or injured, this is either right in line with one strand of moral thinking or, following the other, at least no worse than any other programming. Such a design would be opposed only by someone who argued that the vehicle should be set up to kill more rather than fewer humans; and, barring some quite exceptional circumstances, that, surely, is simply not a plausible moral view.

Secondly, if autonomous vehicles were designed, in Pittsburg or elsewhere, to minimise the number of human deaths or injuries, it is hard to believe that human drivers, acting on their on-the-spot moral judgments, would do any better. Confronted by a trolley scenario—which most drivers encounter as infrequently as a lorry on the Bolivian Death Road—with little or no time to make a moral assessment of the situation, human drivers would act instinctively—and, insofar as human drivers formed any sense of what would be the right thing to do in the particular situation, my guess is that they would generally try to minimise the loss of life.[48] What other defensible response could there be? At all events, we might expect that, with autonomous vehicles, the roads will not only be safer, there will be fewer casualties when there is no option other than the sacrifice of human lives.

Thirdly, even if—at least in the case of autonomous vehicles—there is a reasonably straightforward resolution of the trolley problem, there might well be more difficult cases. Some might think that we face such a case if the choice is between sacrificing innocent passengers in an autonomous vehicle or killing innocent humans outside the vehicle. However, in principle, this does not seem any more difficult: minimising the loss of human life still seems like the appropriate default principle. In the case of the tunnel problem, though, where one life will be lost whichever choice is made, the default principle is not determinative. Secondary defaults might be suggested—for example, it might be suggested that, because the cars present a new and added risk, those who travel in the cars should be sacrificed[49]—but I am ready to concede that this is a case where moralists might reasonably disagree. Moreover, beyond such genuinely difficult cases, there might be some issues where

---

48 Jean-François Bonnefon, Azim Shariff, and Iyad Rahwan, 'The social dilemma of autonomous vehicles' (2016) 352 *Science* (Issue 6293) 1573–1576.

49 Of course, if this is the accepted default, and even if it is accepted that this is the right thing to do, some humans might be reluctant to travel on these terms. And, it might be that the general safety features of the vehicle are such that it simply cannot default to sacrificing its passengers: see e.g. www.dailymail.co.uk/news/article-3837453/Mercedes-Benz-says-driverless-cars-hit-child-street-save-passengers-inside.html (last 5 November 2018).

human agents do have time for moral reflection and where we think that it is important that they form their own view; and there might be cases where there are plausible conflicting options and where moralists want to see this resolved in whatever way aligns with their own moral judgment.

If we make the perhaps optimistic assumption that autonomous vehicles will be programmed in ways that reflect the terms of the social licence agreed by the members of the community in which they will operate, agents will have an opportunity to formulate and express their own moral views as the licence is negotiated and debated. Already, the Open Roboethics Initiative is exploring imaginative ways of crowd-sourcing public views on acceptable behaviour by robots, even in relation to such everyday questions as whether a robot should give way to a human or vice versa.[50] Given such active moral engagement, it cannot be objected that humans are ceding their life-and-death moral judgments to machines. Provided that autonomous vehicles are designed and programmed in accordance with the terms of the social licence that has been agreed for their operation—reflecting not only the community's judgment as to what is an 'acceptable' balance of risk and benefit but also its judgment as to what is morally appropriate—humans are not abdicating their moral responsibilities. Humans, in negotiating the social licence for autonomous vehicles, are debating and judging what is the right thing in relation to the moral programming of the vehicle.

What, though, if it were to be argued that, in order to reinforce the importance of freely doing the right thing and expressing one's human dignity, passengers who enter autonomous vehicles should have the opportunity to override the collectively agreed default? On the face of it, this is a hostage to fortune.[51] Nevertheless, if a moral community is anxious to preserve opportunities for agents to do the wrong (sic) thing, this proposal might be taken seriously.

There is also the argument that vehicles should not be designed in ways that preclude the possibility of doing the right thing (such as stopping to undertake 'Good Samaritan' acts of assistance). This implies that the design needs to allow for a human override to enable the vehicle's passengers to respond to a moral emergency (such as rushing a person who urgently needs medical attention to the nearest hospital). Such contingencies might be provided for by the terms of the social licence, which might then be articulated in a combination of revised rules and measures of technological management.

---

50 See, e.g., AJung Moon, Ergun Calisgan, Camilla Bassani, Fausto Ferreira, Fiorella Operto, Gianmarco Veruggio, Elizabeth A. Croft, and H.F. Machiel Van der Loos, 'The Open Roboethics Initiative and the Elevator-Riding Robot' in Ryan Calo, A. Michael Froomkin, and Ian Kerr (eds), *Robot Law* (Cheltenham: Elgar, 2016) 131–162.
51 Compare, Patrick Lin, 'Here's a Terrible Idea: Robot Cars with Adjustable Ethics Settings' *WIRED*, 18 August 2014: available at www.wired.com/2014/08/heres-a-terrible-idea-robot-cars-with-adjustable-ethics-settings/ (last accessed 5 November 2018).

V The regulatory environment for patient safety: Challenges of complexion and coherence

Hospitals, we know, are dangerous places; even when organisational stresses are low and correct procedures are observed, the safety of patients cannot be taken for granted.[52] As Oliver Quick points out, in the United States alone, it is estimated that there are some 180,000 preventable deaths in hospitals each year, this equating to three jumbo jets crashing every two days.[53] However, numbers (even large numbers) do not always have the salience of specific cases. Hence, following the publication of the report of the public inquiry into the Mid-Staffordshire NHS Foundation Trust (centring on the deaths of patients at Stafford Hospital), no one—not even seasoned observers of the actualities of hospital practice—can think that the regulatory environment for patient safety is fit for purpose.

Introducing the report, Robert Francis QC paints a shocking picture of the suffering of hundreds of patients whose safety was ignored for the sake of corporate self-interest and cost control:

> The evidence at both inquiries disclosed that patients were let down by the Mid Staffordshire NHS Foundation Trust. There was a lack of care, compassion, humanity and leadership. The most basic standards of care were not observed, and fundamental rights to dignity were not respected. Elderly and vulnerable patients were left unwashed, unfed and without fluids. They were deprived of dignity and respect. Some patients had to relieve themselves in their beds when they [were] offered no help to get to the bathroom. Some were left in excrement stained sheets and beds. They had to endure filthy conditions in their wards. There were incidents of callous treatment by ward staff. Patients who could not eat or drink without help did not receive it. Medicines were prescribed but not given. The accident and emergency department as well as some wards had insufficient staff to deliver safe and effective care. Patients were discharged without proper regard for their welfare.[54]

For this deplorable state of affairs, blame and responsibility spread far and wide, from high-level commissions and regulators, to hospital managers and

---

52 See, e.g., Mary Dixon-Woods, 'Why is Patient Safety So Hard? A Selective Review of Ethnographic Studies' (2010) 15 *Journal of Health Services Research and Policy* 11–16.

53 Oliver Quick, *Regulating Patient Safety* (Cambridge: Cambridge University Press, 2017) 1.

54 Robert Francis QC, Press Statement: available at http://webarchive.nationalarchives.gov.uk/20150407084231/http://www.midstaffspublicinquiry.com/report (last accessed 5 November 2018). For vignettes of patient experience at the hospital, see Independent Inquiry into Care Provided by Mid Staffordshire NHS Foundation Trust: January 2005–March 2009 (Volume 2) (HC 375-II) (2010).

individual health-care workers. However, resisting pressure to single out scapegoats or to engage in yet more reorganisation, Francis emphasises that what is needed is 'common values, shared by all, putting patients and their safety first; we need, a commitment by all to serve and protect patients and to support each other in that endeavour, and to make sure that the many committed and caring professionals in the NHS are empowered to root out any poor practice around them'.[55]

Francis is by no means alone in expressing grave concerns about patient safety. Indeed, one book after another, one front-page newspaper story after another, has highlighted the extent to which patients, both in and out of hospitals, are exposed to *unacceptable* risk.[56] Whether we look at the performance of pharmaceutical companies, doctors, nurses, hospital managers, researchers, government ministers, or even the regulatory agencies themselves, it seems that the health of patients is in jeopardy—that is, *unacceptably* and *unnecessarily* in jeopardy. It seems that everyone needs to be reminded (as per Francis) that the first rule of medicine is 'to do no harm' to patients and that, concomitantly, the interests of patients must come first.

While there is much to admire about Francis' strategy in proposing a fundamental cultural change, smart regulators know two things: one is that a desired objective (such as securing the safety of patients) is rarely achieved by relying on a single kind of instrument—rather, the key to effective regulation is applying the optimal combination of instruments;[57] and the other

---

55 Robert Francis QC, Press Statement, p.4: available at: http://webarchive.nationalarchives. gov.uk/20150407084231/http://www.midstaffspublicinquiry.com/report (last accessed 5 November 2018).

56 See, e.g., Emily Jackson, *Law and the Regulation of Medicines* (Hart: Oxford, 2012); and Ben Goldacre, *Bad Pharma* (London: Fourth Estate, 2012). Examples from the press are too numerous to mention but we should also note: Tom Whipple and Michael Savage, 'Cameron hammers labour over failure of NHS Trusts' *The Times* July 17, 2013, p.1 (and similar headlines in pretty much every other newspaper following the publication of Sir Bruce Keogh's post-Francis review of 14 trusts with unusually high mortality rates); and, Chris Smyth, 'Alarm over "high" death rate in English hospitals' *The Times* September 12, 2013, p.1. Even after this, worrying headlines have continued to make it to the front pages: see, e.g., Laura Donnelly, '"Alarming" culture of NHS care' *The Sunday Telegraph*, February 2, 2014, p.1 (reporting the view of David Prior, the chairman of the CQC, that a dysfunctional rift between managers and clinical staff jeopardises the safety of vulnerable patients, that bullying, harassment, and abuse persist, and that whistle-blowers are ostracised); and, more recently, Sarah Boseley, 'Government apologises decades after 450 patients killed in hospital' *The Guardian*, June 21, 2018, p.1 (reporting the historic over-prescribing of life-shortening opioids at Gosport War Memorial hospital, where whistle-blowing nurses were ignored, where the Department of Health buried a report that highlighted routine over-prescription and its fatal effects at the hospital, and where the GMC censured but did not strike off a doctor who was at the heart of the practice).

57 See Neil Gunningham and Peter Grabosky, *Smart Regulation* (Oxford: Clarendon Press, 1998).

is that there are now many kinds of technologies and designs that might be used to support the regulatory objectives.[58] For example, if our concern were not with the safety of patients but the safety of railway passengers, we might rely not only on background health and safety laws but also on the better design of rolling stock[59] together with the installation of surveillance technologies. Similarly, if our concern were with public safety in city centres, we might employ not only various criminal laws but also an array of surveillance (such as CCTV) and identification technologies (such as facial recognition biometrics) to achieve our objectives. Imagine, for example, that with CCTV coverage in operating theatres, surveillance would show (and record) doors frequently opening and shutting 'with staff appearing to treat the operating theatre as a thoroughfare', coupled with a high level of 'case-irrelevant' interruptions and distractions.[60] If we are willing to use smart approaches to secure public safety, and if we think that technologies such as CCTV might contribute to a strengthening of the regulatory environment, the question is whether we should apply a similar strategy to secure patient safety. Why rely only on setting *standards* that require the safety of patients to be the paramount consideration? Why not also introduce *technologies* that channel conduct in this direction or even eliminate altogether unsafe options?[61]

(i) Changing the culture

No one reading the Francis report could doubt that the regulatory environment needs to be much better geared for patient safety. For Francis, the key is to change the hospital culture from one of bullying, target-chasing,

---

58 Seminally, see Lawrence Lessig, *Code and Other Laws of Cyberspace* (New York: Basic Books, 1999). At pp. 93–94, in the context of regulating for the safety of motorists and their passengers, Lessig sets out the range of instruments that are available to regulators in the following terms:

> The government may want citizens to wear seatbelts more often. It could pass a law to require the wearing of seatbelts (law regulating behavior directly). Or it could fund public education campaigns to create a stigma against those who do not wear seatbelts (law regulating social norms as a means to regulating behavior). Or it could subsidize insurance companies to offer reduced rates to seatbelt wearers (law regulating the market as a way of regulating behavior). Finally, the law could mandate automatic seatbelts, or ignition-locking systems (changing the code of the automobile as a means of regulating belting behavior). Each action might be said to have some effect on seatbelt use; each has some cost. The question for the government is how to get the most seatbelt use for the least cost.

59 See, Jonathan Wolff, 'Five Types of Risky Situation' (2010) 2 *Law, Innovation and Technology* 151.

60 See Dixon-Woods (n 52) at 14.

61 For a rare example of this line of thinking, see Jo Bibby, 'Smart sensors and wearable tech: the future for the NHS?' (November 21, 2013) (available at: www.health.org.uk/blog/smart-sensors-and-wearable-tech-future-nhs) (last accessed 5 November 2018).

disengagement, defensiveness, and denial to one that ensures that patients come first. Thus:

> We need a patient centred culture, no tolerance of non compliance with fundamental standards, openness and transparency, candour to patients, strong cultural leadership and caring, compassionate nursing, and useful and accurate information about services.[62]

Translating this into practice, Francis highlights, inter alia, the importance of: clear standards; openness, transparency, and candour; and improved support and training.

First: clear standards. A traditional regulatory approach has three phases: standards are set; inspectors check for compliance; and steps are taken to enforce the standards where non-compliance is detected.[63] If patient safety is to be secured in this way, the right standards must be set, and monitoring and enforcement must be effective and robust. If there are problems in any of these phases, the outcome will be unsatisfactory. Broadly speaking, this is where Francis starts. To secure patient safety, standards (backed by a broad spectrum of support) should be set, compliance monitored, and there should be strong enforcement. It should be made clear to all that the agreed standards are to be observed and that non-compliance simply will not be tolerated.[64] Even if it is not possible to run hospitals on a 'zero-risk' basis, there should at least be a 'zero-tolerance' approach to non-compliance.

Secondly: openness, transparency and candour. If regulators set and enforce the right standards, avoidable harm to patients should be minimised. However, where there are adverse events of one kind or another, it is imperative,

---

62 Chairman's statement (n 55), p.1. Compare, too, Mary Dixon-Woods, 'What we know about how to improve quality and safety in hospitals—and what we need to learn' (February 7, 2013) (available at www.health.org.uk/slideshow-what-we-know-about-how-improve-quality-and-safety-hospitals) (last accessed 5 November 2018).

63 See, e.g., Julia Black, 'What is Regulatory Innovation?' in Julia Black, Martin Lodge, and Mark Thatcher (eds), *Regulatory Innovation* (Cheltenham: Edward Elgar, 2005) 1.

64 Arguably, the reach of the criminal law should extend beyond failures that lead to death or serious injury, to those cases where there is simply wilful neglect or mistreatment of patients. This was a recommendation of the Berwick Report: see National Advisory Group on the Safety of Patients in England, *A Promise to Learn—A Commitment to Act* (London, 2013). In November 2013, the Government announced that it accepted this recommendation and, following a consultation, would implement legislation accordingly: available at www.bbc.co.uk/news/uk-24967230 (November 16, 2013) (last accessed 5 November 2018). The new offence would cover those cases where patients were injured by neglect or mistreatment but also, for example, 'the tolerance of unsanitary conditions, persistent verbal abuse or intimidation, malevolent denial of visiting rights, or summarily discharging sick patients' (ibid). See, too, Karen Yeung and Jeremy Horder, 'How can the criminal law support the provision of quality in health care?' (2014) 23 *BMJ Quality and Safety* 519. For the implementing measures, see ss 20 and 21 of the Criminal Justice and Courts Act 2015.

after the event, that admissions are made. Under-reporting is notorious and defensive practice will do no good. As Francis puts it:

> A duty of candour should be imposed and underpinned by Statute and the deliberate obstruction of this duty should be made a criminal offence.
>
> - Openness means enabling concerns and complaints to be raised freely and fearlessly, and questions to be answered fully and truthfully.
> - Transparency means making accurate and useful information about performance and outcomes available to staff, patients, the public and regulators.
> - Candour means informing any patient who has or may have been avoidably harmed by a healthcare service of that fact and a remedy offered where appropriate, regardless of whether a complaint has been made or a question asked about it.
>
> Every provider trust must be under an obligation to tell the truth to any patient who has or may have been harmed by their care....The deliberate obstruction of the performance of these duties and the deliberate deception of patients in this regard should be criminal offences.[65]

It is hard to gainsay the spirit of this proposal which, by demonstrating a willingness to resort to the criminal law, signals that the new direction is no hollow declaration of intention. However, it is as well to recognise that, where the culture is not unconditionally committed to the paramount importance of patient safety, the call for transparency—running, so to speak, 'across the grain' of health care culture—might meet with resistance.[66] Where there is such a commitment, regulatory reinforcement is not necessary; but where there is no such commitment, there is likely to be resistance to both the primary regulatory objective and the procedural interventions that are designed to support that objective. Indeed, soon after the publication of the Francis Report, a government review led by Ann Clwyd MP uncovered a persisting culture of denial, defensiveness, and delay in relation to patients' complaints.[67]

---

65 Chairman's statement (n 55), p.6.
66 For a seminal commentary on the limits of effective regulatory intervention, see Iredell Jenkins, *Social Order and the Limits of Law* (Princeton NJ: Princeton University Press, 1980).
67 See Nick Triggle, 'NHS complaints revolution "needed"' (available at www.bbc.co.uk/news/health-24669382, October 28, 2013) (last accessed 5 November 2018); and Chris Smyth, 'Days of denying and delaying complaints are over, NHS is told' *The Times*, October 29, p.2; and, more recently, see Laura Donnelly, '"Alarming" culture of NHS care' *The Sunday Telegraph*, February 2, 2014, p.1.

Thirdly, improved support and training. Unless hospitals are to be staffed by robots, duly programmed to act in a caring and compassionate way (a possibility that Francis does not contemplate but which we will consider shortly), health care workers need to be trained and guided. It goes without saying, of course, that even the best-trained nurses will struggle to deliver the right level of care when, relative to the number of patients who require their attention, they are shorthanded. Nevertheless, at a time when public finances are under serious pressure, it is worth emphasising that it may not be possible to put all patients first—if the number of nurses is reduced, some patients will have to come second, some third, and so on.

After the Francis Report, there is a clear view of the purposes to which the regulatory environment should be orientated; patient safety is the focal object of the exercise. However, the question now is not whether the approach to patient safety should be more regulatory but whether it should be more technocratic.

(ii) A technocratic alternative?

Francis makes 290 recommendations designed to change the hospital culture and make sure that patients come first. Some use of IT is contemplated in these recommendations but, for Francis, the key is to change the culture, not to draw on new technologies. Given that computer hardware and software, like the patients and the medical staff that they are intended to serve and to assist, are susceptible to bugs and viruses, and given some of the early adverse episodes with digital health records and prescriptions,[68] we might sympathise with the thought that what we need to be investing in is a new culture of care rather than new technologies. Indeed, we might agree with Harold Thimbleby that 'the reality is that behind the façade of superficial wonder, modern hospital IT is too complicated for its own good, for the good of patients, and for the good of staff.'[69]

Nevertheless, mindful of such caveats, let us suppose that we took more seriously the idea that we might make a more intense use of modern technologies with a view to ensuring that patients are treated decently and with respect for their dignity—and, above all, that they are treated in an environment that respects their need to be safe. If we adopted such a strategy in

---

68 See, e.g., Robert Wachter, *The Digital Doctor: Hope, Hype, and Harm at the Dawn of Medicine's Computer Age* (New York: McGraw Hill Education, 2015).

69 Harold Thimbleby, 'Misunderstanding IT: Hospital Cybersecurity and IT Problems Reach the Courts' (2018) 15 *Digital Evidence and Electronic Signature Law Review* 11, 26 (focusing on allegations of the wilful neglect of patients at the Princess of Wales Hospital in Bridgend, the subsequent trial [*R v Cahill* and *R v Pugh* at Cardiff Crown Court in October 2014], the general tendency to be overconfident about the reliability of computers, and the consequent inappropriate blaming and stigmatisation of nursing staff.

relation to hospitals and health care generally, would this be an advance in our regulatory approach?[70]

In response, we can focus on two types of technological intervention: first the use of various kinds of surveillance and monitoring technologies; and, secondly, the use of various kinds of robotic technologies.

Surveillance technologies

Surveillance technologies might be employed in hospitals or other health-care facilities for many different purposes, some of which will be more controversial than others.[71] For example, Tim Lahey lists the following as some of the less controversial instances:

> Greenwich Hospital in Connecticut conducts video monitoring for patients deemed at risk of falling. Nebraska Medical Center developed an unrecorded video-monitoring system for high-risk patients, such as those on suicide watch. Rhode Island Hospital was required to install recording equipment in operating rooms after multiple mistakes in surgery. And one company, Arrowsight, has developed video-auditing technology to monitor clinician hand-hygiene practices at sinks and safety and cleanliness in operating rooms. These all seem like thoughtful uses.[72]

Nevertheless, the trade-off between privacy and patient safety merits careful attention. This being so, I suggest that, where our interest is in the use of surveillance technologies for the sake of patient safety, we should divide such use into two categories: (i) where surveillance technologies are employed to protect patient, P, against himself; and (ii) where surveillance technologies are used to protect patient, P, against the harmful acts of others.

In the first kind of case, the idea is to maintain surveillance of P lest P should act in ways that are harmful to P's own health and safety. For example, there might be concerns that patients (as in Lahey's reference to the Greenwich Hospital use) might injure themselves by falling or that they will, quite literally, self-harm; or it might be that P suffers from dementia and the technology makes sure that P does not stray beyond the boundaries of a safe

---

70 Compare Karen Yeung and Mary Dixon-Woods, 'Design-Based Regulation and Patient Safety: A Regulatory Studies Perspective' (2010) 71 *Social Science and Medicine* 502–509; and Quick (n 53) who remarks at 180 that 'The model professional should be emotionally intelligent, patient-centred and fully engaged with technological innovation and initiatives in patient safety research.'

71 Tim Lahey, 'A Watchful Eye in Hospitals' *The New York Times* (op ed) (February 16, 2014) (available at www.nytimes.com/2014/02/17/opinion/a-watchful-eye-in-hospitals.html) (last accessed 5 November 2018).

72 Lahey (n 71).

area; or it might be that the technology is in the form of sensors that alert P to various risk factors (such as elevated blood pressure). Quite apart from keeping an eye on P for P's own good, it is possible that some of these technologies might enhance P's freedom because, with surveillance (including RFID and GPS tracking), there is a safety net (e.g., in the case of an elderly P with dementia). Of course, if P is able to consent, then the surveillance measures should be subject to P's consent. However, if the measures cannot be tailored to individual consent, or if the individuals do not have the capacity to consent, then a regulatory approach will have to balance the various interests involved (including making a judgment about the balance between P's interest in freedom and well-being against P's interest in privacy).

The second kind of case is one where surveillance technologies are employed to prevent P being harmed by others (for example, by doctors, nurses, porters, other patients, visitors, trespassers, and so on). In some cases, the harm inflicted might be intentional, in others it might be reckless, and in others unintentional; but, in all cases, the technology would promise some positive ex ante (deterrent) effect and an ex post record. Again, Lahey cites some helpful examples. Although the potential privacy objections that we identified in relation to a paternalistic intervention remain relevant, the fact that the technologies are designed to protect agents against unwilled physical harm caused by others makes the justificatory case (even in the absence of consent) more straightforward. To be sure, the impingement on privacy should not be more than is necessary and should be proportionate; but the shape of the justification here is familiar.

Nevertheless, there are a number of further considerations in this second type of case that take us into much less familiar territory. Where surveillance technologies are in operation and are known to be in operation, they signal that it is not in the interests of those who might harm P to do so because they will be seen. Now, where (as in the Francis Report) the strategy is to educate regulatees into putting the safety of patients first, the prudential signals transmitted by surveillance technologies might interfere with the internalisation of the core regulatory objective. As is well known, one of the effects of introducing CCTV coverage in the cities is to displace at least some crime to zones that are not covered by surveillance. Where surveillance is comprehensive, this is no longer a problem. However, there is an underlying problem: namely, that the reason why the pattern of conduct conforms to the regulatory objective is purely prudential. If patients are now safe, that is an advance; but it is not because the value of prioritising patient safety has been internalised as such.[73]

---

73 For an instructive commentary on a broad range of concerns that can be provoked by the introduction of a camera in a university kitchen (to encourage cleanliness), see Edna Ullmann-Margalit, 'The Case of the Camera in the Kitchen: Surveillance, Privacy, Sanctions,

If surveillance morphs into hard technological management, so that it simply is not possible to harm patients, this achieves the regulatory objective but, as we know by now, it intensifies concerns about the loss of moral signals from the regulatory environments. Hospitals with technological management might represent a better environment for the physical well-being of patients but not necessarily for the moral health of the community.

Caring robots and smart machines

As we have seen, amongst the key elements of the Francis Report are proposals for training, transparency, better information, and so on. Humans need to be educated. Suppose, though, that we could staff our hospitals in all sections, from the kitchens to the front reception, from the wards to the intensive care unit, from accident and emergency to the operating theatre, with robots and smart machines. Moreover, suppose that all hospital robot operatives were programmed (in the spirit of Asimov's laws) to make patient safety their top priority.[74] Is this the way forward?[75]

Already, some will sense that this is the opening chapter of what turns out to be a dystopian script. At 'Robot Hospital', the performance of the robots is less reliable than their designers anticipated; far from being a safe haven, the hospital proves to be even more dangerous than the old hospital, staffed by humans, that it replaced. To be sure, this is not a happy story but it hardly compels us to abandon any plan to replace humans with robot operatives in hospitals. Provided that a precautionary strategy is followed, with robots being introduced slowly to perform low risk functions, and provided that

---

and Governance' (2008) 2 *Regulation and Governance* 425. Significantly, one of the concerns expressed is that the camera might be ineffective and even counter-productive by displacing unclean practices. However, although there is an undercurrent of concern about the negative impact of the camera on the community's culture, it does not quite articulate as a concern about the 'complexion' of the regulatory environment. On CCTV surveillance more generally see Beatrice von Silva-Tarouca Larsen, *Setting the Watch: Privacy and the Ethics of CCTV Surveillance* (Oxford: Hart, 2011).

74 According to the first of Asimov's three laws, 'A robot may not injure a human being or, through inaction, allow a human being to come to harm.' See http://en.wikipedia.org/wiki/Three_Laws_of_Robotics (last accessed 5 November 2018).

75 See Hannah Devlin, 'London hospitals to replace doctors and nurses with AI for some tasks' *The Guardian*, May 21, 2018: available at www.theguardian.com/society/2018/may/21/london-hospitals-to-replace-doctors-and-nurses-with-ai-for-some-tasks (last accessed 5 November 2018). According to this report, a 'three-year partnership between University College London Hospitals (UCLH) and the Alan Turing Institute aims to bring the benefits of the machine learning revolution to the NHS on an unprecedented scale.' The idea is that 'machine learning algorithms can provide new ways of diagnosing disease, identifying people at risk of illness and directing resources. In theory, doctors and nurses could be responsively deployed on wards, like Uber drivers gravitating to locations with the highest demand at certain times of day. But the move will also trigger concerns about privacy, cyber security and the shifting role of health professionals.'

engineers learn from any instances of robot malfunction, we can plausibly imagine a hospital of the future where the operatives are robots. No one, surely, can fail to be impressed by the video of a da Vinci robot suturing a grape.[76]

Nevertheless, is there not something deeply ethically disturbing about the idea of, say, robot nursing care, of Nursebots who can 'do care' but without actually caring about humans? Where we are troubled by a technological development that is not obviously harmful to human health and safety, damaging to the environment, or invasive of human rights (to privacy, expression, and so on), we often appeal to the concept of human dignity to mark our concern.[77] We see this, for example, when debates are opened about the possible use of modern technologies for the purpose of human enhancement.[78] However, as Michael Sandel rightly remarks:

> It is commonly said that enhancement, cloning, and genetic engineering pose a threat to human dignity. This is true enough. But the challenge is to say *how* these practices diminish our humanity.[79]

In just the same way, if we contend that Nursebots pose a threat to human dignity, the challenge is to say precisely how humanity is thereby diminished or compromised.

Recalling the shocking images of the Francis Report, we might respond that the outstanding problem with the level of human nursing care at the hospital was that it was an affront to human dignity. In one sense—by no means an unimportant sense—the circumstances in which some patients were left was 'undignified' or involved an 'indignity'. However, the compromising of human dignity that we are looking for is one in which humans are commodified—that is to say, where humans are treated as things. Where nursing functions are carried out by robots, we might plausibly ask whether, on either side of the Nursebot/human-patient relationship there is some element of commodification.

---

76 www.youtube.com/watch?v=0XdClHUp-rU (last accessed 5 November 2018). For discussion of liability issues should the use of the robot harm the patient, see Ugo Pagallo, *The Laws of Robots* (Dordrecht: Springer, 2013) 88–94 (discussing Roland Mracek's claim against both the producers of the da Vinci surgery robot and the Bryn Mawr hospital in Philadelphia).

77 See Roger Brownsword, 'Human Dignity, Human Rights, and Simply Trying to Do the Right Thing' in Christopher McCrudden (ed), *Understanding Human Dignity* (Proceedings of the British Academy 192) (Oxford: The British Academy and Oxford University Press, 2013) 345.

78 See, e.g., Michael Sandel, *The Case Against Perfection* (Cambridge, Mass.: Harvard University Press, 2007).

79 *Ibid.*, 24.

First, do Nursebots view the human patients for whom they care as mere things? Do they distinguish between, say, the hospital beds and the patients who occupy them? I imagine that the robots can be programmed to respond in suitably different ways—for example, so that they do not attempt to engage the bed in conversation. However, we might then object that programmed caring and compassionate responses are not the same as human responses; and that this is where humanity is diminished. When humans are taught to talk the talk of care and concern, but without really relating to the other, we know that this is not the genuine article. It is not authentic because, quite simply, it is not expressed for the right reason.[80] Moralists, as we have emphasised, are not satisfied with a pattern of right conduct; moralists aspire to do the right thing *for the right reason*. No matter how smart a robot, no matter how professional the learned caring conduct of a human, it falls short of the mark unless it is motivated by the judgment that this is doing the right thing. This is not to deny that, all things considered, we might prefer a motivationally imperfect caring act to the kind of insouciance that sadly characterises much human 'care'; but we should at least be aware of the price we pay in the trade.

We might also detect a more general problem for an aspirant moral community where robots displace humans from caring roles—that is to say, a problem other than that associated with the socially disruptive effects of this loss of jobs. In some ethical communities, there might be a commitment to undertake caring responsibilities for members of one's family or friends; to act in this way is part of what defines one as a member of this particular (caring) community. Indeed, if it is suggested that there is a danger of moral complacency in leaving caring to the Nursebots, we might expect to find this standing on the premise that humans have non-delegable caring responsibilities to those who are in need.

Secondly, how do human patients relate to their robot carers? If, on the one hand, human patients ('mistakenly') view the Nursebots as humans, this might seem to be rewriting the frame of 'humanity', not to say confusing the essentials of human relationships and emotions. Yet, we tolerate humans treating their animal pets as though they are human; and we might even carry this over to some robots. Possibly, there are some special reasons to be concerned in relation to child development;[81] but, otherwise, this is not a major

---

80 Compare Charles Taylor, *The Ethics of Authenticity* (Cambridge, Mass: Harvard University Press, 1991). Taylor argues persuasively that the real issue is not whether we are for or against authenticity, but '*about* it, defining its proper meaning' (at 73); and, for Taylor, authenticity is to be understood as a moral ideal that accords importance to agents fulfilling themselves by being true to themselves—but without this 'excluding unconditional relationships and moral demands beyond the self' (at 72–73).

81 See, Jason Borenstein and Yvette Pearson, 'Companion Robots and the Emotional Development of Children' (2013) 5 *Law, Innovation and Technology* 172.

issue.[82] If, on the other hand, human patients view their robot carers as mere things, this might give rise to quite different ethical concerns—in particular, if it is argued that Nursebots should be recognised as having 'rights' or fundamental interests. Still, unless Nursebots are systematically abused, this seems to be no different to humans mistreating other humans or primates. Obviously, this should not be encouraged, but it does not look like a conclusive reason for declining to develop and use Nursebots.

(iii) Taking stock

There are many difficult questions raised by the vision of hospitals that secure patient safety by technological means. For moral communities, these questions intensify as the regulatory environment progressively hardens. With the rapid development of a range of technologies that might serve for such safety purposes, it is imperative that, in both specialist regulatory circles and in the public square, there is a broad conversation about the acceptability of a technocratic approach.[83]

We also need to debate the ethics of relying on Nursebots. Unlike the use of technologies as regulatory tools, reliance on Nursebots does not present an across-the-board challenge to moral communities. However, in some quarters, it will be argued that Nursebots raise questions about authenticity as well as chipping away at some important moral responsibilities. We might or might not be persuaded by these concerns but they need to be articulated and given serious consideration.

V Conclusion

In his comparative and historical review, Gert Brüggemeier singles out the two general liability clauses of the Russian Civil Code of 1922 as most neatly expressing the 'characteristic paths of liability law of the "risk society".'[84] One path is 'quasi-strict liability for businesses'; and the other is 'strict liability for technical risks'.[85] While each legal system makes its own distinctive response

---

82 See, further, the NESTA 'hot topics' discussion on emotional robots held on July 2, 2013: available at http://www.eventbrite.com/event/6960490013 (last accessed 5 November 2018).

83 See Roger Brownsword, *Rights, Regulation and the Technological Revolution* (Oxford: Oxford University Press, 2008). For one sign of a growing awareness of this concern, see Viktor Mayer-Schönberger and Kenneth Cukier, *Big Data* (London: John Murray, 2013) 162:

> Perhaps with such a [big data predictive] system society would be safer or more efficient, but an essential part of what makes us human—our ability to choose the actions we take and be held accountable for them—would be destroyed. Big data would have become a tool to collectivize human choice and abandon free will in our society.

84 Brüggemeier (n 1) 25.
85 *Ibid.*

to the first technological disruption, adjusting its liability rules in its own way, the general pattern of the movement away from the traditional coherentist model is clear. As new forms of transportation, new machines, and new risks are created, we can expect there to be some continuation of this regulatory-instrumentalist approach. However, we also have to reckon now with the second type of disruption. The rules which are revised and redirected to regulate autonomous vehicles, drones[86], surgical and caring robots, and so on, will operate alongside technological measures that contribute to the desired regulatory effect. Instead of relying ex post on tort law, we will come to rely more and more on preventative designs that serve our regulatory purposes. In just those places where we have traditionally relied on rules that set the standard for human health and safety—whether safety on the roads, in workplaces, or in the hospital—the standard will be embedded in the relevant products, the process will be automated, human operators will be removed from the loop, and tort-like correction will give way to technologically-secured protection.

---

86 Following the drone-related chaos at Gatwick airport in December 2018, there was precisely such a call for a mixed approach, employing both better rules and smarter regulatory technologies. See, e.g., BBC News, 'Gatwick airport: How countries counter the drone threat', December 21, 2018, https://www.bbc.co.uk/news/technology-46639099 (last accessed 21 December 2018).

11

REGULATING TRANSACTIONS
The future of contracts

I Introduction

Writing in 1920, John Maynard Keynes remarked that

> [t]he inhabitant of London could order by telephone, sipping his morning tea in bed, the various products of the whole earth, in such quantity as he might see fit, and reasonably expect their early delivery upon his door-step; he could at the same moment and by the same means adventure his wealth in the natural resources and new enterprises of any quarter of the world ... without exertion or even trouble.[1]

Had he been writing in 2020 (let alone in the year 2061), Keynes would no doubt have marvelled at the development of information and communication technologies that have facilitated the 24/7 ordering and speedy delivery of the various products of the earth—all without significant exertion or trouble. While the possibility of placing an order by telephone did require some clarification and confirmation of the applicable rules of the law of contract, it was neither a technological disruption requiring an urgent international response nor a disturbance of any great magnitude. By contrast, although the initial regulatory response to e-commerce, treating online transactions in just

---

1 John Maynard Keynes, *The Economic Consequences of the Peace* (1920) 10–12, cited in Douglas W. Arner, Jànos Nathan Barberis, Ross P. Buckley, 'The Evolution of Fintech: A New Post-Crisis Paradigm?' [2015] University of Hong Kong Faculty of Law Research Paper No 2015/047, University of New South Wales Law Research Series [2016] UNSWLRS 62: available at papers.ssrn.com/sol3/papers.cfm?abstract_id=2676553 (last accessed 6 November 2018).

the way that their offline equivalents would be treated[2] and confirming the validity of e-transactions and e-signatures,[3] did not suggest any great disturbance to the law of contract, the need to confirm the applicability of the law of contract to digital transactions was seen as a matter of urgency.[4]

While there was some pressure in Europe for a more proactive and comprehensive approach to the legal framework for e-commerce, Directive 2000/31/EC on Electronic Commerce famously failed to clarify and harmonise the relevant formation rules. To date, then, we can say that the law of contract has been gently stirred but not radically shaken up by the technological developments that are associated with the information society. However, as transacting in online environments becomes more the rule than the exception,[5] is this likely to remain the case? Even though, subject to some rule adjustment, contract law has persisted through countless previous technological transformations, is it immune to the rapid developments that we are witnessing in information and communications technologies? Moreover, as the possibilities for the automation and technological management of transactions intensify, will the law of contract continue to be relevant?

We can engage with this overarching question by considering, first (in Part II of this chapter), the shape and direction of the modern law of contract, where, in a development parallel with that in tort law, we find the simple transactionalist idea of being bound by deals to which one has agreed being modified (for the sake of easing the burden on nascent enterprise and innovation) and then being complemented by corrective regulatory interventions (especially in the mass consumer market).[6] Then (in Part III), we can introduce two questions that are provoked by new transactional technologies: one is whether there is a need to refine and revise the 'offline' law of contract where contracting takes place in 'online' environments; and the other is whether, given the automation of transactions, the rules of contract law will be replaced and rendered redundant. With a focus on European e-commerce law, I respond to the former question in Part IV; and, in Part V, I respond to the latter question—a question, of course, that reflects the

---

2 In Europe, the basic provision is Directive 2000/31/EC on Electronic Commerce.

3 On e-transactions, see Directive 2000/31/EC; and, on e-signatures, see Directive 1999/93/EC.

4 See, in particular, the early initiative taken by the UNCITRAL Model Law on Electronic Commerce 1996.

5 For some broader reflections on this shift, see Roger Brownsword, 'The Shaping of Our On-Line Worlds: Getting the Regulatory Environment Right' (2012) 20 *International Journal of Law and Information Technology* 249. See, too, Zygmunt Bauman and David Lyon, *Liquid Surveillance* (Cambridge: Polity Press, 2013).

6 See, Roger Brownsword, 'Regulating Transactions: Good Faith and Fair Dealing' in Geraint Howells and Reiner Schulze (eds), *Modernising and Harmonising Consumer Contract Law* (Munich: Sellier, 2009) 87.

impact of the 'second disruption' as the technological management of transactions becomes an actuality.

My principal conclusions are these. First, the rules which apply to (already heavily regulated) consumer transactions may need some revision in order to correct for unacceptable practices in online environments, Secondly, the development of 'smart' contracts and the like will not render redundant either the traditional rules of contract law or (what is left of) the transactionalist idea of contract. Thirdly, however, as both consumption of goods and services, as well as commercial trading, become progressively more automated, integrated and embedded—with human contractors out of the loop and with no standard contracts in sight—there will need to be further regulatory intervention. Finally, when transactions are so automated, integrated and embedded, traditional coherentist conversations about how these phenomena fit with the law of contract should be set to one side; at this point, it will be the time for a regulatory-instrumentalist conversation.

II The direction of the modern law

The impact of the first technological disruption on contract law, as on tort law, is to provoke changes to the rules. The nature of the correction is two-fold. On the one hand, the liability rules in contract and tort should neither inhibit beneficial innovation nor bankrupt nascent businesses; on the other hand, agents who are vulnerable to major new physical and financial risks (whether as employees or consumers) need to be protected.[7] Thus, when contract law abandons a pure transactional theory of contractual rights and obligations in favour of an objective theory of agreement, this serves to shield carriers against otherwise crippling claims for compensation (when valuable packages are lost or when there are accidents on the railways); and when, in the middle years of the last century, a mass consumer market for new technological products (cars, televisions, kitchen appliances, and so on) develops, a fundamental correction is needed to protect consumers against the small print of suppliers' standard terms and conditions.

The adoption of an objective theory of agreement signifies that the enforcement of contractual rights and obligations is no longer predicated on a party's having actually agreed to the terms and conditions.[8] Rather, the question is whether a co-contractor (or an impartial third party) could say honestly and reasonably that they understood the other party to be agreeing. In this way, there was a decisive modification to transactionalist theory.

---

7 For an interesting debate about how far it is in the interests of employees to abandon the general law of contract in favour of tailored regulation: see, e.g., Gwyneth Pitt, 'Crisis or Stasis in the Contract of Employment?' (2013) 12 *Contemporary Issues in Law* 193.

8 Compare, Morton J. Horwitz, *The Transformation of American Law 1870–1960* (Oxford: Oxford University Press, 1992) 36 et seq.

However, as we noted in an earlier chapter, even in the nineteenth century, there was the hint of an entirely non-transactionalist view that the enforcement of contractual terms and conditions should be judged relative to reasonable business practice. This non-transactional idea took hold in the second half of the last century giving rise to widespread legislative intervention in the consumer marketplace to ensure that the balance between the interests of suppliers and the interests of consumers was fair and reasonable.[9] Today, at any rate in English law, following the Consumer Rights Act 2015, we can say that consumers engage, not so much in contracts, but in regulated transactions; that there is a clear bifurcation between the treatment of commercial (business to business) contracts and consumer contracts; and that this latter branch of the law is dominated by a regulatory-instrumentalist mind-set.

Before we turn to commercial contracts, we can speak briefly to one of the first cases—a case from the early 1970s—where the courts had to deal with a dispute about the terms and conditions of a contract which was automated on the suppliers' side. This was in *Thornton v Shoe Lane Parking Ltd*,[10] where the contract was for the parking of a car at a car park that had an automated entrance system (equipped with traffic lights, a barrier, and a ticket machine but with no human operator in attendance). The question was whether the car park could respond to a claim in negligence brought by Mr Thornton (who had been injured at the car park) by relying on the exclusionary terms and conditions displayed at the car park. The Court of Appeal held that, because Mr Thornton had not been given reasonable notice of the critical exclusion of liability, this term had not been incorporated into the contract and so the defendants could not rely on it. One reason in support of this ruling was that the design of the entrance to the car park meant that motorists were already committed to entering the car park before they had notice of the terms.[11] Applying a standard objective test, unless Mr Thornton was a regular user of the car park, no one could reasonably think that he was agreeing to terms and conditions of which he was unaware. However, the principal reason in support of the ruling was that the particular exclusionary term was either unusual or destructive of ordinary rights and, therefore, needed to be specifically highlighted and drawn to the attention of car park users. This line of thinking hovers between transactionalism and non-transactionalism; but, before the end of the decade, the Unfair Contract Terms Act 1977 had been enacted, providing among other things that exclusionary terms of

---

9 Compare, Hugh Collins, *The European Civil Code* (Cambridge: Cambridge University Press, 2008) 26: 'Instead of talking about *caveat emptor* (let the buyer beware) as in the past, today we speak of ensuing that consumers receive good-quality goods that meet their expectations in return for a fair competitive price.'

10 *Thornton v Shoe Lane Parking* [1971] 2 QB 163.

11 This, we should note, involved a textbook coherentist attempt to place the architecture of the car park and the actions of the motorist in the standard matrix of invitation to treat, offer and acceptance.

the kind at issue in *Thornton* simply could not be relied on (in effect, such terms were black-listed). From that point on, as we have said, we see the consumer marketplace progressively subjected to regulation that marginalises transactionalism.

What, then, of commercial contracts? Here, the position is more complex. The thinking behind some legislative interventions, such as the Arbitration Act 1979, is clearly regulatory-instrumentalist. However, the bulk of the commercial law of contract is in the cases where, for some time, there has been a fierce debate about the right way to approach the interpretation of commercial contracts (it should be recalled that a high percentage of litigated commercial transactions hinge on rival interpretations of contractual terms) as well as the implication of terms.[12] On the one side, minimalists argue in favour of a literal reading of contracts and a high threshold for the implication of terms;[13] on the other, contextualists argue that contracts should be read in their factual matrix and that both interpretation and implication should be sensitive to commercial purpose and good sense.[14] Both sides might claim to be faithful to transactionalist principles, but with differing views as to what counts as best evidence of what the parties agreed, as well as being coherent-ist. Equally, though, we might construct a regulatory-instrumentalist nar-rative in which, while contextualists see the purpose of the law as being to support community norms and self-regulating traders,[15] minimalists see the

---

12 On interpretation, the seminal cases are *Mannai Investments Co Ltd v Eagle Star Life Assur-ance Co Ltd* [1997] 3 All ER 352 and *Investors Compensation Scheme Ltd v West Bromwich Building Society* [1998] 1 All ER 98; with *Arnold v Britton* [2015] UKSC 36 and *Wood v Capita Insurance Services Ltd* [2017] UKSC 24 representing some retrenchment. On implied terms, the leading case on a contextualist approach is *Attorney General of Belize v Belize Telecom Limited* [2009] UKPC 11; and, for a decisive step back, see *Marks and Spencer plc v BNP Paribas Securities Services Trust Company (Jersey) Limited* [2015] UKSC 72. Generally, see Roger Brownsword, 'The Law of Contract: Doctrinal Impulses, External Pressures, Future Directions' (2014) 31 JCL 73 and 'After Brexit: Regulatory Instrumental-ism, Coherentism, and the English Law of Contract (2017) 34 JCL 139.

13 See, Jonathan Morgan, *Contract Law Minimalism* (Cambridge: Cambridge University Press, 2013) 1, arguing that contract law will only return to being fit for purpose (that is, fit to support commercial transactions) when it turns back from its contextualist trajectory and reverts to applying 'a clear and minimal set of hard-edged rules'.

14 See Catherine Mitchell, *Contract Law and Contract Practice* (Oxford: Hart, 2013). In *Total Gas Marketing Ltd v Arco British Ltd* [1998] 2 Lloyd's Rep 209 at 221, Lord Steyn (who, along with Lord Hoffmann has been the principal judicial author of contextualism) sum-marised the essential idea in the following way: '[questions of interpretation] must be con-sidered in the light of the contractual language, the contractual scheme, the commercial context and the reasonable expectations of the parties.'

15 Compare the deference to market-specific custom and practice in *Transfield Shipping Inc of Panama v Mercator Shipping Inc of Morovia, The Achilleas* [2008] UKHL 48. There, the charterers of a single-deck bulk carrier were in breach of contract by being some nine days late in redelivering the vessel to the owners. As a result of this late redelivery, the owners were put in a difficult position in relation to a new fixture of the vessel that had been agreed with another charterer. In order to retain this fixture, the owners agreed to reduce the daily rate by $8,000. The question was whether the owners were entitled to recover the loss of

purpose of contract law as being to provide a particular kind of dispute-resolution service for commercial contractors who cannot otherwise settle their differences.

In the light of all this, we can expect regulatory-instrumentalist thinking to shape the further automation of consumer transactions. For example, if Mr Thornton were now riding in an autonomous vehicle, so that the transaction was fully automated, there would surely be no reference to offers and acceptances or the like; whatever risks there were in this scenario would be covered by an agreed regulatory scheme. Moreover, if a technocratic approach secures the agreed features of that scheme, then it would be natural for that to be implemented as part of the agreed package for the services provided. What will happen in the commercial sector is less clear. Automated transactions are very much part of the landscape. Nevertheless, some contractors might reject technological management of their transactions, preferring to regulate through customary rules; and, others might insist on a clear transactionalist basis for the adoption of new transactional technologies (such as electronic bills of lading).[16]

Before we get to these questions, though, we must take stock of the impact on the law of contract of the development of the Internet as a platform for e-commerce. For, whatever the direction of travel of the offline law of contract, it is clear that the law must respond appropriately to the significant and increasing movement of transactions to online environments.

III Refining, revising, and replacing the law: two questions

Twenty years ago, when UNCITRAL published its Model Law on Electronic Commerce,[17] there was felt to be an urgent need to set out 'internationally

---

$8,000 per day for the duration of the new charter (a sum of $1,364,584.37) or merely the difference between the market rate and the charter rate for the nine days late redelivery (a sum of $158,301.17). The majority of the arbitrators and the lower appeal courts, applying the general rules, held that the owners were entitled to the higher amount: their reasoning was that the charterers must have known that the owners were likely to arrange a new onward charter, that market rates fluctuate, and that late delivery might reduce the profitability of the onward charter (i.e., the forward fixture). However, the House of Lords held unanimously (albeit with some hesitation) that the owners were restricted to the lesser sum, this apparently being in line with the expectation of owners and charterers who deal in the shipping sector.

16 See, e.g., John Livermore and Krailerk Euarjai, 'Electronic Bills of Lading and Functional Equivalence' (1998) (2) *The Journal of Information Law and Technology* (JILT): available at https://warwick.ac.uk/fac/soc/law/elj/jilt/1998_2/livermore/livermore.doc 1 (last accessed 6 November 2018); and, for the current position, UK P&I Club, 'Electronic Bills of Lading' (May 2017), available at www.ukpandi.com/fileadmin/uploads/uk-pi/Documents/2017/Legal_Briefing_e_bill_of_Lading_WEB.pdf. (last accessed 6 November 2018).

17 Available at www.uncitral.org/uncitral/en/uncitral_texts/electronic_commerce/1996Model.html (last accessed 6 November 2018).

acceptable rules aimed at removing legal obstacles and increasing legal predictability for electronic commerce'.[18] In particular, it was thought to be important to provide 'equal treatment to paper-based and electronic information'.[19] In Europe, too, this sense of urgency and importance was shared. If small businesses in Europe were to take advantage of the opportunity afforded by emerging ICTs to access new markets, and if consumers were to be sufficiently confident to order goods and services online, it needed to be absolutely clear that e-transactions (no less than 'traditional' contracts) were recognised as legally valid and enforceable.[20] Accordingly, the regulatory priority was to remove any uncertainty about the legal status of 'online' transactions, Article 9(1) of the e-commerce Directive duly requiring Member States to ensure that, in effect, their national legal frameworks for 'offline' transactions (paradigmatically, for the supply of goods and services) apply also to e-transactions.[21]

Today, in Europe, against the background of the Digital Single Market Strategy,[22] there is a renewed sense of urgency. If the participation (especially, the cross-border participation) of small businesses and consumers in e-commerce is to increase and, with that, if the European economy is to grow, further legal action is required: there are obstacles to be removed and legal predictability needs to be improved. As the Commission has put it, if the potential of e-commerce is to be 'unleashed', there is a *'need to act now on the digital dimension'*.[23]

As is well known, the so-called legislative framework for e-commerce is not found in one comprehensive Directive; and the e-commerce Directive itself is but one element in a group of legal instruments that have been pressed into regulatory service as e-commerce has developed.[24] In the case of those legislative instruments, such as the Directives on consumer sales and

---

18 *Ibid.*, (expressing the purpose of the Model Law).

19 *Ibid.*

20 Roger Brownsword and Geraint Howells, 'When Surfers Start to Shop: Internet Commerce and Contract Law' 19 *Legal Studies* (1999) 287.

21 It was in this context that the idea crystallised that regulators should act on the principle that 'what holds off-line, must hold on-line': see Bert-Jaap Koops, Miriam Lips, Corien Prins, and Maurice Schellekens (eds), *Starting Points for ICT Regulation—Deconstructing Prevalent Policy One-Liners* (The Hague: T.C.M. Asser, 2006).

22 European Commission, Communication from the Commission to the European Parliament, the Council, the European Economic and Social Committee and the Committee of the Regions, 'A Digital Single Market Strategy for Europe' COM(2015) 192 final (Brussels, 6.5.2015).

23 European Commission, Communication from the Commission to the European Parliament, the Council and the European Economic and Social Committee, 'Digital contracts for Europe—Unleashing the potential of e-commerce' COM(2015) 633 final (Brussels, 9.12.2015), p.7 (italics in original).

24 This group currently comprises: Directive 2011/83/EU on consumer rights (amending earlier Directives); Directive 1999/44/EC on consumer sales and guarantees; Directive 93/13/EEC on unfair terms in consumer contracts; and Directive 2000/31/EC on e-commerce.

on unfair terms in consumer contracts, that pre-dated the development of
e-commerce there was no need to consider whether offline environments for
consumer transactions were comparable to online environments. However,
by the time of the e-commerce Directive, this had become one of the central
regulatory questions. In some respects, notably in relation to the potential
liability of ISPs for the transmission, caching, and hosting of information,[25]
the Directive recognises that online environments are different and that some
adjustment of offline law is required. However, so far as contracts are concerned, the Directive reflects an approach that David Post describes as that of
the 'unexceptionalists',[26] this being the view that online transactions require
no special regulatory treatment (other than to confirm their legal enforce-
ability in principle).

The Directive's unexceptionalist view of contracts is underpinned by
three key assumptions.[27] First, it is assumed that online contracts are *functionally* equivalent to offline contracts—for example, the sale of a book at
Amazon online is no different functionally from the sale of a book offline at
a bricks-and-mortar bookshop. Secondly, it is assumed that the *context* for
online transactions is not materially different from the context for offline
transactions: to the contrary, there is a relevant similarity between offline
and online shops. Granted, the architecture of online shopping environments is different from the architecture of offline shopping environments
and it affects the options available to consumers. However, it was assumed
that the differences would not operate in ways that would be prejudicial
to online shoppers. Indeed, there was even reason to think that the online
architecture—at any rate, 'click wrap' assent rather than 'browse wrap'
notice—might prove to be more protective of consumers' interests because
it would make it possible to encourage consumers to read the supplier's
terms and conditions carefully before proceeding with the purchase.[28]
Thirdly, building on the first two assumptions, we have the idea of *normative*

---

25 See, Articles 12, 13 and 14 of the Directive. The fact that the Directive treats the Internet as
'exceptional' relative to, say, the law of defamation and copyright but as 'unexceptional' relative to contract law is not necessarily contradictory: compare Margot E. Kaminski, 'Robots
in the Home: What Will We Have Agreed To?' (2015) 15 *Idaho Law Review* 661, 663; and
Ryan Calo, 'Robotics and the Lessons of Cyberlaw' (2015) 103 *California Law Review*
513, 550–553.

26 David G. Post, *In Search of Jefferson's Moose* (Oxford: Oxford University Press, 2009) 186.

27 See, Roger Brownsword, 'The E-Commerce Directive, Consumer Transactions, and the
Digital Single Market: Questions of Regulatory Fitness, Regulatory Disconnection and Rule
Redirection' in Stefan Grundmann (ed), *European Contract Law in the Digital Age* (Ant-
werp: Intersentia, 2017) 165.

28 Moreover, with burgeoning ex post consumer reviews, one might expect this information
to have some corrective effect on conspicuously unfair terms: see, e.g., Shmuel I. Becher
and Tal Z. Zarsky, 'E-Contract Doctrine 2.0: Standard Form Contracting in an Age of
Online User Participation' (2008) 14 *Michigan Telecommunications and Technology Law
Review* 303.

equivalence, the assumption being that those laws that apply to offline trans-actions should also apply to online transactions. In Europe, this means that the general law of contract, in conjunction with a special body of regulation for consumer transactions, should be copied across from offline transactions to online transactions.

Today, we sense that the distinction between a mature off-line consumer marketplace and an embryonic on-line marketplace—a distinction that is implicit in but central to the e-commerce Directive—is the precursor not only to an onl*ife* marketplace that blurs the online/offline distinction[29] but also to a quite different distinction between consumer transactions that are actively negotiated and performed by human agents and transactions for the supply of consumer goods and services that are processed automatically (by 'mindless' agents) with at best the passive involvement of human agents.

Framing e-commerce in this way, two questions arise. First, even if, at the time of its enactment, the e-commerce Directive was right to treat offline and online transactions as functionally, normatively, and contextually equiva-lent, does that continue to be the appropriate regulatory view? Or, should some special regulatory allowance be made for what now seem to be mate-rial contextual differences between the two transactional environments? Sec-ondly, how should regulators view the distinction between transactions that are made (offline or online) between business suppliers and consumer con-tractors and automated transactions for the supply of goods or services to consumers? Should regulators treat this distinction as immaterial and regard anticipatory and automated transactions as equivalent to non-automated transactions? If, instead of regulating the consumer marketplace by specifying the transactional ground rules for sellers and purchasers, the technological management of transactions takes the regulatory strain, what should we make of this? Rather than asking whether the e-commerce Directive (or the regula-tory environment of which the Directive is a part) needs some modification, should we be asking whether the assumptions that the Directive made about how to regulate the routine consumption of goods and services need to be fundamentally reconsidered?[30]

---

29 See, Mireille Hildebrandt, *Smart Technologies and the End(s) of Law* (Cheltenham: Edward Elgar, 2015).

30 In general, the European Commission's current approach, albeit conspicuously regulatory-instrumentalist, is geared to incremental change and modification rather than a radical re-thinking of the regulatory environment for consumer transactions and routine consumption. See, e.g., the European Commission Staff Working Document (Report on the Fitness Check of EU consumer and marketing law) SWD(2017) 209 final, Brussels, 23.5.2017 (where the substantive provisions are seen as being largely fit for purpose but the enforcement of the law and the remedies for consumers need to be strengthened); and, European Commission, *Communication from the Commission to the European Parliament, the Council and the Euro-pean Economic and Social Committee, A New Deal for Consumers* COM(2018) 183 final, Brussels 11.4.2018. That said, in the latter document it is recognised (at p.9) that there is a

We can tackle the first clutch of questions in the next section of the chapter; and, then, in the following section, we can respond to the second set of questions.

IV Refining and revising the law of contract

As consumers increasingly turn to online environments to order goods and services, we need to consider whether the offline law is fit for purpose in these new contractual settings. We can discuss two particular problems: one, the familiar question of unfair terms; and the other, the new concern about profiling.

(i) Unfair terms

The e-commerce Directive says nothing specifically about unfair terms in online consumer contracts. However, protective measures relating to offline transactions apply; and, if a term will pass muster in an offline contract, the implication is that it will also be treated as 'not unfair' in an online contract. Yet, even if the offline tests are fit for purpose, is it right to apply the same tests to online transactions? Are there not material differences between offline and online contracting environments?

Let us start by noting a couple of almost too-obvious-to-record, yet highly significant and distinctive, features of online contracting. First, there is the striking fact, noted by Mark Lemley, that consumers who make their purchases in offline stores are rarely required to sign up to the suppliers' terms and conditions in the way that is characteristic of online shopping.[31] Thus,

> Merchants and consumers at grocery stores, restaurants, bookstores, clothing stores, and countless other retail outlets seem perfectly able to enter into contracts without a written agreement specifying their rights and obligations. Nonetheless, many of those same retail outlets impose standard form contracts on their online users, probably because it is easier to get someone to click 'I agree' as part of an online transaction than it is to have a clerk obtain a signature on a written form.[32]

Secondly, online transactions are technologically mediated in a way that is quite different from the paradigmatic offline transaction. Consumers are,

---

need to monitor consumer markets where artificial intelligence, the Internet of Things, and mobile e-commerce might present issues that require regulatory attention.
31 Mark A. Lemley, 'Terms of Use' (2006) 91 *Minnesota Law Review* 459.
32 Lemley (n 31) 466.

so to speak, technologically 'mediated consumers', approaching 'the marketplace through technology designed by someone else'[33]—note, first the technological mediation and, secondly, the fact that it is not designed by the consumer. As Ryan Calo memorably puts it, in these environments, an offer is 'always an algorithm away'.[34] If, in these new transactional settings, online suppliers are opportunistically loading their standard forms to the detriment of consumers in a way that does not happen in offline contracting, this immediately raises questions not only about the equivalence of offline and online contracts but also about whether online suppliers are taking unfair advantage of the online environments in which they can now sell to consumers.

Although some commentators believe that the offline controls are perfectly adequate to regulate online business practice,[35] and although a survey of the websites of 500 leading Internet retailers has suggested that, in fact online sellers are not taking advantage of consumers and that 'the contract terms and contracting interfaces of internet retailers are surprisingly benign',[36] the view persists that online suppliers are taking advantage of consumers, that, as a result, consumers are exposed to new vulnerabilities and new kinds of unfairness, and that a new approach is called for.

According to Nancy Kim, for example, efforts to create a 'smooth website experience' for consumers militate against consumers reading terms; and this is compounded by companies who take advantage of consumer behaviour by including 'ever more aggressive and oppressive terms'.[37] Kim argues that the courts should respond, inter alia, by reinvigorating 'reasonable notice' requirements in a way that encourages the design of online environments that force consumers to slow down as well the employment of both visual (e.g. traffic light) and sensory signals to aid consumer navigation of sites.[38]

---

33 Ryan Calo, 'Digital Market Manipulation' (2014) 82 *The George Washington Law Review* 995, 1002.

34 Calo (n 33) 1004.

35 See, e.g., Robert A. Hillman and Jeffrey J. Rachlinski, 'Standard-Form Contracts in the Electronic Age' (2002) NYULR 429.

36 Ronald J. Mann and Travis Siebneicher, 'Just One Click: The Reality of Internet Retail Contracting' (2008) 108 *Columbia Law Review* 984, 1000.

37 Nancy S. Kim, 'Situational Duress and the Aberrance of Electronic Contracts' (2014) 89 *Chicago-Kent LR* 265, 265–266.

38 Nancy S. Kim, 'The Duty to Draft Reasonably and Online Contracts' in Larry A. DiMatteo et al (eds), *Commercial Contract Law: Transatlantic Perspectives* (Cambridge: Cambridge University Press, 2012) 181, 190–200. She also suggests, in Kim (n 37), a limited extension of duress—'situational duress' as she calls it—to regulate those practices where consumers are given no reasonable alternative other than to agree to certain terms and conditions. The use of such design features is important not only for the incorporation of terms but also for the formation of contracts: see Elizabeth Macdonald, 'When is a Contract Formed by the Browse-Wrap Process?' (2011) 19 *International Journal of Law and Information Technology* 285.

While there has been a tendency to assimilate e-contracts to paper contracts, Kim emphasises that the focus should be on their differences rather than their similarities.[39] In a similar vein, Stephen Waddams has argued that online environments tend to exacerbate familiar offline problems with standard forms and unfair terms in consumer contracts.[40]

So far as the United Kingdom is concerned, the principal question is whether the controls on unfair contract terms, now in the Consumer Rights Act 2015, are fit for purpose. The effectiveness of these controls hinges on a number of considerations that reach far beyond the bare terms of the statutory provisions. For example, these controls will be of limited effect unless suppliers understand what is required, unless they have internalised these standards of fair dealing, unless consumers are aware of their rights, and unless regulators are sufficiently resourced to ensure compliance.[41] In all of this, the courts have a limited role to play but the signals given by the appeal courts are not unimportant.

In the years following the enactment of the Unfair Contract Terms Act 1977, one would have been reasonably confident that a consumer who challenged a term as unfair and unreasonable would have succeeded.[42] However, in the present century, such confidence may be misplaced. For, in two major test cases brought by the Office of Fair Trading,[43] the consumer challenges have failed; and, most recently, in *ParkingEye Limited*,[44] the UK Supreme Court again rejected the consumer's challenge.

In *ParkingEye*, the question was about the fairness of a particular term at a retail shopping car park. Notices at the car park clearly indicated that parking, although free, was limited to a maximum of two hours, and that parking for more than the permitted time would result in a charge of £85. In the test case, the motorist left his car parked for 56 minutes more than the permitted time. All seven members of the UK Supreme Court took their lead from

---

39 Kim (n 37) 286.

40 Stephen Waddams, 'Contract Law and the Challenges of Computer Technology' in Roger Brownsword, Eloise Scotford, and Karen Yeung (eds), *The Oxford Handbook of Law, Regulation and Technology* (Oxford: Oxford University Press, 2016) Ch.13. These difficulties are even more pronounced in relation to the terms and conditions that come with smart devices (such as smartphones, cars, homes, fitness monitors, and so on): see Scott R. Peppet, 'Regulating the Internet of Things: First Steps Toward Managing Discrimination, Privacy, Security and Consent' (2014) 93 *Texas Law Review* 85, 139–147.

41 Compare the comments in SWD(2017) 209 final (n 30) at p.74 et seq (on the lack of progress in relation to traders' compliance with consumer protection rules and the challenges to the effectiveness of the law).

42 See John N. Adams and Roger Brownsword, 'The Unfair Contract Terms Act: A Decade of Discretion' (1988) 104 *Law Quarterly Review* 94.

43 *Director-General of Fair Trading v First National Bank plc* [2000] 1 All ER 240; and *Office of Fair Trading v Abbey National plc* [2009] UKSC 6.

44 *Cavendish Square Holding BV v Talal El Makdessi; ParkingEye Limited v Beavis* [2015] UKSC 67.

the CJEU in *Aziz*[45]—where it was emphasised that the test of unfairness (in Article 3.1 of the Unfair Terms in Consumer Contracts Directive[46]) is very general in nature and leaves much to the national courts who are in a position to provide the context and particular circumstances for the test's application. However, in holding by 6 to 1 that the over-parking term was not unfair, while the majority highlighted the legitimate interest of the company, the one minority judge, Lord Toulson, highlighted the view of the consumer.

Against the background of Advocate General Kokott's opinion in *Aziz*, the majority (led by Lord Neuberger) reasoned, first, that the charge did operate to the detriment of the consumer because, under the ordinary law, the unlicensed parking of the car would have amounted to a trespass (damages for which would have been limited to the occupation value of the space). Secondly, with regard to the question of the parking operators' good faith, the question—so the majority held—is not whether the particular motorist 'would in fact have agreed to the term imposing the £85 charge in a negotiation, but whether a reasonable motorist in his position would have done so'.[47] Thirdly, having taken into account the importance to both the freeholders and the lessees of the retail outlets as well as to the shoppers of ensuring that parking space was available (and not clogged up by commuters using the adjacent railway station or by other long-term users), the majority were satisfied that 'a hypothetical reasonable motorist would have agreed to objectively reasonable terms, and these terms are objectively reasonable.'[48]

By contrast, Lord Toulson adopted a protective test of good faith such that the burden is on the supplier to show that it could be fairly assumed 'that the consumer would have agreed [to the term] in individual negotiations on level terms'.[49] For a number of reasons—largely, common sense reasons why, in special circumstances, an £85 charge should not be applied— Lord Toulson declared himself not satisfied that the car park operators could discharge the burden of showing that they could reasonably assume that 'a party who was in a position to bargain individually, and who was advised by a competent lawyer, would have agreed to the [term that imposed the charge].'[50]

Returning to online consumer environments, there will be many examples of sites (like the offline car park) that are free to access (in the sense that no fee has to be paid) but that make entry and use subject to non-negotiable terms and conditions.[51] If a consumer wishes to challenge one of those terms

45 *Mohamed Aziz v Caixa d'Estalvis de Catalunya, Tarragona I Manresa (Catalunyacaixa)* Case C-415/11.
46 Directive 93/13/EEC.
47 *Cavendish Square Holding BV v Talal El Makdessi; ParkingEye Limited v Beavis* [2015] UKSC 67, [108].
48 *Ibid.*, [109].
49 *Ibid.*, [309].
50 *Ibid.*, [314].
51 For discussion of the fairness of 'give data to get service' deals, see Natali Helberger, 'Profiling and Targeting Consumers in the Internet of Things—A New Challenge for Consumer

as unfair, *Aziz* suggests a reasonably high level of protection. However, if national courts, as Lord Toulson puts it, apply *Aziz* in a way that 'waters down the test'[52]—for example, by prioritising a not unreasonable business perspective over a not unreasonable consumer perspective, or even by recalling a notion of 'freedom of contract' that treats consumers as bound by terms of which they have been made aware despite having no freedom to negotiate them—then the application of the offline rules to online environments will leave consumers vulnerable.

Summing up, there are several questions, both of principle and of an empirical nature, threaded together in these remarks about potentially unfair terms in online consumer contracts. Perhaps the first question is whether there is evidence of significant use of unfair terms in online consumer transactions—that is to say, terms that either would not appear in equivalent offline consumer contracts or that would not be used in any offline contracts.[53] If so, the next question is whether consumers can be plausibly said, at best, to be consenting to these terms or, at least, to be proceeding with reasonable notice of such terms. If not, the question then is whether the Directive on unfair terms in consumer contracts is fit to correct such unfairness. If not, the question is what kind of further protective legal rules are required—or perhaps the time is right for a more technocratic approach.[54] Imagine, for example, that regulators or consumers (just like legal practitioners) were able to rely on smart machines to scan online terms and conditions to see which, if any, were arguably unfair; or imagine that a supplier's standard terms and conditions had to be displayed in a format that would not permit anything other than clearly fair terms. While, in the former case, a technocratic intervention

---

Law' in R. Schulze and D. Staudenmayer (eds), *Digital Revolution: Challenges for Contract Law in Practice* (Baden-Baden: Nomos, 2016) 135, 142–151.

52 *Cavendish Square Holding BV v Talal El Makdessi; ParkingEye Limited v Beavis* [2015] UKSC 67, [315].

53 Compare, too, European Commission, Communication from the Commission to the European Parliament, the Council, the European Economic and Social Committee and the Committee of the Regions, 'Online Platforms and the Digital Single Market: Opportunities and Challenges for Europe' COM(2016) 288/2, Brussels, May 25, 2016, at 12–13, where the Commission recognises that there needs to be further investigation into whether, in B2B relations, on-line platforms are imposing unfair terms and conditions on SMEs and micro-enterprises.

54 For an interesting and highly relevant application of new technological tools to the regulation of unfair terms in consumer contracts, see Hans-W Micklitz, Przemyslaw Palka, and Yannis Panagis, 'The Empire Strikes Back: Digital Control of Unfair Terms of Online Services' (2017) 40 *Journal of Consumer Policy* 367. Similarly, but with machine learning applied to flag up provisions in on-line privacy policies that are problematic relative to the GDPR, see Giuseppe Contissa, Koen Docter, Francesca Lagioia, Marco Lippi, Hans-W Micklitz, Przemysław Palka, Giovanni Sartor and Paolo Torroni, 'Claudette Meets GDPR: Automating the Evaluation of Privacy Policies Using Artificial Intelligence' (July 2, 2018). Available at SSRN: ssrn.com/abstract=3208596 (last accessed 6 November 2018).

might support a transactionalist view, in the latter it might achieve a more acceptable balance of interests.

(ii) Profiling

Writing at the turn of the century, in a prescient article, Richard Ford imagines an online environment in which a person's data is used by suppliers and others to develop a profile of the-consumer-that-is-me to the point that they 'will suggest books I'll enjoy reading, recordings I'll like to listen to, restaurants I'll be glad I tried, even if I wouldn't have chosen any of them on my own'.[55] While some of these suggestions will coincide with the consumer's likes, others will not. However, the more that a person's preferences are revealed and used to refine the profile, the more that person will like them and the more reliant on them they will become. Before too long, there will be little or no distance between the person and their profile with the profilers seeming to know the individual better than they know themselves.

Yet, if online suppliers know me better than I know myself, how far does this depart from the world of offline consumption? Possibly, there will be some marketplaces in which traders really do know their customers but, in many offline environments, each consumer is simply another number to add to the footfall. Does such a departure give rise to regulatory concern with regard to the balance of power in online environments, to the transparency of the dealing, and to the autonomy of consumers? As with the concern about unfair terms, are there features of online consumer environments that cumulatively differentiate them in a material way from offline marketplaces?

It is in the nature of consumer profiling, particularly when powered by machine-learning, that it will become more 'discriminating' as more data is acquired relating to the preferences of individual consumers or groups of consumers. For present purposes, let me assume that whatever discrimination is exercised it is used in a way that is lawful.[56] Nevertheless, this does not exhaust concerns about the vulnerabilities of consumers in online environments.

---

55 Richard T. Ford, 'Save the Robots: Cyber Profiling and Your So-Called Life' (2000) 52 *Stanford Law Review* 1576, 1576.

56 Compare the protective decision in *CCN Systems Ltd v Data Protection Registrar* (Case DA/90 25/4/9, judgment delivered 25 February 1991), where the question was whether it was fair to use 'third-party' address-based information to make a decision where a first-party (who might have no relationship with the third party) applies for credit. Even though it was accepted that such third-party information might have general predictive value and utility, its use was held to be unfair to the individual.

Articulating concerns of this kind, Eliza Mik has written:

> Online businesses influence consumer behaviour by means of a wide range of technologies that determine *what* information is displayed and *how* and *when* it is displayed. This creates an unprecedented power imbalance between the transacting parties, raising questions not only about the permissible levels of procedural exploitation in contract law, together with the adequacy of existing consumer protections but also, on a broader level, about the impact of technology on the autonomy of the consumer.[57]

Mik does not claim that online consumer environments are wholly different in kind to their offline equivalents; but she does argue that there are elements in the former that render them significantly different. Thus:

> Malls, shops and the packaging of goods have always been designed to influence our purchasing decisions. Malls, shops and vending machines do not, however, change their layout and contents in real-time to match specific individuals at crucial moments related to their psychological or physiological states. Traditional transacting environments cannot detect and exploit individual vulnerabilities. There is fairness in the fact that everybody is presented with the same range of choices in the same way. Every sales person faces limits as to what she can directly observe from the interaction with the consumer. She does not know the consumer's prior purchasing history with *other* shops, his circle of friends, his medical correspondence and his cognitive preferences. She can adapt the sales strategy to a limited range of consumer characteristics but, leaving aside extreme examples, is unable to detect idiosyncratic vulnerabilities. In recommending specific products, the sales person relies on her intuition and experience—not on the combined experience derived from billions of transactions. Some shops pay higher rent for space in high-traffic areas, but once the spaces have been allocated mall operators cannot promote certain shops by making others more difficult to access.
>
> … Unquestionably, commercial websites can be compared to shops or malls. We can focus on the similarities: in both instances a range of goods is presented for sale. We can also focus on the differences. Apart from the obvious ones, such as automation and lack of direct human contact, the cognitive and sensory experience of navigating web interfaces differs from walking around supermarkets. Browsing the Amazon

---

57 Eliza Mik, 'The Erosion of Autonomy in Online Consumer Transactions' (2016) 8 *Law, Innovation and Technology* 1, 2.

app differs from browsing a physical bookstore. Amazon is not a shop but a symbolic abstraction of a shop ...The web must be recognized as a discrete transacting environment, where consumers process and interact with information differently than offline, and where they are subject to different, more powerful influences.

... In sum, there is nothing new in the fact that businesses try to influence consumption choices. It is the differences in the available tools and in the characteristics of the online environment that warrant legal attention.[58]

How, then, should we respond to these different tools and characteristics? Should we have recourse to the protective (transactionalist) principles of the general law of contract, or should we rely on a more regulatory approach?

Within each legal system, there will be doctrines (such as, in English law, the doctrine of undue influence) that seek to protect parties whose transactional autonomy is compromised.[59] Typically, the context is one in which A is so reliant on B for advice and assistance that A routinely follows B's advice. This does not presuppose any active or conscious wrongdoing on B's part; the problem is that A no longer exercises any independent transactional judgment. In many ways, we might find such a relationship of dependence in the dealings between consumers and their online suppliers. However, the kind of profiling and personalised service that is under question is not only widespread in online consumer environments, it is presented as a service to better inform consumers. As Ford notes, the suggestions that come in might be for goods or services that we have not thought about buying, and which, having acted on the recommendations, we found we liked. There does not seem to be too much to object to in that. However, if particular consumers find that they are unable to put any distance between such recommendations and their shopping choices, they might argue that the online transactions into which they enter are no longer 'their own'; and it might be possible to invoke protective general doctrines. That said, because the question raised here is pervasive—and because, as Natali Helberger rightly remarks, there is a 'fine line between informing, nudging and outright manipulation', all this presenting 'one of the main challenges for consumer law and policy in the context of

---

58 Mik (n 57) 22–24. In similar terms, see Calo (n 32).
59 In the leading case of *Royal Bank of Scotland plc v Etridge (no. 2)* [2001] UKHL 44, [11], Lord Nicholls remarked that 'there is no single touchstone for determining whether the principle [of undue influence] is applicable. Several expressions have been used in an endeavour to encapsulate the essence: trust and confidence, reliance, dependence or vulnerability on the one hand and ascendancy, domination or control on the other. None of these descriptions is perfect. None is all embracing. Each has its proper place.'

profiling and targeting in the Internet of Things'[60]—perhaps a regulatory response is more appropriate. This, indeed, is the route that Mik takes and her view is that the best regulatory resources currently in place in Europe are in Directive 2005/29/EC on unfair commercial practices (the UCPD).

As is well known, the UCPD prohibits unfair commercial practices—particularly misleading and aggressive commercial practices—that distort the economic behaviour of consumers. At the time of the UCPD's enactment, the notions of 'nudging' and of the 'choice architecture' were not yet part of everyday regulatory discourse. Today, it is a very different story. For example, Article 22 of the Consumer Rights Directive recognises that there is a problem where the choice architecture of a website is such that pre-ticked boxes default the consumer to additional 'options', such as insurance. Here, the regulatory response is to require an express opt-in by the consumer, failing which any payment for such additional services is to be reimbursed. Hence, if we think that the problem with profiling and personalisation is that it involves an unacceptable 'nudge'—the supplier nudging the consumer towards a purchase that (i) is in the economic interest of the supplier but (ii) is not necessarily in the economic interest of the consumer (even though the supplier can plead in good faith that it reasonably believes that the subject matter of the purchase might be one in which the consumer is interested)—the question is whether the spirit and intent of the UCPD can be brought to bear on these online commercial practices.

The general test of unfairness, in Article 5(2) of the UCPD, provides that a commercial practice is unfair if (a) it is contrary to the requirements of professional diligence and (b) 'it materially distorts or is likely to materially distort the economic behaviour with regard to the product of the average consumer whom it reaches or to whom it is addressed, or of the average member of the group when a commercial practice is directed to a particular group of consumers.' According to Article 2(e), a practice materially distorts the economic behaviour of consumers if its use is such as 'to appreciably impair the consumer's ability to make an informed (sic) decision, thereby causing the consumer to take a transactional decision that he would not have taken otherwise'. Given that the two classes of unfair commercial practices at which the Directive is principally aimed are those that are (i) misleading and (ii) aggressive, we need to read the definition of material distortion in Article 2(e) as saying 'to appreciably impair the consumer's ability to make a decision *that is both free and informed*, thereby causing the consumer to take a transactional decision that he would not have taken otherwise'. Employing this definition, we can then treat material distortion as occurring where, because of misleading commercial practices, the consumer's decision is not

---

60 Natali Helberger (n 51) 152.

properly *informed* but also where, because of aggressive commercial practices, the consumer's decision is not *free*.

Applying this test to the use of nudges by online sellers, there are two senses in which we might suggest that this practice materially distorts the transactional decisions made by the average consumer. First, we might suggest that nudged transactional decisions are not free; secondly, we might suggest that they are under-informed or mis-informed—and, in both cases, we would argue that consumers make a decision to purchase that they otherwise would not have made.

In response to the first suggestion, it might be objected that transactional decisions are never free; and that, insofar as the Directive prohibits aggressive commercial practices, it is contemplating a much more overt impingement on the freedom of consumers. That said, those who advocate nudges do not shrink from the first kind of objection. To the contrary, such advocates regard it as a virtue of nudging that it is fully compatible with the autonomy of those who are nudged because there is always the opportunity to opt out.[61] In practice, of course, where the norm is to go with the nudge, the defaults might become somewhat 'sticky' and the opportunity to opt out is more formal than real; but, where a supplier is simply recommending a product that seems to fit well with the consumer's profile, the choice architecture requires the latter to opt in—the nudge is a gentle one. Moreover, even if one accepts the nudge, the consumer has the right to withdraw within 14 days.[62] By contrast, if the default were set for 'buy' and, to avoid the purchase, it required the consumer to opt out, this would be a much more questionable—perhaps even 'aggressive'—nudge, analogous to offline inertia selling that has long been regulated.[63]

With regard to the second suggestion, the obvious response is that suppliers are endeavouring to ensure that their consumer customers are fully informed. Again, advocates of online selling will claim that there is so much more information available to consumers than in offline shopping environments. However, this might be thought to miss the point. The objection is not that the consumer was misled in relation to the characteristics of the products that were purchased or recommended but that the supplier 'manipulated' the consumer without disclosing the process that enabled such targeted and 'accurate' recommendations to be made. While such an objection would seem to point to an example of a failure to meet the requirements of

---

61 Richard H. Thaler and Cass R. Sunstein, *Nudge: Improving Decisions About Health, Wealth and Happiness* (New Haven: Yale University Press, 2008).

62 See Article 9(1) of Directive 2011/83/EU on consumer rights.

63 Compare, too, Larry A. DiMatteo, 'Regulation of Share Economy: A Consistently Changing Environment' in Schulze and Staudenmayer (n 51) 89, 100 (for a non-existent 'unsubscribe option' in Lyft's terms of service with customers).

'professional diligence' (specifically by being contrary to good faith), it is not so clear how this practice results in a material distortion of the average consumer's transactional decisions.

Within the Directive, there might be a peg for this kind of objection in the notion of 'undue influence', which is defined in Article 2(j) as follows:

> 'undue influence' means exploiting a position of power in relation to the consumer so as to apply pressure, even without using or threatening to use physical force, in a way which significantly limits the consumer's ability to make an informed decision.

As with the general definition of material distortion, we surely have to read this definition as saying 'in a way which significantly limits the consumer's ability to make a decision *that is both free and informed*—not least because, within the overall scheme of the Directive, the idea of undue influence is given, alongside harassment, coercion and the use of physical force as a feature of aggressive commercial practices. Reading the Directive this way, it might be argued that a personalised and profiled nudge amounts to aggression by undue influence: because the nature of the nudge is not disclosed, there is a lack of information about the extent to which the transactional decision of the consumer is being influenced. Or, to put this more straightforwardly, there is a lack of transparency. Indeed, here, we should note the Commission's more recent recognition that '[g]reater transparency is ... needed for users [of online platforms] to understand how the information presented to them is filtered, shaped or personalised, especially when this information forms the basis for purchasing decisions ...'.[64] That said, the benchmark in the Directive is the average consumer, this being a consumer 'who is reasonably well-informed and reasonably observant and circumspect, taking into account social, cultural and linguistic factors'.[65] No doubt, there will be a good deal of variation in the understanding of consumers about how the choice architecture of their online shopping environments structures and influences their transactional decisions; and it is not clear that the offline benchmark works so well in online environments. However, the more that consumers understand about the way that online environments manage and channel their behaviour, the less plausible it will be to plead that nudging amounts to undue influence and an unfair commercial practice.[66] At this stage, if there is to be an argument against nudging, it might be, as Cass Sunstein has more recently suggested, one that hinges on

---

64   European Commission (n 53) 10–11.
65   Recital 18 of the Directive.
66   Compare Helberger (n 51) 157: 'One important factor in this context could be the knowledge of the user of the fact that she is being persuaded, based on her persuasion profile. This is because only with such knowledge is she actually able to mobilise her defence strategy, should she wish to do so. Another important factor could be her ability to opt out of

the importance of active choosing when 'learning, authenticity, responsibility, and the development of values and preferences are [at stake]'.[67] However, the idea that routine consumer choices fall into this special category is not terribly plausible.

Although there is nothing in the Directive to support the idea that nudging itself should be categorically considered unfair, at some time in the future, one might imagine an argument against nudging and profiling that raises the question of whether there is a risk that consumers might become trapped in their own (machine learning-assisted) personal 'echo-chambers', and whether this risk needs to be managed.[68] The thought is that those consumers who simply want to experiment or change their profiles might find it difficult to do so. Might it be that consumers of this kind—consumers, of course, who are not in this case assisted by their right to withdraw within the cooling off period—are 'vulnerable' in a sense recognised by the Directive?

In fact, Article 5(3) of the Directive does make special allowance for 'consumers who are particularly vulnerable to the practice or the underlying product because of their mental or physical infirmity, age or credulity in a way which the trader could reasonably be expected to foresee'. Where a consumer is within such an identifiable group, the benchmark is adjusted to that of the average member of that group. However, if we try to align this with the complaint that a group of consumers find themselves trapped in their own profiles, there is an obvious difficulty in saying that the reason for their alleged vulnerability is their mental or physical infirmity, or their age or credulity. There might also be a difficulty in treating a practice that steers a consumer away from trying new and different kinds of product options as a material distortion of the consumer's transactional decision. To be sure, informal guidance on the Directive suggests that the concept of a 'transactional decision' should be read broadly to include decisions both to purchase and not to purchase.[69] However, the example that is given in the guidance is an offline one: a consumer sees a promotional offer for a digital camera;

---

targeted messages, as a means to restore the imbalance in control power between user and trader.'

67 Cass R. Sunstein, *Choosing Not To Choose* (Oxford: Oxford University Press, 2015) 119–120.
68 See, e.g., European Data Protection Supervisor, *Towards a New Digital Ethics* (Opinion 4/2015) 11 September, 2015, 13:

> Profiles used to predict people's behaviour risk stigmatisation, reinforcing collective stereotypes, social and cultural segregation and exclusion, with such 'collective intelligence' subverting individual choice and equal opportunities. Such 'filter bubbles' or 'personal echo-chambers' could end up stifling the very creativity, innovation and freedoms of expression and association which have enabled digital technologies to flourish.

69 European Commission, *Guidance on the Implementation/Application of Directive 2005/29/ EC on Unfair Commercial Practices* (Commission Staff Working Document) Brussels, 3 December 2009 SEC(2009) 1666.

the consumer goes to the shop intending to buy the camera; but, the consumer then learns that the price advertised applies only if the purchaser also subscribes to a photography course; so, the consumer decides not to buy the camera.[70] Although this sequence of events ends with a decision not to purchase, it is a long way from the kind of situation that we are discussing; and, even if these difficulties can be overcome, it is far from clear whether a creative interpretation of the Directive, or a bespoke regulatory response, is the better way of handling this question. As with the control of unfair terms, it might be that the smartest response to profiling involves some element of technocratic intervention.

With this thought, we can move on to the second cluster of questions identified in Part III, questions that arise when, further to a radical transformation of our routine consumption practices, humans are pretty much taken out of the transactional loop, consumer contractors becoming mere consumers.

V Technological management and transactions

Anticipating the creation of 'an economy which ... is planned not by bureaucrats or CEOs but by the technostructure',[71] Ariel Ezrachi and Maurice Stucke question the fitness of traditional competition law to regulate novel forms of behavioural discrimination and collusion. We might ask the same question about the law of contract—and, indeed, we might wonder whether the rules of contract law have any place in a techno-managed sphere of transactions.[72] Arguably, they do not. After all, if transactions are mechanically formed and performed with humans out of the loop, the continuing relevance or role of rules that are addressed to flesh and blood (and potentially defaulting) agents is less than clear. Moreover, we might also sense that there are deeper implications for the traditional transactionalist idea of contract when smart machines and connected devices do all the work. Indeed, according to Shoshana Zuboff, a new form of (surveillance) capitalism may install a new kind of sovereign power,[73] where a big data corporation

> may sell access to an insurance company, and this company purchases the right to intervene in an information loop in your car or your kitchen

---

70 *Guidance* (n 69) 22 23.

71 Ariel Ezrachi and Maurice E. Stucke, *Virtual Competition* (Cambridge, Mass: Harvard University Press, 2016) 32–33.

72 Compare, Ian R. Kerr, 'Bots, Babes and the Californication of Commerce' (2004) 1 *University of Ottawa Law and Technology Journal* 284, 288–289 (arguing that it has become disingenuous to characterise many online transactions as giving rise to contracts in the traditional sense).

73 Shoshana Zuboff, 'Big Other: Surveillance Capitalism and the Prospects of an Information Civilization' (2015) 30 *Journal of Information Technology* 75.

in order to increase its revenues or reduce its costs. It may shut off your car, because you are driving too fast. It may lock your fridge when you put yourself at risk of heart disease or diabetes by eating too much ice cream. You might then face the prospect of either higher premiums or loss of coverage. ... [S]uch possibilities ... represent *the end of contracts.* ... [They replace] the rule of law and the necessity of social trust as the basis for human communities with a new life-world of rewards and punishments, stimulus and response.[74]

Yet, is this right? When transactions are technologically managed, does this represent the end of *contracts*? And is this the end of the line for the *law* of contracts?

We can explore these questions by considering, first, the significance of the development of so-called 'smart contract' applications and then we can contemplate whether there is any sense of contract that might persist in a world where the provision of consumer goods and services is automated, integrated and embedded and in which commerce is conducted by smart machines.

(i) Smart contracts

If we define smart contracts as 'a set of promises, specified in digital form, including protocols within which the parties perform on these promises',[75] then this implies that the promises and their performance are encoded in some way; and, although this definition long pre-dates the development of blockchain technology (and cryptocurrencies such as Bitcoin), it is in the context of this particular technological development that so-called 'smart contract' applications are currently envisaged.[76]

About such applications there are many preliminary questions: in particular, there are questions about the concept of a smart contract (whether this is simply the encoding of some transactional function or a self-executing transaction that would be recognised as a legally binding 'fiat' contract), about the extent of the technological capabilities, and about the reception of such

---

74 *Ibid.*, 85–86 (emphasis added).
75 See, e.g., the Chamber of Digital Commerce, 'Smart Contracts: 12 Use Cases for Business & Beyond A Technology, Legal & Regulatory Introduction—Foreword by Nick Szabo', available at www.digitalchamber.org/smartcontracts.html (last accessed 6 November 2018) p.8 (citing Nick Szabo).
76 See, e.g., Aaron Wright and Primavera De Filippi, 'Decentralized Blockchain Technology and the Rise of *Lex Cryptographia*', papers.ssrn.com/sol3/papers.cfm?abstract_id=2580664 (March 10, 2015) (last accessed 6 November 2018); D. Tapscott and A. Tapscott, *Blockchain Revolution*, (London: Portfolio Penguin, 2016) 101–103; Riikka Koulu, 'Blockchains and Online Dispute Resolution: Smart Contracts as an Alternative to Enforcement' (2016) 13.1 *SCRIPTed* 40; and Phillip Paech, 'The Governance of Blockchain Financial Networks', LSE Law, Society and Economy Working Papers 16/2017 (available at papers.ssrn.com/sol3/papers.cfm?abstract_id=2875487) (last accessed 6 November 2018).

transactional technologies in both commercial and consumer marketplaces.[77] There is also, of course, the recurrent question whether we should frame our legal questions in a coherentist or a regulatory-instrumentalist way.

In its most recently announced programme of work, where the Law Commission states that it will be looking at smart contracts,[78] we might detect signs of coherentist thinking. This is because the Commission, mindful that blockchain transactional records cannot be rectified or reversed, wonders (in an apparently coherentist way) 'how this feature would interact with contract law concepts such as implied terms or contracts which are held to have been void from the outset'.[79] However, the Commission's initial thinking is more complex: its coherentist questions are posed alongside a background regulatory-instrumentalist concern that, if English courts and law are to remain a competitive choice for business contractors, then there 'is a compelling case for reviewing the current English legal and regulatory framework to ensure that it facilitates the use of smart contracts.'[80]

Let me start by giving a simple example that should ease some of the perplexity that surrounds the question whether smart contracts fit the template for fiat contracts and, concomitantly, how the law of contract might be applied (and might continue to be relevant) where smart contracts are used. The stock example of a smart contract takes the form of a coded instruction for a 'payment' to be made by A to B conditional on the occurrence of some event (if x, then y). While bets and wagers fit this specification, they are hardly typical of everyday contracts for the supply of goods or services. I suggest that a better example of a payment contract (although not a contract for the supply of goods and services) can be found in some familiar nineteenth-century jurisprudence in which one of the leading cases is *Tweddle v Atkinson*.[81]

Essentially, the form of the dispute in *Tweddle v Atkinson* was as follows. It was agreed between A and B that they would each pay a sum of money to C (who was the son of A) on the marriage of C to D (D being the daughter of B). C duly married D. However, B died before his portion of the agreed sum had been paid to C. Although A and B had expressly agreed that C should

---

77 For some very sound preliminary cautions, see ISDA Linklaters, *Whitepaper: Smart Contracts and Distributed Ledger—A Legal Perspective* (August, 2017); for insightful discussion of terminology, technical limitations and connection with the real world, see Eliza Mik, 'Smart Contracts: Terminology, Technical Limitations and Real World Complexity' (2017) 9 *Law, Innovation and Technology* 269; and for the reception of smart contracts by commercial contractors, see the excellent discussion in Karen E.C. Levy, 'Book-Smart, Not Street-Smart: Blockchain-Based Smart Contracts and The Social Workings of Law' (2017) 3 *Engaging Science, Technology, and Society* 1.

78 Law Commission, *Thirteenth Programme of Law Reform* (Law Com No 377) (HC 640, December 13, 2017).

79 *Ibid.*, p.20 (para.2,39).

80 *Ibid.*, p.20 (para 2,39).

81 (1861) 25 JP 517, 1 B & S 393, 30 LJQB 265.

be entitled to enforce the contract, the court declined to enforce the agreement at the suit of C. Whatever the precise doctrinal reason for treating C as ineligible to enforce the contract between A and B—whether because C gave no consideration for the promises made by A and B, or because C was not a party to the agreement between A and B—the fact of the matter was that the court would not assist C. Given this restriction in the law, had smart contract technologies been available at the time, we might think that this was just the kind of case in which their use might have appealed to A and B. For the smart contract would simply need to be coded for payment of the agreed sum in the event of the marriage taking place: that is to say, the technology would ensure that, once the marriage of C and D had taken place, then the agreed sums would be paid to C.

There are, however, several reasons that might have dissuaded A and B from using a smart contract. First, the context in which A and B agreed to pay sums to C was one of a potentially closer relationship between the two families. Even if the families had not previously 'got on'—indeed, even if A and B did not trust one another—this might have been exactly the wrong time to suggest the use of smart contracts and signal this lack of trust. Secondly, unless the smart contracts were so smart as to use fiat currency, the inconvenience of obtaining, or subscribing to, the particular cryptocurrency might have deterred A and B from making use of the technological option. Moreover, even if this inconvenience did not deter A and B, the cryptocurrency might not be welcomed by C which, in turn, might have steered A and B away from this option. Thirdly, if there were concerns about 'bugs' in smart contract coding or if there was any doubt about the reliability of off-chain sources (that notify the occurrence of the relevant event), A and B might have preferred not to take the risk.

Fast forwarding to the present century, English law is no longer so restrictive of third-party claims, one of the principal purposes of the Contracts (Rights Against Third Parties) Act, 1999, being to give third-party beneficiaries, such as C, a right to enforce in precisely such a case as *Tweddle v Atkinson* (where this was clearly the intention of A and B).[82] Accordingly, in a similar post-Act scenario, the calculation for A and B would be different. The risk would be not so much that C might be left without a legal remedy but that the practical cost of obtaining a remedy might be an issue for C. Hence,

---

82 For commentary, see John N. Adams, Deryck Beyleveld, and Roger Brownsword, 'Privity of Contract—the Benefits and the Burdens of Law Reform' (1997) 60 *Modern Law Review* 238. To the extent that the reform of the privity rule required a legislative intervention, this suggests a regulatory-instrumentalist approach. However, to the extent that the reform was based on classical contractual principles (giving effect to the intentions of the contracting parties), rather than addressing the commercial inconvenience of the rule, this was more like a coherentist exercise.

for A and B, the question would be whether, all things considered, their preferred option is to commit the contract to the blockchain.

In our example, both parties perform by paying a sum of money to C. However, like bets and wagers, this is also atypical and we should note that the utility of smart contracts will be limited so long as they are restricted to making a transfer of value—that is to say, so long as they cannot undertake the counter-performance (the supply of [non-digital] goods or services) that is still characteristic of fiat contracts. Another limitation is that, for some time, coders are likely to be challenged by those many contractual principles and terms that turn on vague concepts such as 'reasonableness', or 'good faith, or the use of 'best endeavours', and so on.[83] Yet another limitation might be that smart contracts can present a risk of commercially sensitive or confidential information being disclosed.[84] And, a further limitation might arise where a buyer is concerned that a *seller* might become insolvent; here, a smart contract for payment to the seller will work for the buyer only if payment is contingent on actual performance by the seller.[85]

It follows that, even where we can find a gap in legal support for transactions, there might be reasons—arising from the context or the nature of the performance, or because of doubts about the counter-party's performance (and possible insolvency), or simply for reasons of net disutility—that militate against reliance on smart contracts. Hence, assuming that there is a viable infrastructure for arbitrating and enforcing transactions, fiat contracts, governed by the law of contract, might still be the preferred option.

Now, we can take up the question of whether smart contracts meet the criteria for a standard fiat contract.[86] Whereas standard fiat contracts for, say, the supply of goods or services are formed by an 'offer and acceptance' between the supplier and the customer and they are represented by an exchange of 'consideration' (on the one side, the goods or services to be supplied and, on the other, the price to be paid),[87] both the process for creating a smart con-

---

83 See, e.g., Primavera De Filippi and Aaron Wright, *Blockchain and the Law* (Cambridge, Mass.: Harvard University Press, 2018) 77.

84 See, e.g., De Filippi and Wright (n 83) 83–84.

85 To be sure, in principle, a smart contract might be coded or 'architected' in just the way that the buyer needs; but the practical question is whether the seller will agree to such an arrangement. For further issues relating to insolvency, see Scott Farrell, Heidi Machin, and Roslyn Hinchliffe, 'Lost and Found in Smart Contract Translation—Considerations in Transitioning to Automation in Legal Architecture' in UNCITRAL, *Modernizing International Trade Law to Support Innovation and Sustainable Development: Proceedings of the Congress of the United Nations Commission on International Trade Law* (Volume 4), pp 95–104 (www.uncitral.org/pdf/english/congress/17-06/83_ebook.pdf) (last accessed 6 November 2018).

86 For example, see Kevin Werbach and Nicolas Cornell, 'Contracts *Ex Machina*' (2017) 67 *Duke Law Journal* 313, 338 et seq.; and Farrell, Machin, and Hinchliffe (n 85), p.96 et seq.

87 See, e.g., John N. Adams and Roger Brownsword, *Understanding Contract Law* 5th edn (London: Sweet and Maxwell, 2007).

tract and the lines of computer coding that then represent that contract look rather different.[88] As Primavera De Filippi and Aaron Wright put it:

> Where traditional legal agreements and smart contracts begin to differ is in the ability of smart contracts to enforce obligations by using autonomous code. With smart contracts, performance obligations are not written in standard legal prose. Rather, these obligations are memorialized in the code…using a strict and formal programming language (like Ethereum's Solidity).[89]

Accordingly, if we focus narrowly on the coded instructions that express the smart contract, we may well wonder whether what we are looking at does meet the criteria for a fiat contract. There are several points to make in relation to this question about the characterisation of smart contracts.

First, let us suppose that we translate the computer coding into a natural language. If, once the coded instructions are translated back into a natural language, they do not reveal a 'deal' (an exchange of values), then this particular smart contract will still not look like a fiat contract—but, of course, this follows from the fact that there is no deal, not from the fact that this is a smart contract. By contrast, if the translated smart contract does disclose a deal, then (other things being equal[90]) this has the makings of a fiat contract. So, we should not jump to the conclusion that smart contracts can never be treated as standard fiat contracts simply because some examples of smart contracts do not have the basic elements for a standard fiat contract.

Secondly, and most importantly, where a smart contract is being used as a tool to give effect to a standard fiat contract, the law of contract will engage with the latter and only indirectly with the former. To explain, in our example (based on *Tweddle v Atkinson*), suppose that C and D marry but the Oracle (the off-chain trusted third-party information provider) does not transmit that the relevant event has taken place. As a result, the transfer to C is not triggered. Setting aside the possibility that C might have a legal claim against the Oracle, C's obvious remedy is to sue A and B for breach of their (standard fiat) contract. To be sure, C's complaint is that the smart contract has not worked as intended; but, if we set aside the possible liability of others, C is arguing that the responsibility for this failure lies with A and B. At no point does C's claim against A and B—a claim for breach of contract against

---

88 Compare, for example, the 'tutorial' on creating a smart contract at medium.com/@ConsenSys/a-101-noob-intro-to-programming-smart-contracts-on-ethereum-695d15c1dab4 (accessed 6 November 2018).

89 De Filippi and Wright (n 83) 74.

90 For example, even if there is a 'deal', it will not be treated as a fiat contract unless it is supported by an intention to create legal relations: see *Balfour v Balfour* [1919] 2 KB 571.

A and B—seem to hinge on whether the smart contract itself fits the standard specification for a fiat contract.

What, though, if the Oracle incorrectly advises that the marriage between C and D has taken place, so that the payment is made when it should not have been? In such a scenario, quite apart from any claim against the Oracle, there may be a restitutionary mechanism to reverse the payment without having to decide whether the smart contract is a contract as such;[91] or, it may be that the background agreement between A and B provides for a risk of this kind (allocating the risk to one of the parties and, possibly, providing also for insurance against the risk). Accordingly, to pick up the Law Commission's apparent concerns, if there are questions about implied terms (for example, about whose risk it is if the smart contract does not work as intended), this will be a question about the background fiat contract, engaging the usual rules of contract law, not a question about implying terms into a smart contract as such; and, as for the supposed irreversibility of the blockchain record, even if this were correct (which it is not because, where there are 'errors', the blockchain might be rectified by 'forking'[92]), this does not preclude the court ordering a restitutionary payment.

For good measure, we might add that, for the same reason, there is really no puzzle about the application of doctrines such as duress or undue influence. For example, if A has applied unfair pressure to B to commit to the deal, B might take court action in an attempt to avoid the transaction before it has been committed to the blockchain; if B tries to avoid the transaction when it is on the blockchain but not yet executed, it may not be possible to prevent the payment being made but, in principle, a court might order C to restore the value received. Nor is there a problem if it is argued that the smart contract is void for illegality or the like. It may seem strange that the law of contract plays a residual public policy function but that is as far as any peculiarity goes. What is contrary to public policy is the purpose of the background fiat contract; and it will be this contract that is declared to be void.

So, what does all this signify for fiat contracts and the rules of contract law? First, so long as smart contracts are simply an optional tool, then contractors (whether in families or in business) might prefer to use fiat contracts. For example, international traders who are habituated to using fiat contracts, letters of credit, banks, and courts may prefer to continue to do business this way. Secondly, so long as smart contracts are simply being used as tools to give effect to identifiable background fiat contracts, the law of contract

---

91 In English law, the decision of the UK Supreme Court in *Patel v Mirza* [2016] UKSC 42 has significantly raised the profile of a possible restitutionary response to claims arising in connection with illegal contracts.

92 See, e.g., Kelvin F. Low and Ernie G.S. Teo, 'Bitcoins and other Cryptocurrencies as Property?' (2017) 9 *Law, Innovation and Technology* 235 (concerning the responses to famous hacks at Mt Gox and of the DAO).

continues to be relevant and applicable in the various ways that have been indicated. Accordingly, it may be premature to be declaring either that, after smart contracts, the law of contract will be irrelevant or that this is the end of trust in transactions.

Nevertheless, we should not conclude that, intensified and pervasive technological management notwithstanding, it will be business as usual for the law of contract and for the transactionalist idea of contract. When what we have in the background is not an identifiable fiat contract but a networked environment of smart machines, when smart (or algorithmic) contracts have a life of their own detached from fiat contracts or club rules, and with human contractors fully out of the loop, we have a very different situation.[93] At that point, we should set aside our coherentist questions and engage in a regulatory-instrumentalist conversation.

(ii) Smart machines and automated transactions

Whatever the future of smart contracts, we need to think in a technology-neutral way about smart machines, imagining a world where commerce is, so to speak, a 'conversation conducted entirely among machines'.[94] Imagine, as Michal Gal and Niva Elkin-Koren foresee it, that

> [y]our automated car makes independent decisions on where to purchase fuel, when to drive itself to a service station, from which garage to order a spare part, or whether to rent itself out to other passengers, all without even once consulting with you.[95]

In that world, humans have been taken out of the transactional loop, leaving it to the technology to make decisions that humans would otherwise be responsible for making. The relevant humans include not only the primary contractors but also trusted third parties (such as banks that administer letters of credit or other documents in international trade).[96] To this extent, smart transactional technologies are in the same category as autonomous vehicles, albeit that the critical decisions that are made by the former will typically have positive or negative *economic* consequences for the humans who are

---

93 For some helpful reflections along such lines, see Lauren Henry Scholz, 'Algorithmic Contracts' (2017) 20 *Stanford Technology Law Review* 128.

94 Per W. Brian Arthur, 'The Second Economy', *McKinsey Quarterly* (October 2011), quoted in Nicholas Carr, *The Glass Cage* (London: Vintage, 2015) 197.

95 Michal S. Gal and Niva Elkin-Koren, 'Algorithmic Consumers' (2017) 30 *Harvard Journal of Law and Technology* 309, 309–310.

96 For discussion of trust in banks as opposed to trust in other payment systems, see Iris H.-Y. Chiu, 'A New Era in Fintech Payment Innovations? A Perspective from the Institutions and Regulation of Payment Systems' (2017) 9 *Law, Innovation and Technology* 190.

connected to the transaction rather than involving matters of human health and safety.

In the consumer marketplace, we can already anticipate a world where our routine needs are taken care of by cyberbutlers (or digital assistants), connected devices, the Internet of Things, and a succession of smart environments. Precisely how this will work, both technologically and practically, remains to be seen. However, let us assume that, at least in the initial stages of this transactional revolution, consumers are faced with a choice of service providers, each provider offering a package of terms and conditions. Some of these packages will be relatively narrow and specialised, others will be far broader. Within these broader packages, defaults will be set for supplying needs that are widely regarded as basic, repeat, and routine; but then there will be goods and services that some individuals will choose to treat as routine while others will want to treat them as non-routine (meaning that they will be ordered only as and when they are specifically requested); and, perhaps, there will be some goods and services which will never be ordered without the specific request of the consumer.[97] So, contract is the basis on which consumers sign up to a particular package; and, within the package, specific authorisation is sometimes required for non-routine purchases. The critical point is that, so far as the package contract is concerned, the ideal-typical contractual (transactional) conditions must be satisfied; or, as Margaret Radin says, 'If the future of contract makes it ever more clear that the only point of choice is whether or not to buy the product-plus-terms, we could focus our attention on making that choice really a choice.'[98] However, so far as consent and authorisation within the package are concerned, the requirements may be specified by the package or by background regulation.

If the anchoring point for the package is to be transactional, then this is where we need to find what Ian Kerr terms a 'jural relation that is founded upon agreement' and the 'manifestation of a mutual concordance between [two] parties as to the existence, nature and scope of their rights and duties';[99] this is where the consumer has to give a free and informed consent to the package and its terms. While everyone can agree that coercive and fraudulent actions by the supplier are incompatible with transactionalism, beyond this, as we consider duress, pressure and influence, or notice of terms, we enter

---

97 Compare Gal and Elkin-Koren (n 95) at 317, suggesting that, even those consumers who enjoy shopping (especially for personal items such as a 'piece of jewelry') might prefer automated (and algorithmically managed) supply of 'more mundane products, like pet food'. And, some may 'prefer to have algorithms make all of their consumption decisions'.

98 Margaret Jane Radin, 'Humans, Computers, and Binding Commitments' (2000) 75 Ind. LJ 1125, 1161. If the choice is simply too complex for consumers, the alternative is to leave it to regulators to black-list some terms, to require others to be set aside for separate explicit agreement, and to require others to be conspicuously displayed.

99 Kerr (n 72).

into deeply contested territory where more and less protective conceptions of 'being free' and 'being informed' are articulated. Still, until automated supply is the norm, the significance of entering into a package needs to be highlighted: the context in which packages are offered needs to be such that the consumer is free to say 'no' to *any* package (not just 'no' to some packages but eventually 'yes' to one); and the consumer needs to understand that declining any package is an option. Moreover, the consumer needs to understand that, if he signs up for a package, many of his routine needs will be automatically serviced. This invites a regulatory framework that takes this one transaction out of the background and puts it into the spotlight—and all this *a fortiori* if the founding contract is itself presented in an online environment.

In the evolving world of automated consumption, as I have said, there may well be a significant number and range of transactions that require the authorisation of the consumer. However, the degree of delegated authority enjoyed by cyberbutlers may vary from one consumer to another.[100] Where the delegation is only of an advisory nature, this seems relatively unproblematic; but where the delegating human charges the cyberbutler with making fairly open-ended executive decisions ('order what I need') there are obvious risks that the former's needs and preferences will be misinterpreted. Should the cyberbutler surprise or disappoint the consumer whose needs it services, that is one thing; but should the cyberbutler exceed its authority, that raises questions about liability and compensation. While coherentists will probably look for guidance to traditional (or modified) offline agency rules (with the acts of the 'agent' being 'attributed' to the consumer), those with a regulatory mind-set will more likely favour a regulated scheme of insurance or indemnification.[101] There is also, of course, the possibility that the design of the technology is so smart that, even if a malfunction occurs, it will be detected and corrected before any loss occurs.

All this said, such a precautionary and consumer-protective regulatory intervention might soon seem anachronistic. As the balance changes between automated and non-automated consumption, there will probably be defaults

---

100 Compare Roger Brownsword, 'Autonomy, Delegation, and Responsibility: Agents in Autonomic Computing Environments' in Mireille Hildebrandt and Antoinette Rouvroy (eds), *Autonomic Computing and Transformations of Human Agency* (London: Routledge, 2011) 64.

101 Compare Emily M. Weitzenboeck, 'Electronic Agents and the Formation of Contracts' (2001) 9 *International Journal of Law and Information Technology* 204; Silvia Feliu, 'Intelligent Agents and Consumer Protection' (2001) 9 *International Journal of Law and Information Technology* 235; and Gunther Teubner, 'Digital Personhood: The Status of Autonomous Software Agents in Private Law' *Ancilla Juris* (2018)107. For agency applied to algorithms, see Lauren Henry Scholz (n 93) at 165 ('The law should treat the intent and knowledge level of companies or individuals who use algorithms for contracting in the same way as the law would treat the intent and knowledge level of a principal in agency law'): available at ssrn.com/abstract=2747701 (last accessed 7 November 2018).

and nudges towards automated packages; and this might well be viewed as acceptable when it is reasonable to assume that this is what most consumers want. In time, consumers who want to consume in the old-fashioned way might find that this is no longer a practical option. In time, the justification for what is in effect imposed automation is not that it rests on consensually-agreed packages but that it serves the general welfare. Instead of contract and transactionalism, the justification is the general utility based on an acceptable distribution of benefit and risk.[102]

So much for the automated supply of consumer goods and services. Beyond that, however, in business dealing, when commerce is largely conducted by smart machines, what sort of transactional tasks and functions might they undertake, what sort of transactional decisions might they make? Following Ian Macneil, we can grant that there is a sense in which all transactions between humans are relational.[103] However, we can place some transactions, such as the one-off exchange between one-time contractors, at one end of the relational spectrum of duration and complexity and other transactions, such as major construction contracts, at the opposite end of the spectrum. Once smart machines run the simple exchange at one end of the relational spectrum, they will take over relatively simple formation and performance functions—executing payments is the paradigmatic example; and, to this extent, not only will the humans be taken out of the loop, the transaction will be isolated from the web of background social relationships and expectations that Macneil rightly insists give context to human contracting. At the other end of the spectrum, there will be a range of in-contract decisions to be made, for example, involving variation, renegotiation, exercises of discretion, serving notices, and so on.[104] To what extent such functions can be encoded or automated is moot; and, whether the functions are simple or complex, it remains to be seen whether humans will want to be taken out of these transactional loops.

Assuming that a community—whether the community at large or, say, a particular business community—sees benefits (such as simple convenience or efficiency, maximising one's time, reducing court enforcement costs or designing in compensatory arrangements that are a workaround relative to restrictions in the law of contract [such as third-party claims]) in entrusting transactions to smart technologies, then regulatory-instrumentalists will view

---

102 See Roger Brownsword, 'The Theoretical Foundations of European Private Law: A Time to Stand and Stare' in Roger Brownsword, Hans-W Micklitz, Leone Niglia, and Stephen Weatherill (eds), *The Foundations of European Private Law* (Oxford: Hart, 2011) 159 at 161–164.

103 See, especially, Ian R. Macneil, *The New Social Contract* (New Haven: Yale University Press, 1980).

104 Compare Paech (n 76) 26 (on the possibility that, while some parts of the transaction might, as it were, 'go smart', others might be treated as 'non-smart').

the governance of such transactions as an exercise in putting in place proportionate and acceptable measures of risk management.

In general, there are at least five kinds of risk that might be anticipated as
follows:

(i) there might be a malfunction in the technology; this means that the
regulatory framework must provide for the correction of the malfunction
(for example, completion of performance or setting aside a transaction
that should not have been formed); the required corrective measures
might include rescission or rectification of transactions on which the
malfunction has impacted as well as arrangements for compensation and
insurance;

(ii) the integrity of the technology might be compromised by unauthorised
third-party acts; these acts will need to be covered by the criminal law or
tort law or both, and there will need to be provision for the adjustment
of losses so occasioned;[105]

(iii) the way that the technology functions (not malfunctions) might mean
that some decisions are made that do not align with the preferences,
interests, or intentions of a party who is relevantly connected to the
transaction—for example, as when Knight Capital Group's automated
programme 'flooded exchanges with unauthorised and irrational orders,
trading \$2.6 million worth of stocks every second';[106] this suggests that
regulators need to find ways of bringing humans back into the loop for
critical decisions, or capping the financial loss that can be generated by
the machine, or putting humans in control (just like principals) of the
terms and conditions of the machine's authorisation;

(iv) as the famous flash crash of May 6, 2010 reminds us, fully automated
algorithmic trading can involve systemic risks to whole markets and
not just to individual contractors;[107] which implies that the regulatory
framework needs to make provision for 'an "emergency stop" function,
enabling [supervisors] to halt the automatic termination of contracts
recorded in a blockchain financial network'[108]—that is, more generally,
provision for human oversight of markets along with powers to intervene
and suspend trading and transactions; and,

(v) although the technologies seem to comply with general legal requirements, the fact that some have 'black-box' elements might seem unduly

---

105 See discussion in Low and Teo (n 92).
106 Nicholas Carr (n 94) 156. In all, there were some \$7billion in errant trades and the com-
pany lost almost half a billion dollars.
107 See Frank Pasquale and Glyn Cashwell, 'Four Futures of Legal Automation' (2015) 63
UCLA Review Discourse 26, 38–39.
108 Paech (n 76) 21.

risky[109]—particularly, for example, if there is a concern that they might be operating in unlawful discriminatory ways;[110] this implies that regulators have to either grant a degree of immunity to these technologies (reflecting how far the risk of unlawful practice can be tolerated) or introduce measures to constrain such practices.

These short remarks do no more than scratch the surface of the questions that are already being addressed as smart transactional technologies emerge.[111] For example, what view will be taken of arguably disproportionate use of technological force to secure compliance—the stock example is that of a lessor of vehicles who is able to immobilise the vehicle should a lessee fail to make the payment on time? Or, where business contractors use technological management to achieve an outcome that the law of contract does not assist (but which is not otherwise legally prohibited), will this lack of congruence be a ground for challenge?[112] As such questions are sharpened and debated, we can expect that regulatory-instrumental responses will challenge coherentism, with the rights and responsibilities of at least some connected human parties being governed by the bespoke provisions of a regulatory regime rather than being 'transactionally' determined as in traditional contract law. That said, to the extent that those downstream rights and obligations that are associated with the technological management of transactions are rooted in a founding authorisation or master agreement, there will need to be careful regulatory oversight of the protocols that tie in the connected parties to the transactions that are executed in their names.[113]

---

109 Generally, see Frank Pasquale, *The Black Box Society* (Cambridge, Mass: Harvard University Press, 2015).

110 See, e.g., Cathy O'Neil, *Weapons of Math Destruction* (London: Allen Lane, 2016).

111 See, further, Roger Brownsword, 'Smart Contracts: Coding the Transaction, Decoding the Legal Debates' (OUP/UCL, 2019) (forthcoming in Phillip Hacker ed); 'Regulatory Fitness: Fintech, Funny Money and Smart Contracts' (2019) *European Business Organization Law Review* (forthcoming); and 'Smart Transactional Technologies, Legal Disruption, and the Case of Network Contracts' (CUP, 2019) (forthcoming in Larry di Matteo ed).

112 For example, see the discussion in Werbach and Cornell (n 84) 338 et seq. See, too, Wright and De Filippi (n 76) 26. Here, having raised the question of 'what is legally versus technically binding', and having noted that there may be some divergence between what the law of contract will enforce or invalidate and what smart contracts, operating 'within their own closed technological framework', will enforce or invalidate, the authors conclude that while 'implementing basic contractual safeguards and consumer protection provisions into smart contracts is theoretically possible, it may prove difficult given the formalized and deterministic character of code'. However, beyond the question of technical possibility, there is the question of how the State manages the conflict between the law of contract, consumer and commercial, and the actual operation of smart contracts. In other words, is there an implicit requirement that the latter should be congruent with the former?

113 Compare Brownsword (nn 27 and 111).

VI Conclusion

The kinds of conversation that we need to have about the future of contract law seem to be increasingly regulatory and of a new coherentist kind. The traditional principles and doctrines of the law of contract are being overtaken by the next generation of transactional technologies.

In the field of consumer transactions, which is already heavily regulated, there may need to be a rebalancing of the interests of suppliers and consumers, leading to a degree of rule revision together with some innovative uses of novel consumer-protective technologies. In particular, consumers seem to be at a distinct disadvantage in online environments where they can be bounced into accepting unfair terms and conditions, profiled, prodded, and subjected to dynamic pricing. There may also need to be some regulatory interventions to support and to stabilise an acceptable balance of interests between the parties who trade in the share economy.

In the field of commercial transactions, it will be largely the self-regulatory choices made by trading communities that will determine the future of the rules of contract law. Some trading communities may resist and reject technological management; the rules of contract law will continue to apply. Even where commercial parties decide to buy into the technological management of transactions, there may be background anchoring agreements or club rules that set the protocols for the use of these technologies; and, to this extent, traces of contract and transactionalism will persist.

All this said, the direction of travel is likely to be towards not just smart contracts (which do not, as tools to be used by contractors, displace the law of contract) but towards the integrated and embedded automation of transactions. In this world, online contracting (where humans continue to function as active contractors) will be overtaken by automated consumption (where humans become passive contractors); humans will be taken out of the transactional loop; and even the residual traces of contract and transactionalism will be gone. When we find that we simply cannot answer our traditional coherentist questions, this is a sign that these are no longer the right questions to be asking. In such a transactional world, we need to have not only (regulatory-instrumentalist) conversations about acceptable balances of interests but also (new coherentist) conversations about the kind of community that we want to be.

12

REGULATING THE INFORMATION SOCIETY

The future of privacy, data protection law, and consent

I Introduction

There are many questions that we might ask about both the regulatory challenges and the regulatory opportunities that are presented by what is commonly referred to as the 'information society'—this reflecting the ubiquitous mediation of modern technologies in the accessing, acquisition, transmission, exchange, and processing of information, the breadth and depth of our reliance on these technologies, and the extent to which all of this contributes to the generation of new information (for example, in genomics).[1] In this context, we might ask: What kind of information society does our particular community want to be? What is the future for traditional legal rules that protect our interests in privacy and confidentiality and, concomitantly, for the transactionalist idea of consent by reference to which agents authorise, and others justify, acts that would otherwise infringe these interests? How do these rules lie alongside those legal frameworks, reflecting a regulatory-instrumentalist

---

1 For my purposes, I need not join the debate about whether the concept of the information society captures the emergence of a new type of society: on which, see Frank Webster, *Theories of the Information Society*, 3rd edn (London: Routledge, 2006). Unquestionably, the information society speaks to a time and place in which technological developments disrupt both earlier (pre-Internet) rules of law and invite a radically different (technocratic) regulatory mind-set. Trying to quantify or measure the disruptive impacts on the economy, on occupational patterns, on culture, and the like is another matter; but, there is no doubt that the impacts of the principal technologies reach far and wide. See, further, the report by the Secretary-General of the UN to the Commission on Science and Technology for Development concerning the implementation of the outcomes of the World Summit on the Information Society (Geneva, May 2018), document A/73/66–E/2018/10: available at unctad.org/en/PublicationsLibrary/a73d66_en.pdf (last accessed 9 November 2018).

approach, that aim to allow data to flow while securing fair, proportionate, transparent, and secure processing of our personal information? Finally,
where does technological management fit into this picture? Should regulators
be looking for technological fixes to the problems created by Big Data?[2]

In this chapter, I will respond to these questions in four stages. First (in
Part II), I sketch the changing landscape of our informational interests.
While some of these interests (and their covering rights) have applications in
both traditional offline and modern online environments for transactions and
interactions, others have been elaborated specifically for online environments;
and, while some of these interests (for example, the interests in privacy and
confidentiality) concern control of the *outward* flow of information about a
particular person, others (such as the general right to be informed, the right
to know, and the right not to know) concern the *inward* flow of information.
Clearly, for any community, it is important to decide which of these interests
should be privileged and protected—in other words, what kind of information society does the community want to be? Secondly (in Part III), I will
seek to clarify some questions about informed consent, broad consent, and
informed choice. In each case, the questions concern the kind of information that is required. However, the information that is relevant to a consent
being 'informed' is analytically distinct from the information that is at issue
in relation to the informational rights themselves. Unless this distinction is
made, there is scope for confusion with regard to such matters as so-called
'broad consent' and the making of 'informed choices'.[3] Thirdly (in Part IV),
in the particular context of brain imaging technologies, I explore the way in
which the privileging of the privacy interest might be justified; and, I suggest
that, if we conceive of privacy as akin to a property interest, this will encourage a much more careful approach to informed consent. Finally (in Part V),
with the focus on Big Data, and the pressure to let data flow for the sake of
realising the benefits of machine-learning and AI, I consider how coherentist,
regulatory-instrumentalist and technocratic mind-sets might shape a community's thinking about data governance.

II The landscape of informational interests

What informational interests do we have? In traditional offline environments,
there are many claims made to informational rights as well as allegations

---

2 See, e.g., Viktor Mayer-Schönberger and Kenneth Cukier, *Big Data* (London: John Murray,
   2013).
3 For an earlier attempt to dispel such confusion, see Roger Brownsword, 'Informed Consent:
   To Whom it May Concern' (2007) 15 *Jahrbuch für Recht und Ethik* 267, and 'Big Biobanks:
   Three Major Governance Challenges and Some Mini-Constitutional Solutions' in Daniel
   Stretch and Marcel Mertz (eds), *Ethics and Governance of Biomedical Research—Theory and
   Practice* (Switzerland: Springer, 2016) 175.

of informational wrongs. However, our interest is primarily in the privacy and confidentiality of information. It is not the case that such informational interests have no application to modern online environments and are found exclusively in offline settings; rather, interests that have been crafted in offline environments now need to be translated and applied to online environments. Alongside these interests, there are, however, also some informational interests that have been specifically articulated and elaborated in the context of new information and communication technologies. In this latter category, the various interests in fair, transparent, proportionate, and secure processing of our personal data are the most obvious examples. We can start by identifying the key informational interests that we still claim in offline environments before turning to the interests that are emerging in online environments.

(i) Traditional offline informational interests

The interests that we traditionally recognise fall into two broad categories: one category comprises our interests in controlling access to information that we hold concerning ourselves; and the other category comprises our interest in receiving, or in not having disclosed to us, information that relates to us. For convenience, we can characterise the first category of interests as concerning the outward flow of information and the second as concerning the inward flow.[4]

Interests relating to the outward flow of information

In traditional offline environments, we have an interest in controlling the outward flow of information that touches and concerns us in some material respect. Sometimes, we want to prevent any release or circulation of the information; paradigmatically, this is the function of our 'privacy' interest. At other times, we are willing to permit a limited release of the information but we want to control the extent of its circulation; paradigmatically, this is the function of our interest in 'confidentiality'. To mark a file 'private' is to signal that, without proper authorisation, it should not be read; to mark a reference 'confidential' is to signal that the information is exclusively for the use of the recipient.

Although our interests in informational privacy and confidentiality have been crafted in offline environments, it bears repetition that, with the development of modern online environments, these interests not only retain their offline significance but also have an online presence. Moreover, one of the most challenging questions for today's information societies is to work out

---

4 Seminally, see, Samuel Warren and Louis Brandeis, 'The Right to Privacy' (1890) 4 *Harvard Law Review* 193; and Charles Fried, 'Privacy' (1968) 77 *Yale Law Journal* 475.

how rights (and rules) relating to privacy and confidentiality fit into the regulatory frameworks for the processing of data.

Privacy

Famously, it was the potentially unwarranted intrusion of the camera and photography that prompted Samuel Warren and Louis Brandeis to argue for recognition of a right to privacy.[5] In the years that followed, while a number of strands of privacy took root in the US jurisprudence,[6] the United Kingdom was notoriously reluctant to recognise an explicit right to privacy.[7] Even following the enactment of the Human Rights Act 1998, the right was not explicitly recognised.[8]

Be that as it may, in both the common law and civilian world, privacy is the most protean of concepts, with (to use Graeme Laurie's terminology) both spatial and informational articulations.[9] However, even if we have anchored privacy to the idea of an informational interest, there is still a good deal of analytical work to be done.[10] As a first step, from a background of very broad usage of the term, we need to bring into focus in the foreground a narrow, more particular, informational interest in privacy.

Sometimes, we use privacy as an umbrella term that covers a number of our informational interests, including not only our interests in privacy and confidentiality as sketched above but also a number of interests relating to the collection and fair processing of our personal data that are gathered together under the data protection principles.[11] Generically, we sometimes refer to this bundle of claims, both offline and online, as our 'privacy' rights.

Privacy in this broad sense is both conceptually unsatisfactory and risky in practice. As to the former, this broad usage fails to respect the difference

---

5   (n 4).
6   See, William L. Prosser, 'Privacy' (1960) 48 *California Law Review* 383.
7   For example, in *Kaye v Robertson* [1990] EWCA Civ 21, the Court of Appeal upheld a claim for a blatant breach of privacy by invoking the tort of malicious falsehood (the wrong being not the infringement of privacy but the implication that the claimant had consented to the acts in question).
8   See, e.g., Rachael Mulheron, 'A New Framework for Privacy? A Reply to Hello!' (2006) 69 *Modern Law Review* 679.
9   Graeme Laurie, *Genetic Privacy* (Cambridge: Cambridge University Press, 2002).
10   See, e.g., Tjerk Timan, Bryce Clayton Newell, and Bert-Jaap Koops (eds), *Privacy in Public Space* (Cheltenham: Edward Elgar, 2017).
11   In Europe, the standard reference points for such principles are: the Council of Europe's Convention for the Protection of Individuals with Regard to the Automatic Processing of Personal Data, 1981; and the General Data Protection Regulation (Regulation (EU) 2016/679). Stated shortly, the principles require that personal data should be processed fairly, lawfully, and in accordance with individuals' rights; that processing should be for limited purposes, adequate, relevant and proportionate; that retained data should be accurate; that retention should be secure and for no longer than necessary; and, that data transfers to other countries should be subject to equivalent protection.

between information that we want to keep entirely to ourselves and information that we are prepared to see having a limited release; and, once data protection is brought under the privacy umbrella, there is a tendency to confuse interests in opacity (that is, an agent's interest in keeping certain information to itself—an interest in others not knowing) with those in transparency (that is, an agent's interest in knowing when, and for what purpose, information is being collected—an interest in openness).[12] As to the latter, the 'superficial conflation of the fundamental right to privacy and data protection'[13] is risky, indeed doubly risky: on the one hand, if data protection is subsumed under privacy, it might weaken the former; but, on the other hand, if data protection subsumes privacy, this might weaken the latter.

In its narrow usage, privacy refers only to the interest that we have in controlling access to, and sharing, information about ourselves.[14] When, in this sense, we claim that some information is private, we mean, at minimum, that we have no obligation to supply this information to others; if we want to treat the information as purely our own business, we are entitled to do so. However, as a corollary, we might assert that it is wrong, too, for others to try to obtain access to that information. On this view, where A's information is private, not only does B have no right to know (A, the right-holder is under no requirement to disclose), it is also wrong for B and others to try to access the information (against the will of A). Where there is a breach of privacy in this sense, the wrongdoer might compound the infringement by passing on the information to third parties; but it is the wrongful accessing of the information, rather than its further circulation, that is the paradigmatic violation of privacy (in the narrow sense).

It is far from clear how we determine whether information of a particular kind or in a particular place is protected under our privacy right—that is to say, it is not clear how we decide whether our informational privacy right is engaged. Commonly, we ask whether there is a legitimate or a reasonable

---

12 Serge Gutwirth and Paul de Hert, 'Privacy, Data Protection and Law Enforcement. Opacity of the Individual and Transparency of Power' in E. Claes, A. Duff, and S. Gutwirth, (eds), *Privacy and the Criminal Law*, (Antwerp and Oxford: Intersentia, 2006) 61. Compare, too, Hielke Hijmans and Herke Krannenborg, 'Data Protection Anno 2014: How to Restore Trust? An Introduction' in Hielke Hijmans and Herke Kranenborg (eds), *Data Protection Anno 2014: How to Restore Trust?* (Cambridge: Intersentia, 2014) 3, 15:

[Constitutional values] give guidance to all stakeholders in how to act in this highly dynamic environment with constant, and often fundamental, innovations. In this context, it seems more and more important to distinguish between the value of privacy, the 'traditional' fundamental right...that needs to be respected, and the claim individuals have that their data are protected, according to transparent rules of the game.

13 Juliane Kokott and Christoph Sobotta, 'The Distinction between Privacy and Data Protection in the Jurisprudence of the CJEU and the ECtHR' in Hielke Hijmans and Herke Krannenborg (eds) (n 12) 83

14 See Charles Fried, 'Privacy' (1968) 77 *Yale Law Journal* 475.

expectation that the information should be treated as private.[15] This is a matter to which I will return in Part IV of this chapter. However, at this point, it is worth noting that new information and communication technologies play an important role in agitating whatever understanding we have of the scope of privacy protection. For example, in *United States v Antoine Jones*, when the US Supreme Court heard arguments on the question whether a GPS tracking device fixed to a motor car (imagine a device the size of a credit card fixed to the back of the car licence plate) engaged the search and seizure provisions of the Constitution, it was evident that the boundaries of privacy are seen as being always subject to reappraisal.[16] As Justice Alito, having remarked that '[t]echnology is changing people's expectations of privacy',[17] asked:

> Suppose we look forward 10 years, and maybe 10 years from now 90 percent of the population will be using social networking sites and they will have on average 500 friends and they will have allowed their friends to monitor their location 24 hours a day, 365 days a year, through the use of their cell phones. Then … what would the expectation of privacy be then?[18]

In fact, Justice Alito need not have imagined a possible future: for, at the time of the Supreme Court hearing, Uber already had 'a live map revealing the locations of its cars and their passengers, by name. Uber was not only tracking its cars' movements, it was tracking peoples' movements.'[19] It was

---

15 Compare Timan et al (n 10).
16 *United States v Antoine Jones* (arguments heard on November 8, 2011) (transcript on file with author). See, too, the exchange at pp.56–57, where Justice Kagan doubts that we are ready to accept that privacy would not be engaged by 'a little robotic device following you around 24 hours a day anyplace you go that's not your home, reporting in all your movements to the police, to investigative authorities …'. On the significance of *Jones*, see Margaret Hu, 'Orwell's *1984* and a Fourth Amendment Cybersurveillance Nonintrusion Test' (2017) 92 *Washington Law Review* 1819, 1903–1904:

> The cybersurveillance nonintrusion test implicitly suggested by [*Jones*] first shifts the vantage point of the Fourth Amendment analysis from an individual-based tangible harm inquiry to an inquiry of a society-wide intangible harm. Next, the cybersurveillance nonintrusion test shifts the burden from an individual citizen to the government. Under the current privacy test, an individual must first establish a subjective reasonable expectation of privacy. The cybersurveillance nonintrusion test instead requires the government to justify the intrusion of the surveillance on society. The Supreme Court and post-Snowden federal courts appear to be using the Orwellian trope both to establish the society-wide intangible harm: 1984-type scenarios that violate established privacy customs in a democratic society; and to engage the cybersurveillance nonintrusion test to ask the government to overcome the 1984 problem presented by contemporary surveillance methods.

17 *Ibid.*, 43.
18 *Ibid.*
19 Michael J. Casey and Paul Vigna, *The Truth Machine* (London: Harper Collins, 2018) 37.

only three years later, following an outcry about privacy violation, that Uber agreed to encrypt the names of passengers and the geolocation data.[20]

While Justice Alito rightly picks up on the interaction between emerging technologies and our social expectations, it is worth teasing out the slippery steps in the erosion of privacy, namely: (i) we start with certain expectations that are thought to be reasonable and that are protected by rights to privacy and confidentiality; (ii) with the development of social networking sites, and the like, we show that we are willing to give our consent to what otherwise would be an infringement of our privacy and confidentiality rights; (iii) the fact that we are willing so to consent is taken to imply that we are less concerned about our privacy and confidentiality; which (iv) implies a change in our expectations; which (v) entails that our previous expectations are no longer reasonable; and which (vi) translates as a privacy right with a revised (narrowed) range of protection. However, is step (iii) well-founded? Does the fact that we are willing to authorise an act (which would involve a wrongdoing without our authorisation) imply that we are willing to give up having the authorising say? Does the fact that we say 'yes', even that we routinely say 'yes', imply that we are ready to surrender the option of saying 'no'? If not—and the Uber case certainly should give us pause—then we should be taking a much harder look at this process and its supposed outcome.

For the moment, let me simply put down a marker in relation to this harder look. First, if privacy reaches through to the interests that agents necessarily have in the commons' conditions, particularly in the conditions for self-development and agency, it is neither rational nor reasonable for agents, individually or collectively, to authorise acts that compromise these conditions (unless they do so in order to protect some more important condition of the commons). As Maria Brincker expresses this point:

> Agents act in relation not to singular affordances but to affordance spaces: choices are always situated calibrations of multiple interests and purposes given the perceived opportunities. To assess the values and risks of potential actions we need to have expectations regarding the consequences of those actions.[21]

It follows, argues Brincker, that without some degree of privacy 'our very ability to act as autonomous and purposive agents' might be compromised.[22] Secondly, if respect for privacy is a fundamental value of a particular community, then regulators should insulate it against casual corrosion and specious erosion. Thirdly, if privacy is simply a legitimate informational interest that

---

20 *Ibid.*, 37–38.
21 Maria Brincker, 'Privacy in Public and the Contextual Conditions of Agency' in Timan et al (n 10) 64, 88. Similarly, see Hu (n 16).
22 Brincker (n 21) 64.

has to be weighed in an all things considered balance of legitimate interests, then we should recognise that what each community will recognise as a privacy interest and as an acceptable balance of interests might well change over time. To this extent, our reasonable expectations of privacy might be both 'contextual'[23] and contingent on social practices.[24]

As I have indicated, this does no more than scratch the surface of privacy; and, we will pick up some of these threads later in the chapter. However, for the moment, it suffices to be aware of privacy (in the narrow sense) as not only one of our recognised informational interests but, arguably, as more than a mere interest (as an interest that engages special regulatory responsibilities).

Confidentiality

Sometimes A will make information available to B but this will be subject to A's interest in the confidentiality of the information. Where confidentiality is engaged, this means that B should not disclose the information to anyone else—at any rate, this is so unless the context implies that there are others who are on the circulation list. Confidentiality might be engaged in at least three ways.

First, and most flexibly, the confidence might arise by virtue of an agreement between A and B. In such a case, B is bound to respect A's interest in confidentiality because B has undertaken to do so. In principle, the scope of B's undertaking might fall anywhere between (and including) maximally restrictive (B should tell no one) or minimally restrictive (B may tell anyone except Z—for example, anyone other than Z, the subject of a confidential reference). To the extent that B's obligations (to respect the confidence) arise from B's express or implied agreement to receive the information on A's terms, then the confidence is as wide or as narrow, as flexible or as inflexible, as A wants to make it.

Secondly, the confidence might arise by virtue of the nature of the information: quite simply, some information is always to be treated as confidential. Here, it is not necessary for A to put B on notice that the information is confidential; B is assumed to understand that he is the recipient of this kind of information and, thus, subject to a duty of confidence. Provided that everyone understands what kind of information is to be treated as confidential, that is fine. However, where there is uncertainty about this matter, there will be problems (as with privacy itself) in relying on a reasonable expectations test.

---

23 Compare, for example, Daniel J. Solove, *Understanding Privacy* (Cambridge, Mass.: Harvard University Press, 2008), and Helen Nissenbaum, *Privacy in Context* (Stanford: Stanford University Press, 2010).

24 Compare the insightful analysis in Bert-Jaap Koops and Ronald Leenes, ' "Code" and the Slow Erosion of Privacy' (2005) 12 *Michigan Telecommunications and Technology Law Review* 115.

Thirdly, the confidence might inhere in the nature of the relationship between A and B, as is the case, for example, with information given in client/lawyer, confessor/priest, and patient/doctor relationships, and so on. Where the information disclosed by A in such a relationship with B, touches and concerns the vital interests of C, there is deep uncertainty about the circumstances in which B may legitimately break the confidence.[25] Which right has priority, A's right that B should respect the confidence or C's claimed right to be informed about a matter that impacts on C's vital interests?

Interests relating to the inward flow of information

Modern developments in human genetics have generated much debate about a person's right to know (about their genetic profile) and their right not to know (if they prefer to remain ignorant of their genetic profile).[26] Where a person exercises the right to know in order to obtain such information, they might then guard it aggressively (by invoking their privacy rights) against third parties such as employers and insurers. So much is a relatively recent development in the scheme of informational interests. However, we can start with a much more general (positive) right to be informed.

The right to be informed

Arguably, there is a general right not to be misinformed, whether in the lead-up to a transaction or in routine interactions. Such a negative right would shield us against fraud and misrepresentation, and the like. However, it is far from clear—especially, given the interest in privacy—that there is a general (so to speak, all purpose) right to be informed. Possibly, we have some positive informational responsibilities in emergencies, such as to warn of life-threatening hazards (for example, of fires and floods, of unsafe bridges, of dangerous people and wild animals at large); but, aside from such cases, it is not clear that there is a responsibility to keep others informed. Nevertheless, even if there is a very limited background right to be informed, there might be a right to be informed that arises in more particular contexts and circumstances.

Taking our lead from a common approach to privacy and confidentiality, one way of identifying the contexts and circumstances that give rise to a right

---

25 For a classic discussion, see *Tarasoff v Regents of the University of California* 551 P2d 334 (Cal. 1976).

26 See, e.g. Ruth Chadwick, Mairi Levitt, and Darren Shickle (eds), *The Right to Know and the Right Not to Know* (second edition) (Cambridge: Cambridge University Press, 2014); and Roger Brownsword, 'New Genetic Tests, New Research Findings: Do Patients and Participants Have a Right to Know—and Do They Have a Right Not to Know?' (2016) 8 *Law, Innovation and Technology* 247.

to be informed is to ask whether the case is one in which it would be reasonable to expect to be informed. Now, any attempt to base a theory of particular informational rights on the notion of reasonable expectations is slippery because there is no settled reference point to determine what expectations are reasonable.[27] However, where the focus is on particular circumstances and contexts, the reasonableness of an expectation will hinge on the signals given by the relevant agents or on the standard pattern of information-giving or on some background regulatory principle or rule. So, if B explicitly or implicitly undertakes to inform A, or 'to keep A informed', then A's claim to a right to be informed is based on a reasonable expectation. Similarly, if the context is one where the settled practice is for B to inform A (for example, about the risks of a medical procedure that is an option for A, or about some incidental findings in a scanning trial), then once again A has a reasonable expectation that underpins a claimed right to be informed. Finally, if there is a background rule or principle, whether in legislation, case law, or a code of practice, or the like, that requires B to inform A (for example, a rule that requires a prospective insured to make full disclosure to the insurer) then there is a plausible reference point for A's expectation and claimed right to be informed.

For each community, the extent of any such right to be informed will be subject to the variables of personal interaction, custom and practice, regulatory codes, and background philosophy. In communities where the culture is individualistic, agents will be expected to rely on their own resources if they are to be informed; in communities that give more weight to such matters as solidarity, cooperation, and the welfare of others, there will be more support for the right to be informed.

The right to know

The right to know has applications in relation to both public and personal information. So far as access to *public* information is concerned, over one hundred countries worldwide have freedom of information laws.[28] The general purpose of these laws, implicitly recognising a right to know, is to enable citizens to access official information and documents. This is designed to enable citizens to play a more effective role in the public sphere and to hold public officials to account. In some spheres, notably in the sphere of environmental law, the right to know is specifically underpinned by legal provisions—as for example, in the Aarhus Convention, the Preamble to which provides

---

27 Compare Roger Brownsword, 'The Ancillary Care Responsibilities of Researchers: Reasonable but Not Great Expectations' (2007) 35 *Journal of Law, Medicine and Ethics* 679.

28 See, en.wikipedia.org/wiki/Freedom_of_information_laws_by_country (last accessed 9 November 2018).

that 'citizens must have access to information, be entitled to participate in decision-making and have access to justice in environmental matters …'.[29]

Turning to access to *personal* information, as we have said, developments in human genetics have prompted claimed rights both to know and not to know in relation to information about one's own genetic profile or particulars. For example, children who are adopted or whose parents made use of assisted conception technologies with third-party donors of sperm or egg cells, might claim the right to know the identity of their genetic parents. Similarly, the information arising from genetic tests on A might be relevant to the health and well-being of other members of A's family. If A declines to pass on the results of the test, A's family members might complain that this infringes their right to know.

Conceptually, the so-called right to know is a particular application of the right to be informed. In the absence of a regulatory requirement, developed custom and practice, or a personal undertaking, on what basis do children or family members assert that they have a reasonable expectation to be given the genetic information that they wish to have? Unless they rest their case on a (philosophically contestable) natural right to be informed, their best argument might be that their expectation is reasonable because of the particular and special nature of the genetic information. Genetic information, it has often been claimed, is exceptional; and if the right to be informed is indeed exceptional, then this is why it is engaged in these cases. Of course, the more that custom and practice supports this kind of claim, the easier it becomes to plead a reasonable expectation; and it is precisely because practice is still unsettled that, in many places, the right to know is controversial in its application.

Although questions of this kind have been debated for some years, two developments have served to concentrate the mind once again on these matters. One is the development of non-invasive prenatal testing (NIPT)—which is now being piloted within the UK national screening pathway for Down's syndrome.[30] The attraction of this simple blood test is that it promises to reduce the need for an invasive amniocentesis test or chorionic villus sampling and, with that, to reduce the number of babies lost during pregnancies.[31] However, because NIPT presents an opportunity to provide information about the fetus that goes beyond the trisomies,[32] even to the point of full

---

29 UNECE Aarhus Convention Access to Information, Public Participation in Decision-Making and Access to Justice in Environmental Matters (1998), 2161 UNTS 447.
30 See J. Gallagher, 'Safer Down's test backed for NHS use' (2016) available at www.bbc.co.uk/news/health-35311578 (last accessed 9 November 2018).
31 For a successful trial led by Professor Lyn Chitty at Great Ormond Street Hospital, see www.rapid.nhs.uk/about-rapid/evaluation-study-nipt-for-down-syndrome (last accessed 9 November 2018).
32 For example, Sequenom's MaterniT 21 PLUS 'can tell you if you are having a boy or a girl, and screens for both common and rare chromosomal abnormalities. The test screens for trisomy 21 (Down syndrome), trisomy 18 (Edwards syndrome), trisomy 13 (Patau syndrome),

genomic profiling, as well as returning information about the mother,[33] it invites questions about how broad the test should be and how far the mother's right to know might extend.[34] The other development is the emergence of big biobanks (comprising large collections of biosamples and participant data) all being curated for the benefit of health researchers. For some time, one of the most problematic aspects of biobank governance has been whether researchers who access a biobank have any responsibility to return potentially clinically significant results to individual (identifiable) participants.[35] At UK Biobank,[36] where the general rule is that there will be no individual feedback of research findings, but where the biosamples of all 500,000 participants have now been genotyped, and where a major imaging sub-study aims to enrol 100,000 participants on the basis that potentially clinically significant findings will be returned, there are some complex questions of both principle and practice.[37]

The right not to know

Let us suppose that A has undergone genetic tests which show that she has markers indicating a predisposition to breast cancer. A is minded to tell her sister, B; but A knows that B does not want to have this information. If, nevertheless, A insists on telling B, B might claim that this violates her right

and many others that can affect your baby's health': see sequenom.com/tests/reproductive-health/maternit21-plus#patient-overview (last accessed 9 November 2018).

33 See, e.g., K. Oswald, 'Prenatal blood test detects cancer in mothers-to-be', *Bionews* 739 (2015) at www.bionews.org.uk/page_503998.asp. (last accessed 9 November 2018).

34 For discussion, see Roger Brownsword and Jeff Wale, 'The Development of Non-Invasive Prenatal Testing: Some Legal and Ethical Questions' (2016) 24 *Jahrbuch für Recht und Ethik* 31, and 'The Right to Know and the Right Not to Know Revisited' (two parts) (2017) 9 *Asian Bioethics Review* 3. See, further, Roger Brownsword and Jeff Wale, Testing Times Ahead: Non-Invasive Prenatal Testing and the Kind of Community that We Want to Be' (2018) 81 *Modern Law Review* 646.

35 See, e.g., Catherine Heeney and Michael Parker, 'Ethics and the Governance of Biobanks' in Jane Kaye, Susan M.C. Gibbons, Catherine Heeney, Michael Parker and Andrew Smart, *Governing Biobanks* (Oxford: Hart, 2012) 282; Deryck Beyleveld and Roger Brownsword, 'Research Participants and the Right to be Informed', in Pamela R. Ferguson and Graeme T. Laurie (eds), *Inspiring a Medico-Legal Revolution* (Essays in Honour of Sheila McLean) (Farnham: Ashgate, 2015) 173; and Roger Brownsword, 'Big Biobanks, Big Data, Big Questions' in Regina Ammicht Quinn and Thomas Potthast (eds), *Ethik in den Wissenschaften* (Tubingen: IZEW, 2015) 247.

36 See www.ukbiobank.ac.uk/ (last accessed 9 November 2018).

37 On the questions of principle, see Roger Brownsword (n 26); and on some practical questions about minimising the return of false positive findings, see Lorna M. Gibson, Thomas J. Littlejohns, Ligia Adamska, Steve Garratt, Nicola Doherty, Joanna M. Wardlow, Giles Maskell, Michael Parker, Roger Brownsword, Paul M. Matthews, Rory Collins, Naomi E. Allen, Jonathan Sellors, and Cathie L.M. Sudlow, 'Impact of detecting potentially serious incidental findings during multi-modal imaging' [version 2]. *Wellcome Open Res* 2017, 2:114, available at wellcomeopenresearch.org/articles/2–114/v2 (last accessed 9 November 2018).

not to know. Again, whether or not this is a plausible claim will depend upon whether B has a reasonable expectation that she has such a right (to resist the inward flow of information). The same analysis would apply, too, if it were C, A's physician, who insisted on informing B (where B had made it known that she did not wish to know); and, of course, C's wrong would be compounded if, in informing B, C was acting directly in breach of A's insistence that the results of her tests should be treated as confidential. Or, suppose that A tests positive for Huntington's disease. It might be claimed that A's close relatives (also A's prospective employers or insurers) have a right to know about A's test results;[38] but, equally, might it be claimed that those who do not wish to know about these results have a right not to know?[39] When genetic sequencing is increasingly affordable, when genetic information—thanks to big data sets and machine learning—promises to be increasingly interpretable, and when genetic information is readily available in a global marketplace,[40] there are many stakeholders debating these rights. In particular, there are patients and research participants (some wanting to know, others preferring not to know), clinicians and medical researchers (some wanting to disclose, others preferring not to), and various commercial interests and lobbyists (some supporting the rights, others opposing them).

The relationship between the right not to know and the right to know bears further analysis. For example, there are questions to be asked about whether these rights really are discrete and whether they share a common root human interest or need. Without attempting to settle such questions, let me make just one point of clarification. In my view, the right not to know should not be treated as simply an indication that an agent, who has the right to know, has opted not to know. To this extent, the two rights are conceptually distinct. To

---

38 On which, see *ABC v St George's Healthcare NHS Trust & Ors* [2015] EWHC 1394 (QB); [2017] EWCA Civ 336. In this case, the claimant, who was pregnant at the relevant time, sued the defendants, complaining that they had failed to inform her that her father had been diagnosed with Huntington's disease. Had the claimant been so informed, she would have known that she was at risk of having the disease and, knowing that her children would also be at risk, she would have terminated the pregnancy. In the High Court, the claim was struck out (as unarguable) on the ground that, because the defendants obtained the information about the father's health status in confidence, and because the father was emphatic that he did not want his daughter to be told, it would not be fair, just, and reasonable to impose on them a duty to inform the daughter. For comment, see Victoria Chico, 'Non-disclosure of Genetic Risks: The Case for Developing Legal Wrongs' (2016) 16(1–2) *Medical Law International* 3; and, Michael Fay, 'Negligence, Genetics and Families: A Duty to Disclose Actionable Risks' (2016) 16(3–4) *Medical Law International* 115. Not altogether surprisingly, the Court of Appeal has now reversed this decision and remitted the case for trial.

39 See, e.g., the Nuffield Council on Bioethics, *Genetic Screening: Ethical Issues* (London, 1993); and Ruth Chadwick, Mairi Levitt, and Darren Shickle (n 26).

40 NB Gina Kolata, 'FDA Will Allow 23andMe to Sell Genetic Tests for Disease Risk to Consumers' *The New York Times* (April 6, 2017): see www.nytimes.com/2017/04/06/health/fda-genetic-tests-23andme.html?_r=0 (last accessed 9 November 2018).

illustrate, let us suppose that A, having undergone genetic tests, has information that is relevant to B. This is how the two rights map onto the situation:

- If B has the right to know, A is required to inform B; but (as we will see in due course) B may signal that the requirement is waived—that is, B may give an informed consent that authorises non-disclosure by A.
- If B has the right not to know, A is required not to inform B; but B may signal that the requirement is waived—that is, B may give an informed consent that authorises disclosure by A.

In practice, then, there are two *legitimate* possibilities: the information is disclosed (whether pursuant to B's right to know, or under the authorisation of B's informed consent [relative to the right not to know]); or the information is not disclosed (whether pursuant to B's right not to know, or under the authorisation of B's informed consent [relative to the right to know]).

(ii) Modern online informational interests

With the development of modern information technologies, where data are machine readable and digitally processable, we need to review the range and adequacy of our informational interests; but, this does not mean that the traditional informational interests no longer persist. Quite rightly, the European Charter of Fundamental Rights[41] recognises that it may be misleading to ground data protection in privacy; privacy interests (protected by Article 7 of the Charter) and data protection interests (protected by Article 8) are not identical.[42] Nevertheless, privacy (along with the other traditional interests) is still relevant in online contexts. For example, even if we are happy to share much information on social networking sites, there might still be some matters that we regard as private and that we do not wish to be collected or disclosed; and, in relation to the information that we freely post on such sites, there might still be questions of confidentiality (that is, questions about the extent of the circulation).

Data protection rights

There is rough agreement on the kinds of informational interests that are distinctively online interests. Copying some of the data processing principles

---

41 See Charter of Fundamental Rights of the European Union (2000/C 364/01) (18.12.2000).
42 Article 7 provides: 'Everyone has the right to respect for his or her private and family life, home and communications.' Article 8(1) provides: 'Everyone has the right to the protection of personal data concerning him or her.'

set out in the Data Protection Directive (now the General Data Protection Regulation),[43] Article 8(2) of the Charter provides:

> Such [personal] data must be processed fairly for specified purposes and on the basis of the consent of the person concerned or some other legitimate basis laid down by law. Everyone has the right of access to data which has been collected concerning him or her, and the right to have it rectified.

So, even if our privacy interest is not engaged, there is still a set of interests relating to the way in which our personal data is collected and processed in online environments,[44] an interest in the purposes for which it is used and circulated, an interest in its accuracy, an interest in its secure storage,[45] and so on. However, in a rapidly changing information society, this still leaves many questions to be answered.

In a timely move, in November 2010, the European Commission launched a consultation on a fresh attempt to regulate for the effective and comprehensive protection of personal data.[46] Right at the start of the consultation document, the Commission recognised that 'rapid technological developments and globalisation have profoundly changed the world around us, and brought new challenges for the protection of personal data'.[47] Recognising the importance of both clarity and sustainability, the Commission opened the concluding remarks to its consultation in the following terms:

> Like technology, the way our personal data is used and shared in our society is changing all the time. The challenge this poses to legislators is to establish a legislative framework that will stand the test of time. At the end of the reform process, Europe's data protection rules should continue to guarantee a high level of protection and provide legal certainty to individuals, public administrations and businesses in the

---

43 Directive 95/46/EC; Regulation (EU) 2016/679.

44 The importance of the interest in transparency is underlined by Articles 10 and 11 of the Data Protection Directive; according to these provisions, the data subject has a right to be informed that his or her personal data are being processed, as well as being told the purpose of such processing.

45 After the loss, in October 2007, of the personal records of some 25 million people, including dates of birth, addresses, bank account and national insurance numbers, we can take it that secure use is likely to be recognised as one of the more important aspects of this interest. See Esther Addley, 'Two Discs, 25m Names and a Lot of Questions' *The Guardian*, November 24, 2007 www.guardian.co.uk/uk_news/story/0,,2216251,00.html (last accessed 9 November 2018).

46 European Commission, Communication (A comprehensive approach on personal data protection in the European Union), Brussels, 4.11.2010, COM(2010)609 final.

47 *Ibid.*, p.2.

internal market alike for several generations. No matter how complex the situation or how sophisticated the technology, clarity must exist on [sic] the applicable rules and standards that national authorities have to enforce and that businesses and technology developers must comply with. Individuals should also have clarity about the rights they enjoy.[48]

Eight years later, after a marathon legislative process, the General Data Protection Regulation (GDPR) came into effect.[49]

The challenge for the Regulation, like the Data Protection Directive that it has replaced, is partly that its key concepts—such as 'personal data', 'processing', 'data controller', 'pseudonymisation' and so on—have to be applied to rapidly changing technological formats, applications, and uses not all of which can be anticipated. However, the main problem is that the legal framework expresses both a transactionalist view—indeed, one of the salient features of the GDPR is to attempt to strengthen the consent provisions—and a regulatory-instrumentalist view that tries to balance legitimate interests in collecting and processing data with the interests that we each have in the acceptable use of our personal data. If the transactionalist provisions applied only to data that are 'private' and the other provisions to data that are 'personal but not private', we might be able to make sense of the law. Arguably, the special categories of data picked out by Article 9.1 of the GDPR—namely, 'personal data revealing racial or ethnic origin, political opinions, religious or philosophical beliefs, or trade union membership, and the processing of genetic data, biometric data for the purpose of uniquely identifying a natural person, data concerning health or data concerning a natural person's sex life or sexual orientation'— appropriately constitute the classes of personal data that are covered by the privacy right. However, even if it is conceded that this list correctly identifies information that should be respected as intrinsically private (and confidential), it remains for discussion whether this list is exhaustive or merely indicative, and whether what is private offline is equally private online. So long as the scope of the privacy interest is contested and contingent, we have problems.

From a regulatory-instrumentalist perspective, it is important that data governance does not over-regulate data collection and processing, lest the benefits of big data are lost, but neither should it under-regulate lest there are infringements of our privacy interest or right (whatever its precise scope) or there are unacceptable uses of personal data. In the well-known *Lindqvist* case,[50] as in its subsequent case law,[51] the European Court

---

48 *Ibid.*, p.18.
49 Regulation (EU) 2016/679.
50 Case C-101/01 (*Bodil Lindqvist*).
51 See, e.g., Case C-582/14 *Patrick Breyer v Bundesrepublik Deutschland* (2016) (on dynamic IP addresses); and Case C-434/16 *Peter Nowak v Data Protection Commissioner* (2017) (on comments made by an examiner on an examination script).

of Justice has interpreted the concept of 'personal data' very broadly. In *Lindqvist* itself, personal data were held to include commonplace information such as the names, telephone numbers, hobbies, and so on, of Lindqvist's colleagues (which one might well think of as examples of one's 'personal details'); but, by treating information that 'relates to' the data subject in the extended sense that it might have some effect on that person, information that one would simply not think of as being part of one's personal details are also treated as personal data. So, in the recent *Nowak* case, an examiner's comments on the data subject's examination script were treated as personal data (even though, in this context, the data subject's right to check the accuracy of the data would make no sense).[52] While such a broad reading means that the law is engaged, it also means that data should flow only in accordance with the principles for lawful collection and processing.

As we saw in Chapter Four, the complex relationship between new technological developments and our uncertainty about our informational interests is exemplified by the *Google Spain* case.[53] First, there is the widespread confusion involved in failing to differentiate between those rights (privacy being the paradigm) that protect an interest in opacity and those rights (fair collection and processing of personal data being the paradigm) that relate to an interest in transparency. Secondly, as we highlighted in our earlier discussion of the case, the balancing exercise undertaken by the court mixes fundamental rights with non-fundamental rights or legitimate interests.[54] Although the GDPR gestures in the direction of the right to be forgotten in Article 17, where it elaborates the data subject's right to erasure, we are no nearer to settling the scope and strength of this right.[55]

In short, even though we live in information societies, where we attach huge importance to information and rely heavily on modern technologies, the landscape of our informational interests lacks definition as well as normative settlement.

---

52 *Nowak* (n 51). At [35], the CJEU confirms that information 'relates to' the data subject where 'by reason of its content, purpose or effect, it is linked to a particular person'.

53 Case C-131/12, *Google Spain SL, Google Inc v Agencia Española de Protection de Datos (AEPD), Mario Costeja González* [2014] available at curia.europa.eu/juris/document/document_print.jsf?doclang=EN&docid=152065 (last accessed 9 November 2018).

54 On which, see Eleni Frantziou, 'Further Developments in the Right to be Forgotten: The European Court of Justice's Judgment in Case C-131/12, *Google Spain SL, Google Inc v Agencia Española de Proteccion de Datos*' (2014) 14 *Human Rights Law Review* 761, esp 768–769.

55 For insightful commentary, see Orla Lynskey, 'Control over Personal Data in a Digital Age: *Google Spain v AEPD and Mario Costeja Gonzalez*' (2015) 78(3) *Modern Law Review* 522 (arguing, inter alia, that the labelling of the right as a 'right to be forgotten' is misleading).

III Informed consent and informational rights

In this part of the chapter, I clarify the role of informed consent in relation to the set of informational interests sketched. To the extent that these interests are covered by a right (under a choice-theory of rights), then rights-holders may authorise acts that would otherwise involve a violation of their rights. Such authorisation is given through a process of informed consent. So far so good. However, we should not confuse (i) the information that the rights-holder must have to give a valid (informed) consent with (ii) the information that is specified for a consent within a particular regulatory regime, or (iii) with the information that the same rights-holding agent should have pursuant to exercising particular informational rights; and we need to bear in mind the distinction between a consent being invalid (because it is not sufficiently informed) and an authorisation failing to cover some particular acts because it is not expressed in terms that are sufficiently specific and explicit.

We can proceed in three stages: (i) clarifying the informational conditions for a valid informed consent; (ii) discussing the potential validity of broad consent in the context of big biobanking projects; and (iii) clarifying the informational requirements for making an informed choice.

(i) The informational conditions for a valid informed consent

It is commonly agreed that a consent will not operate as a valid authorisation in relation to any right (informational or otherwise) unless it is both freely given and properly informed. Both these conditions are problematic.[56] For present purposes, we need not try to resolve the difficulties concerning the former condition. Our question is: what information must the rights-holder have before a purported consent is properly informed? I suggest that we should answer this question by reminding ourselves that the information relates directly to the giving of an authorisation with reference to a particular right. Suppose, for example, the background right is not informational as such; suppose it is a right to physical integrity or to property. The information that bears on the validity of the consent must, or so I suggest, concern the meaning and significance of consent being given. Accordingly, I propose that, where A consents to act x by B, the consent will be properly informed provided that the rights-holder, A, understands:

- that he or she has the particular right in question (to physical integrity, or property, or [to reintroduce an informational right] privacy, or whatever)
- that he or she has the option of giving or withholding consent

---

56 For discussion, see Deryck Beyleveld and Roger Brownsword, *Consent in the Law* (Oxford: Hart, 2007).

- that, if consent is withheld, there is no penalty
- that, if consent is given, an act, x, is being authorised that, without the consent, would involve a violation of the right, and
- that, if consent is given to B, then provided that B acts within the scope of the authorisation, B (by doing x) does no wrong to A—that is, A understands that he or she is precluded (by the consent) from complaining about B's doing of x.

If A understands all of this, then I suggest that, for the purposes of determining whether A has given a valid consent to B, A's consent is sufficiently informed.

To avoid any misunderstanding, I should add that this thin account of *informed* consent does not, of course, entail that A has no protection against being misinformed or duped by B, or others. If A's consent is procured by fraud or misrepresentation or the like, A will have independent recourse and remedies. In such circumstances, we may want to say that here, too, the reason why A's consent is not valid is that it was not a properly informed consent. However, I would prefer to differentiate between A being underinformed (in the way that the thin account has it) and A being misinformed.

(ii) The potential validity of a broad consent

For many years, medical researchers have relied on 'biobanks' (that is, collections of biological samples and tissues that are curated and used for particular research purposes). However, we are entering an era of 'big biobanks' in which the collection of biological samples is complemented by various kinds of personal data (such as data concerning lifestyles) as well as medical records, all linked to further databases that are health-related (such as cancer registries and hospital episode statistics). The hope that encourages investment in such big biobanks is that, by interrogating the data, researchers will understand much more about the causes of prevalent diseases—in particular, by understanding the causal significance of a person's genetic profile, their lifestyle, and their exposures, and so on—as well as finding responses that work more effectively for patients.

In at least three respects, today's big biobanks can be contrasted with yesterday's biobanks. First, there is the scale and size of the new generation of biobanks. As the name implies, these biobanks are 'big', recruiting large populations of participants, not just a small group of people who happen to have a particular disease. For example, UK Biobank, which is an exemplar of a big biobank, has 500,000 participants; Genomics England is aiming to sequence 100,000 genomes; in the United States, the Precision Medicine Initiative is seeking to recruit more than one million volunteers; and, albeit a rather different kind of biobank, 23andMe already has a very large collection of biosamples supplied by its customers. Secondly, big biobanks are created, not for research into one particular disease (such as a cancer or stroke or coronary disease) but as a resource to be available for research into any and all diseases (again, to refer to UK Biobank, the resource is open for use by bona fide researchers for health-related research purposes). Thirdly, in big biobanks, when the biological samples are genotyped and sequenced, further data are added to the resource. Moreover, as data linkages intensify, and as biobanks themselves are increasingly linked to one another, it becomes clear that this is where big biobanking and big data meet.[57]

Suppose, for example, that a researcher is interested in finding out more about the causes of heart disease. If a biobank can provide the researcher with the genetic profile, general health data, environmental and life-style data in relation to 25,000 participants who have a diagnosis for heart disease and the same data in relation to 25,000 participants who do not have that diagnosis, it might be possible by trawling through the data to identify relevant differences between the groups. For example, if it were not already known that smoking tobacco is a risk factor for various diseases, this might be a factor that would show up; or it might be that the analysis will highlight some differences in the genetics or even, for example, whether the participants who have the diagnosis have dogs (which need to be exercised, which then impacts on the participant's weight, which in turn has implications for coronary health). It is a common view, however, that the consent given by participants when they enrol as participants in such biobanking projects is suspect because each participant is asked to consent to the use of their biosamples and data for a broad range of not fully specified health research purposes. As we have said, there might be biobanks and research projects that are dedicated to investigating just one disease or condition. However, in big biobanks, we can take it that the participants will not have been asked to consent to the use of their biosamples and data for research into coronary problems or for any other specific condition; rather they will have been asked to agree to use for health research generally, possibly with heart disease getting a mention as one of many prevalent conditions in which researchers might be interested. Is it correct to infer that, because the authorisation given is broad, participants cannot be giving an informed consent to the use of their biosamples and data for research?

A few years ago, in a case involving the biobank, Lifegene, the Director of Sweden's Data Inspection Agency ruled that the gathering of personal information for 'future research' is in breach of the (Swedish) Personal Data Act. The problem identified by the Agency was that Lifegene's expressed research

---

57 See, e.g., Roger Brownsword, 'Regulating Biobanks: Another Triple Bottom Line' in Giovanni Pascuzzi, Umberto Izzo, and Matteo Macilotti (eds), *Comparative Issues in the Governance of Research Biobanks* (Heidelberg: Springer, 2013) 41; and 'Big Biobanks, Big Data, Big Questions' in Regina Ammicht Quinn and Thomas Potthast (eds), *Ethik in den Wissenschaften* (Tubingen: IZEW, 2015) 247.

purposes were too general to satisfy section 9c of the Act, which provides that the collection and processing of personal information must be for 'specific, explicitly stated and justified purposes ...'.[58] In neighbouring Finland, the legislature tried to remove any doubt about the legality of broad consent by enacting a Biobank Act (Act 688/2012) that makes it clear that a participant may consent to a broad range of purposes.[59] According to some commentators, the authorisation may 'include research into health-promoting activities, causes of disease, and disease prevention and treatment, as well as research and development projects that serve healthcare'.[60] However, to the extent that biobanking of this kind is subject to European data protection legislation, Article 4 of the GDPR seems to be more in line with the Agency's view in the Lifegene case than with that of the Finnish Act, the GDPR defining consent in the following terms:

> 'consent' of the data subject means any freely given, specific, informed and unambiguous indication of the data subject's wishes by which he or she, by a statement or by a clear affirmative action, signifies agreement to the processing of personal data relating to him or her.

What should we make of this? First, it depends whether the proposition is (i) that a broad consent can never be a valid consent or (ii) that a broad consent can be a valid consent and can authorise acts that clearly fall within its terms but there might be some acts that are not covered by a broad authorisation. Secondly, what we make of these propositions depends on how we characterise the relationship between the biobank and its participants—whether we view a participant's consent as a waiver of the benefit of a right, or as a contract, or as a signal of assent within a regulated scheme.[61]

The first possibility is that we view participants as negotiating waivers of the benefit of their rights to privacy and confidentiality (which normally attach to sensitive medical information). Such a consent will be a valid waiver only if it is free and informed. However, I have suggested that, although the requirement of being informed is heavily contested, it is enough that the rights-holder understands the significance of the waiver. On this view, there is no reason why a broad consent should not be adequately informed.

58 I am indebted to Adrienne Hunt for drawing my attention to this case and to Søren Holm for translating and summarising the case. See ethicsblog.crb.uu.se/2011/12/20/the-swedish-data-inspection-board-stops-large-biobank/ (last accessed 9 November 2018).
59 See Sirpa Soini, 'Finland on a Road towards a Modern Legal Biobanking Infrastructure' (2013) *European Journal of Health Law* 289.
60 See Joanna Stjernschantz Forsberg and Sirpa Soini, 'A big step for Finnish biobanking' (2014) 15 *Nature Reviews Genetics* 6.
61 Compare, Roger Brownsword, 'Big Biobanks: Three Major Governance Challenges and Some Mini-Constitutional Solutions' in Daniel Stretch and Marcel Mertz (eds), *Ethics and Governance of Biomedical Research—Theory and Practice* (Switzerland: Springer, 2016) 175.

Nevertheless, if the rights-holder subsequently complains that some research use—or commercial exploitation of the research—at the biobank falls outside the scope of the authorisation, there might be a question, not so much about the general validity of the consent but about the scope of the authorisation.

Taking a similar approach, we might see little problem with a person freely consenting to the terms for participation in the Personal Genome Project-UK (PGP-UK), even though those terms are likely to be extremely unattractive to some potential participants.[62] Indeed, it will take a special kind of person to sign up for a project that explicitly says that, while participants 'are not likely to benefit in any way as a result of [their] participation',[63] there are numerous potential risks and discomforts arising from participation.[64] The consent to be given by participants is 'open', in the sense that all medical information attached to a person's record will be made available online; and, while participants' names and addresses will not be advertised, participants are warned explicitly that they might quite easily be identified and their privacy cannot be guaranteed.[65] Moreover, prospective participants are put on notice that PGP-UK 'cannot predict all of the risks, or the severity of the risks, that the public availability of [participant] information may pose to [participants] and [their] relatives'.[66] Clearly, this is not for everyone, and especially not for the risk-averse. However, if a participant signs up for these terms on a free and informed basis, then their open consent, just like a broad consent, authorises certain research activities; and they have accepted the risks that are within the scope of the authorisation.

The second possibility is that we view the relationship between a biobank and its participants as contractual. Even if, according to English law, the agreement between the biobank and its participants would not be treated as a legally enforceable contract (because it is not backed by the requisite intention),[67] the relationship is nevertheless 'contractual' in the sense that the participants sign up to a package of terms and conditions. Moreover, on

---

62 Ian Sample, 'Critics urge caution as UK genome project hunts for volunteers' *The Guardian*, November 7, 2013 (available at: www.theguardian.com/science/2013/nov/07/personal-genome-project-uk-launch (last accessed 9 November 2018)). See, Article 7.1 of the informed consent document: available at

63 www.personalgenomes.org.uk/static/docs/uk/PGP-UK_FullConsent_06Jun13_with_ amend.pdf (last accessed 9 November 2018).

64 See *ibid.*, Article 6, running to no fewer than six pages.

65 See *ibid.*, Article 6.1(a)(iv).

66 See *ibid.*, Article 6.1(a)(vi).

67 In the English law of contract, an independent requirement of an 'intention to create legal [contractual] relations' was introduced by the Court of Appeal in *Balfour v Balfour* [1919] 2 KB 571. This requirement translates into a rebuttable presumption that domestic and social agreements are not backed by the requisite intention and thus should not be treated as legally enforceable contracts; by contrast, the presumption is that business agreements are backed by an intention to create legal relations and are enforceable. It is not clear how an otherwise contractual relationship between researchers and participants would be classified;

this view, the justification for the biobank's use of the participants' biosamples and data is 'transactional'. On some interpretations of transactionalism, a failure to be at one in relation to all permitted research uses might invalidate the consent, but the more likely view is that this kind of failure simply goes to the scope of the permission and the interpretation of the authorising terms and conditions. The question then might be whether the biobank and the researchers who access the biobank samples and data could reasonably rely on the participants' consent or whether, as relationalists or contextualists might view it, the particular use at issue was in line with the parties' honest and reasonable expectations given the context in which they were operating.

The third possibility is that governance of biobanks is taken to be a matter of striking an acceptable balance of the different interests of stakeholders (particularly participants, researchers, funders, and potential beneficiaries of the research) in the use of the participants' biosamples and data. If such a balance requires uses to be specifically and explicitly authorised, in the way that the GDPR has it, then consent that falls short of that standard will not do—or, at any rate, it will not do unless a special research exemption is adopted. However, this is not something that is intrinsic to the idea of informed consent; this is a particular take on consent within a particular regulatory-instrumentalist view.

(iii) Informed choice

How do these remarks about consent bear on the turn in modern medical law, away from a paternalistic culture (in which doctors decided what was in a patient's best interests) to one in which doctors are expected to assist patients to make informed choices about their treatments?

As we saw in Chapter Ten, in *Montgomery v Lanarkshire Health Board*,[68] the UK Supreme Court, resoundingly rejecting the applicability of the *Bolam* test to the case at hand, held that the relationship between clinicians and patients must be rights-respecting rather than paternalistic and that patients have a right to be informed about their options (together with their relative benefits and risks).

If, after *Montgomery*, patients are recognised as having a right to make an informed choice as to their treatment, and if this is a right such as the right to privacy and confidentiality, then there is a distinction between patients being asked to waive the benefit of that right and patients seeking to exercise their right. In the former case, the waiver should be subject to an informed consent of the kind that I have specified above. In the latter case, the emphasis is

---

but, my guess is that the presumption would be that the parties do not intend to create a legally enforceable agreement.

68 [2015] UKSC 11.

not so much on their consent but on the range of information that they are entitled to have under the right. As the caveats in *Montgomery* indicate, there may be some information in relation to which, or some contexts in which, the right runs out (for example, as where the so-called therapeutic privilege applies).

If, however, the relationship between doctors and their patients is specified in a regulatory-instrumentalist way, the community's idea of an acceptable balance of interests will be used to determine the limits of the right and the conditions in which a waiver is sufficiently informed to relieve doctors of their informational responsibilities.

No doubt, there is more to be said about the different ways in which consent figures in waivers of the benefits of a right, in a contractual transaction, and in assenting to and authorising acts.[69] However, suffice it to say that these complexities persist in communities where information is transferred into online environments. In those environments, we find that both privacy and consent are sometimes viewed as anachronistic. In particular, according to a widely held view, if individual agents are to have control over access to their private information, and if access is to be granted only where those individuals give their consent, this impedes the collective interest and gives too much power to individual gatekeepers.[70] To justify such a restriction, privacy must be really special. In that spirit, we can proceed to ask in the next section what might warrant such a privileging of privacy.

IV Privileging privacy

Having differentiated between informational rights and informed consent, in this part of the chapter I will consider one case in which the way that we understand a particular informational right might indeed impact on how we operationalise informed consent relative to that right. The right in question is privacy; and the context is brain imaging where imagers are routinely shielded by the informed consent given by patients and research participants.

Modern developments in neurosciences, particularly the development of powerful imaging technologies, have given rise to a concern that the inner sanctum of privacy (the human brain) is under threat.[71] How are we to

---

69 See further Beyleveld and Brownsword (n 56).
70 Compare the discussion in Roger Brownsword, 'Rights, Responsibility and Stewardship: Beyond Consent' in Heather Widdows and Caroline Mullen (eds), *The Governance of Genetic Information: Who Decides?* (Cambridge: Cambridge University Press, 2009) 99.
71 Here, I am drawing on Roger Brownsword, 'Regulating Brain Imaging: Questions of Privacy and Informed Consent' in Sarah J.L. Edwards, Sarah Richmond, and Geraint Rees (eds), *I Know What You Are Thinking: Brain Imaging and Mental Privacy* (Oxford: Oxford University Press, 2012) 223.

address this concern? In this section, I introduce a distinction that might be helpful, namely that between a fixed (privileged) and a flexible conception of this informational entitlement. While the flexible conception will recognise a prima facie infringement of privacy where a balance of reasonableness so indicates, the fixed conception is engaged purely by the private nature of the information and without any kind of balancing exercise—here, infringement (just like trespass to land) is per se. If the practical difference between the two conceptions is to be measured by reference to their respective scope and strength, then current experience suggests that the protective sweep of the flexible conception (which, by and large, is the dominant conception nowadays) will be wide but weak, whereas (depending upon how precisely we anchor it) the fixed conception will produce a more narrowly targeted but stronger protective effect.

To start with the fixed conception, the basic idea is that an agent may be related to certain information in such a way that the agent's privacy interest is engaged. Just as an agent may claim a controlling interest over an object by asserting, 'This is my property; this is mine', so an agent may claim a controlling interest over information by asserting, 'This is my information; it is just my business; it is private'. On this fixed conception, a privacy claim is analogous to a property claim; in both cases, the reasonableness of the claim is irrelevant—it matters not one jot, for example, that some third party might make better use of the property or information, or that the third party has greater needs than the agent with the controlling interest. If the information is private, it is mine to control.

Because the fixed conception of privacy follows the form of a proprietary claim, we might think that there is actually an identity between privacy and property. At some level, this might indeed be the case. However, to get the relationship between property and privacy clearly and fully into focus would require a major detour. In the present context, let me simply make one comment on this issue. Imagine that the development of imaging technology imitates that of information technology, becoming much smaller, much cheaper, more portable, more widely distributed, and more powerful. Imagine that brains can be scanned and imaged remotely, without the knowledge of the agent being scanned and without any physical contact. Even if privacy claims have the shape of property claims, even if they sometimes need property as their platform,[72] the interest that invites articulation in such a remote scanning scenario surely is one of privacy. No doubt, there is much more that could be said about this; but, for present purposes, let me proceed on the

---

72 Compare the case law on claimant landowners who plead a property-based tort, such as trespass or nuisance, to protect their privacy—e.g., *Baron Bernstein of Leigh v Skyviews and General Ltd* [1978] QB 479 (trespass) and *Victoria Park Racing and Recreation Grounds Co Ltd v Taylor* (1938) 58 CLR 479 (nuisance).

assumption either that the fixed conception of privacy is not reducible to property or that, if it is reducible, privacy is a distinct interest within the bundle of proprietary entitlements.

If some information is to be protected in the way that the fixed conception of privacy envisages, then what is it that justifies such privileged treatment? One possibility is that it is the location of the information that matters—if the information is in the public domain, it is not protected; while, if the information is in a private zone, it is protected. However, even if we can draw a sharp and workable distinction between those zones that are public and those that are private, the fact that the information is in a private zone does not, in itself, seem like a sufficient reason for treating it in the privileged protective way that the fixed conception of privacy proposes.[73] After all, the information might be entirely trivial—in which case, if we insist that it is to be protected as private, this can only be explained as an indirect effect of the zoning rather than by way of a direct concern for the special nature of the information as such. Even if there are good reasons for the zoning, we do not yet have good reasons for according privileged protection to information in private zones. Moreover, if we try to limit the protection to *special* or *sensitive* information that is located in a private zone, we are back to our question: what is it that could possibly make information special or sensitive in the way that the fixed conception of privacy seems to presuppose? Another possibility is that it is the personal signalling or marking of the information that is crucial. If an agent marks a file 'top secret', the information therein remains private even if the file is carelessly left in a public place. But, again, the information so marked might be trivial. Hence, unless agents exercise their prescriptive power in a selectively justifiable way, this does not appeal as a compelling reason for adopting a fixed concept of informational privacy.

What seems to be missing in both the zonal and the personal signalling accounts is a lack of interest in the character of the information that is protected. Yet, the obvious reason for taking a strong line on informational privacy is that there is something about the information itself that is critical. But, what are these critical characteristics?

A promising line of thinking is suggested by Christian Halliburton who, writing with reference to US constitutional jurisprudence, argues that we should recognise an interest in 'personal informational property'.[74] Distinctively, this interest would target 'information which is closely bound up with identity, or necessary to the development of the fully realized person,

---

73 Compare Timan (n 10), doubting the former and agreeing with the latter.
74 Christian M. Halliburton, 'How Privacy Killed *Katz*: A Tale of Cognitive Freedom and the Property of Personhood as Fourth Amendment Norm' (2009) 42 *Akron Law Review* 803.

[and which] like certain types of property, is deserving of the most stringent protection'.[75] Elaborating this idea, Halliburton says:

> I think it is easy to see (and rather difficult to dispute) that our thoughts, our internal mental processes, and the cognitive landscape of our ideas and intentions are so closely bound up with the self that they are essential to our ongoing existence and manifestation of a fully developed personal identity. As such, they are inherently and uncontrovertibly personal information property deserving absolutist protections because any interference with these informational assets cannot be tolerated by the individual. Many would therefore argue that capturing thoughts, spying on mental processes, and invading cognitive landscapes with [brain imaging technologies] deprive the individual not only of property related to personhood, but of personhood altogether.[76]

There is much to ponder in these remarks, where the already complex mingling of property with privacy is joined by the difficult concept of personal identity. Suffice it to say that, if we are to support the fixed conception of privacy, Halliburton gives us, at the very least, some important ideas about how we might justify our position—as indeed, does Brincker, whose linking of privacy to the context for agency and self-development we noted previously.[77]

Turning from the fixed to the flexible conception of privacy, we find that we are on more familiar ground. Here, an infringement is recognised only where a balance of reasonableness so indicates. In the common law world, this conception is expressed by asking whether the complainant has a 'reasonable expectation' of privacy[78]—as Lord Nicholls put it in the Naomi Campbell case, '[e]ssentially the touchstone of private life is whether in respect of the disclosed facts the person in question had a reasonable expectation of privacy.'[79] Typically, in the case-law, this will involve some balancing of the interests of a celebrity complainant against the interests of the media in publishing some story and pictures of the celebrity. Thus, in the J.K. Rowling case,[80] Sir Anthony Clarke MR said:

> As we see it, the question whether there is a reasonable expectation of privacy is a broad one, which takes account of all the circumstances of

---

75 *Ibid.*, 864.

76 *Ibid.*, 868.

77 See text to n 21. Compare, too, Hu (n 16).

78 In the seminal case of *Katz v United States* 389 US 347 (1967) at 361, Justice Harlan set out a famous two-part test: first, the complainant must have exhibited a subjective expectation of privacy; and, secondly, the complainant's expectation must be one that society is prepared to recognise as reasonable.

79 *Campbell v Mirror Group Newspapers Limited* [2004] UKHL 22, [21].

80 *Murray v Express Newspapers plc* [2007] EWHC 1908 (Ch); [2008] EWCA Civ 446 (reversing the trial court decision).

the case. They include the attributes of the claimant, the nature of the activity in which the claimant was engaged, the place at which it was happening, the nature and purpose of the intrusion, the absence of consent and whether it was known or could be inferred, the effect on the claimant and the circumstances in which and the purposes for which the information came into the hands of the publisher.[81]

Although high-profile disputes of this kind are determined very much on a case-by-case basis, it is important to keep an eye on the benchmark or reference point for a judgment that a particular expectation of privacy is reasonable. To recall our remarks earlier in the chapter, the judgments that are made often take their lead from what seems to be reasonable in the light of prevailing custom and practice. However, practice is a shifting scene; and particularly so where new technologies not only make possible ever more remote and undetectable observation but also encourage netizens to be carefree about their personal data. Somewhat bizarrely, if we apply the flexible conception in such conditions we find that the more that there is pressure to push back the line of privacy, the less that it is infringed—because our reasonable expectation has been adjusted (i.e., lowered) by the practice.[82]

Without filling in the substantive detail of the fixed conception, we cannot specify its precise scope. For the sake of argument, however, let us suppose that it would operate with a more restrictive range than the ubiquitous flexible conception. Nevertheless, the thought is that, where the fixed conception was engaged, its protective effect would be stronger than that given by the flexible conception. Why so? Quite simply, because we would take a more stringent view of consent where the authorisation concerned a right, albeit an informational right, that had proprietary characteristics.

To elaborate on this idea: whichever version of privacy we espouse, where the privacy right is engaged there will be a prima face infringement unless there is a free and informed consent that authorises the act that otherwise would be an infringement. As we have said, what constitutes a 'free' and an 'informed' consent is deeply problematic; and the same applies to what constitutes a sufficient signalling of consent (hence the endless debates about

---

81 [2008] EWCA Civ 446, [36].
82 For the best general elaboration of this point, see Bert-Jaap Koops and Ronald Leenes (n 24). And, for a telling example, see Aimee Jodoi Lum, 'Don't Smile, Your Image has just been Recorded on a Camera-Phone: The Need for Privacy in the Public Sphere' (2005) 27 *University of Hawai'i Law Review* 377, 386:

> Many of the same social conditions exist today as they did in the 1990s, but the explosion of technological advances has made individuals far more susceptible to invasions of privacy than ever before. America's voyeuristic tendencies and obsession with reality TV further exacerbates the problem because behaviors that might otherwise be considered unacceptable become normalized.

whether consent has to be by opt-in or by opt-out).[83] For those who want to defend the privacy interest, these requirements will be applied in a demanding way such that, where it is claimed that the right-holder has consented to a privacy-infringing act, it needs to be absolutely clear that the right-holder has authorised the act in question. However, for those who want to lower the barriers, the requirements will be applied in a much less demanding way—for example, instead of consent having to be given explicitly, implicit indications will suffice.[84] In some theoretical accounts, the fixed conception of privacy might actually entail the demanding interpretation of consent. However, for present purposes, let me put the relationship between the rival conceptions of privacy and the different views of consent less in terms of logic than in terms of life. In the current context, where privacy is protected widely but weakly, privacy advocates who are drawn to the fixed conception surely will want to bolster it with a demanding requirement of consent; by contrast, opponents of privacy will tend to argue for the flexible conception in conjunction with a less demanding requirement of consent. Whether practice actually maps in this way is, of course, an empirical question. Nevertheless, it would be surprising if it did not do so.

In the information society, there seem to be three ways in which we might try to take these ideas forward. One option is to articulate a fixed conception of informational privacy that applies to a restricted range of cases but, where it applies, it has strong protective effect. A second option is to stick with, or adjust, a flexible conception of the kind we currently employ. Subject to adjustment, such a conception can range broadly across our interactions and transactions but its protective effect is relatively weak. A third option might be to try to combine the two conceptions so that the fixed conception protects, as it were, the inner sanctum of informational interests but, beyond that, the flexible conception offers a degree of protection that follows the contours of convention and custom and practice. At all events, if we are concerned about a loss of informational privacy, if we are not yet ready to accept that privacy is dead, we need something like the fixed conception—rooted in the commons conditions or recognised as a distinctive value—to make the protective regime more robust by anchoring the ubiquitous idea of a reasonable expectation.

---

83 For comprehensive discussion of each of the elements of an adequate consent, see Beyleveld and Brownsword (n 56).

84 See, Roger Brownsword, *Rights, Regulation and the Technological Revolution* (Oxford: Oxford University Press, 2008) Ch.3. In relation to the variable implementation of the consent provisions in European data protection law, see European Commission, *Comparative Study on Different Approaches to New Privacy Challenges, in Particular in the Light of Technological Developments: Final Report* (January 20, 2010), para.54 et seq.

V Big data: regulatory responsibilities, community framings, and transnational challenges

Big data is big news, big business, big money, and a big cause for concern[85]— as Frank Webster has put it, in the general context of the information society, we like to have '[b]ig terms for big issues.'[86] The consensus is that the collection and use of personal data needs governance and that big data-sets (interrogated by state of the art algorithmic tools) need it a fortiori; but there is no agreement as to what might be the appropriate terms and conditions for the collection, processing and use of personal data or how to govern these matters. In the light of what we have previously said, we should remind ourselves of the responsibilities of regulators, then we can sketch the traditional coherentist, new coherentist, regulatory-instrumental, and technocratic ways of framing the debates and, finally, we can note the transnational nature of the challenge.

(i) The responsibilities of regulators

The paramount responsibility for regulators is to protect and maintain the commons' conditions. Arguably, the collection of big data sets is key to the intelligence activities that are responsible for the security of the State. On the other hand, big data, in both the public and the private sector, can be a major inhibition on the lifestyle choices of agents—even leading individuals to engage defensively in deliberate acts of obfuscation and misinformation to fool their profilers.

Introducing his book on the Snowden revelations, Glenn Greenwald[87] reminds his readers that snooping and suspicionless surveillance is nothing new. Nor, of course, is it just the NSA in the United States that acts, or that has acted, as an agent of surveillance. Famously, three decades ago, the revelations made by Peter Wright (an ex member of the British intelligence services) in his book, *Spycatcher*, caused a furore—on one side, about the publication of confidential information[88] and, on the other, about the bugging and eavesdropping routinely carried out by MI5.

With each new wave of technologies, the focal points for surveillance change—for example, from the opening of mail to wire-tapping. With the development of the Internet, and with our online activities becoming the focus for surveillance, the threat to privacy is amplified. For, as Greenwald

---

85 For a general overview, see Viktor Mayer-Schönberger and Kenneth Cukier (n 2).
86 (n 1) 1.
87 Glenn Greenwald, *No Place to Hide* (London: Penguin, 2014).
88 For the government's unsuccessful attempt to suppress publication, see *Attorney-General v Guardian Newspapers (No 2) ('Spycatcher')* [1990] 1 AC 109.

rightly points out, the Internet is much more than our post office or telephone: the Internet 'is the epicentre of our world ... It is where friends are made, where books and films are chosen, where political activism is organized, where the most private data is created and stored. It is where we develop and express our very personality and sense of self.'[89]

Faced with this threat, we can hear echoes of Al Gore's critique of the information ecosystem[90] in Greenwald's suggestion that we stand at an historic crossroads. The question, as Greenwald, puts it, is this:

> Will the digital age usher in the individual liberation and political freedoms that the Internet is uniquely capable of unleashing? Or will it bring about a system of omnipresent monitoring and control, beyond the dreams of even the greatest tyrants of the past? Right now either path is possible. Our actions will determine where we end up.[91]

The question, then, is simply this: what kind or kinds of information society are incompatible with respect for the commons? In both the public and the private sector data are being gathered on an unprecedented scale;[92] and, if these data are then used to train smart machines that sift and sort citizens (as mooted by the Chinese social credit system)[93] this could be the precursor to a truly dystopian 'system of omnipresent monitoring and control'. If privacy does not stand between us and such a compromising of the commons, then regulators need to find another protective strategy.

---

89 Greenwald (n 87) 5–6.
90 Al Gore, *The Assault on Reason* (updated edition) (London: Bloomsbury, 2017) 294, where the challenge is said to be to restore 'a healthy information ecosystem that invites and supports the ... essential processes of self-government in the age of the Internet so that [communities] can start making good decisions again'.
91 Greenwald (n 87) 6.
92 With reference to the private sector, compare Siva Vaidhyanathan, *The Googlization of Everything (And Why We Should Worry)* (Oakland: University of California Press, 2011).
93 See Yongxi Chen and Anne S.Y. Cheung, 'The Transparent Self Under Big Data Profiling: Privacy and Chinese Legislation on the Social Credit System' (2017) 12 *The Journal of Comparative Law* 356; and Rogier Creemers, 'China's Social Credit System: An Evolving Practice of Control' (May 9, 2018) (available at SSRN: ssrn.com/abstract=3175792) (last accessed 9 November 2018). Creemers points out that the 2014 Plan for the construction of a social credit system does not actually make any reference to 'the sort of ... big data analytics that foreign observers have ascribed to the SCS' (p.13). Nevertheless, the combination of unique identifiers, a lack of anonymity, and increasingly connected data points suggest an evolution of the SCS in the direction anticipated by many observers. We should also note the practice of alternative credit scoring where traditional financial indicators are combined with many non-financial data points (concerning, e.g., where one shops, who is in one's social network, how long one takes to read terms and conditions, and so on) in order to assess creditworthiness and then to make a decision: for a critical commentary, see Mikella Hurley and Julius Adebayo, 'Credit Scoring in the Era of Big Data' (2016) 18 *Yale Journal of Law and Technology* 148.

Once regulators have secured the commons, their next responsibility is to ensure that the distinctive values of the community are respected. In the age of information it is important for societies to have a clear view of the particular kind of information society that they want to be. As we have seen, in Europe, communities are wrestling with the tension between free-flowing data and respect for privacy. If one set of values is prioritised, the tension is removed. But, in Europe, we do not really know what kind of information society we want to be—and this has knock-on effects as we have seen for questions about not only particular rights like the right to be forgotten but more generally for consent. While such substantive matters are still unsettled, European regulators are thinking in more technocratic ways but there has been no debate about the kind of regulatory society that Europe wants to be—one that relies on rules or one that accepts measures of technological management.

Beyond these leading responsibilities, the day-to-day regulatory responsibility is to adjust the balance between a plurality of legitimate interests and stakeholders so that the regulatory framework is generally regarded as acceptable. With new technologies and new applications constantly disturbing this balance, regulators will be kept busy tinkering with the rules; and, it is in this context that the flexible notion of privacy makes some sense.

(ii) Four framings of the issues

When a community opens a debate about big data and the regulatory environment, it might approach it in coherentist, regulatory-instrumentalist, or technocratic ways.

For traditional coherentists, the question is whether big data is consistent with ideas of respect for privacy and confidentiality. These traditional ideas give considerable weight to the transactionalist idea of individual agents authorising data collection by giving their free and informed consent. As transactionalism weakens, what counts as respect might well come down to what the community regards as 'reasonable' (and, concomitantly, what members of the community believe that they can reasonably expect in relation to access to information that is private and confidential).

For new coherentists, there will be questions about the compatibility of big data with (i) the commons conditions, (ii) the distinctive values of the community (including questions about leaving decisions to smart machines and whether decision-making concerning humans should mimic human reasoning processes)[94] and (iii) the understanding of when rules rather than technological management should be used, and vice versa.

---

94 Compare Fleur Johns, 'Data Mining as Global Governance' in Roger Brownsword, Eloise Scotford, and Karen Yeung (eds), *Oxford Handbook on Law, Regulation and Technology* (Oxford: Oxford University Press, 2017) 776.

For regulatory-instrumentalists, the question is one of finding an acceptable balance between, on the one hand, the interest in innovation and capturing benefits that are predicated on big data and, on the other, the interest of citizens in being shielded against any risks that accompany big data.[95] Some of the risks might seem relatively trivial, no more than an inconvenience, but others are more serious. The latter risks are varied—for example, health data might lead to unfair discrimination in insurance and employment; lifestyle data might lead to unfair discrimination in employment but also in access to health care, credit, and so on; the more that we commit financial data to online environments, the more we become potential targets for cybercrime of one kind or another; and, as the remarkable rise and fall of Cambridge Analytica has highlighted, big data might actually threaten the pillars of democracy.[96] Moreover, where data are used to predict whether particular individuals present a risk to others (as in the criminal justice or immigration systems), there is always the risk that we find ourselves being not just a statistical, but an all too real, false positive.

Finally, for technocrats, the question is whether technological measures can be employed to assuage whatever risks there are or other concerns (such as traditional concerns about privacy or modern concerns about data sharing).[97] This might well be the way ahead but, in Europe, even if regulators could do it, we have had no serious debate about whether they should do it.

(iii) The transnational nature of the challenge

Suppose that Europe, or the United Kingdom post-Brexit, had a very clear idea about the particular kind of information society that it wanted to be.

---

95 Some regulatory cultures will be more pro-innovation than others (where the emphasis is on risk minimisation. See Meg Leta Jones, 'The Internet of Other People's Things' in Timan et al (n 10) 242.

96 See, e.g., 'Cambridge Analytica closing after Facebook data harvesting scandal' *The Guardian* (May 2, 2018), available at www.theguardian.com/uk-news/2018/may/02/cambridge-analytica-closing-down-after-facebook-row-reports-say (last accessed 9 November 2018). Generally, on the undermining of democracy, see Jamie Bartlett, *The People vs Tech* (London: Ebury Press, 2018).

97 For example, it has been argued that the best response to concerns about unfairness in algorithms is to be found in the computer scientists' toolkits, see Joshua A. Kroll, Joanna Huey, Solon Barocas, Edward W. Felten, Joel R. Reidenberg, David G. Robinson, and Harlan Yu, 'Accountable Algorithms' (2017) 165 *University of Pennsylvania Law Review* 633. See, too, an excellent discussion in Philipp Hacker, 'Teaching Fairness to Artificial Intelligence: Existing and Novel Strategies against Algorithmic Discrimination under EU Law' (2018) 55 CMLR 1143. Here, in response to the perceived lack of regulatory fitness of EU anti-discrimination law relative to the operation of algorithmic decisions, an approach that draws on the enforcement tools introduced by the GDPR (such as Data protection impact assessments) together with bias-detecting and bias-minimising tools developed by computer scientists is proposed.

That would be an important advance but it would not be the end of the matter. For, a community might be very clear about its governance approach to big data and yet find itself unable to take effective regulatory measures because the processing takes place outside the jurisdiction.

Right from the first days of the Internet, there has been a debate about whether national regulators will be able to control big data operations, particularly those operations that are hosted outside the jurisdiction.[98] The conventional wisdom now is that national regulators are relatively powerless but not completely so. For example, all might not be lost if they have some leverage over the provider's assets or personnel or reputation within the home territory (recall how, in *LICRA v Yahoo!* (2000),[99] the Paris court managed to exert some influence over Yahoo!); or if they can persuade internet service providers, or other online intermediaries, to act as 'chokepoints';[100] or, if they can take advantage of whatever opportunities there might be for various kinds of reciprocal cross-border enforcement or other forms of cooperation between national regulators[101]—for example, there might be opportunities to develop cooperative global state engagement by agreeing guidelines in relation to the leading issues. While it might seem outrageous that so little can be done, it is not the case that regulators can do absolutely nothing.[102] The cyberlibertarians were not entirely correct. Nevertheless, if national regulators are unable to get their act together, the cyberlibertarians might as well have been right.[103]

VI Conclusion

Information societies, big data sets, and sophisticated data analytics do not necessarily signal the end of privacy and confidentiality, the end of informed consent, the end of transactionalism—and, to the extent that such values and ideas are important for the preservation of the commons, regulators should not let them slide. However, with the commons secured, regulators face enormous challenges in articulating a regulatory environment that is

---

98 Seminally, see David R. Johnson and David G. Post, 'Law and Borders—the Rise of Law in Cyberspace' (1996) 48 *Stanford Law Review* 1367.

99 en.wikipedia.org/wiki/LICRA_v._Yahoo! (last accessed 9 November 2018). (last accessed 9 November 2018).

100 Natasha Tusikov, *Chokepoints: Global Private Regulation on the Internet* (Oakland: University of California Press, 2016).

101 See, e.g. Lawrence Lessig, *Code Version 2.0* (New York: Basic Books, 2006) Ch.15.

102 See, too, Mark Leiser and Andrew Murray, 'The Role of Non-State Actors in the Governance of New and Emerging Digital Technologies', in Roger Brownsword, Eloise Scotford, and Karen Yeung (eds), (n 94) 670.

103 Compare Hielke Hijmans and Herke Kranenborg (n 12) 14: '[I]n our view governments have to show that the matter [of data protection] is not growing out of hand and that they should reclaim control by ensuring that the issue is fully covered by the rule of law.'

congruent with the community's uncertain vision of the kind of information society that it wants to be. The turbulence in informational rights and interests is simply a symptom of this malaise. Coherentism is on the back foot; regulatory-instrumentalists sense that the old rules will not suffice but they are unclear about what rules would be fit for purpose (because the purposes themselves are not settled); and while technocrats see possible fixes for concerns about privacy and data governance, they are in a vanguard without knowing whether their communities are fully behind them. New coherentists can and should hold regulators to account in relation to both the maintenance of the commons and respect for the distinctive values of the community; but, if the community does not know what kind of information society it wants to be, regulators either have to take a lead or be patient. In this light, it seems reasonable to conclude that the rules relating to our informational interests will continue to evolve.

Epilogue

13

IN THE YEAR 2161

In Nick Harkaway's dystopian novel, *Gnomon*,[1] we are invited to imagine a United Kingdom where, on the one hand, governance takes place through 'the System' (an ongoing plebiscite) and, on the other, order is maintained by 'the Witness' (a super-surveillance State, 'taking information from everywhere' which is reviewed by 'self-teaching algorithms' all designed to ensure public safety).[2] Citizens are encouraged to participate in the System, casting their votes for this or that policy, for this or that decision. In the latest voter briefing, there is a somewhat contentious issue. In a draft Monitoring Bill, it is proposed that permanent remote access should be installed in the skulls of recidivists or compulsive criminals. The case against the Bill is as follows:

> [It] is a considerable conceptual and legal step to go from external surveillance to the direct constant observation of the brain; it pre-empts a future crime rather than preventing crime in progress, and this involves an element of prejudging the subject; once deployed in this way the technology will inevitably spread to other uses ...; and finally and most significantly, such a device entails the possibility of real-time correction of recidivist brain function, and this being arguably a form of mind control is ethically repellent to many.[3]

Now, some of this has been discussed in previous chapters (notably, the movement towards actuarial and algorithmic criminal justice). However, the book has proceeded on the assumption that the regulatory environment is largely

---

1 Nick Harkaway, *Gnomon* (London: William Heinemann, 2017).
2 (n 1) 11.
3 (n 1) 28.

an *external* signalling environment. To be sure, the argument of the book has been that we should re-imagine that regulatory environment so that techno-logical management is included as a strategy for ordering human behaviour. However, the striking feature of the Monitoring Bill in Harkaway's dystopia is precisely that it makes a conceptual leap by asking us to imagine a regula-tory environment that shifts the regulatory burden from external to internal signals. While this may be a prospect that is both too awful to contemplate and anyway too remote to worry about, in this Epilogue we can very briefly ponder that possibility.

It might be said that a movement from the external to the internal is already underway. Many of the devices upon which we rely (such as smart-phones and various quantified-self wearables) are only just external, being extensions to ourselves (much like prosthetic enhancements). Without our necessarily realising it, these devices involve 24/7 surveillance of some parts of our lives—and, might it be, as Franklin Foer suggests, that wearables such as 'Google Glass and the Apple Watch prefigure the day when these companies implant their artificial intelligence within our bodies'?[4] Then, of course, there are experiments with novel neurotechnologies, including brain-computer interfaces, which also problematise the boundary between external and internal regulatory signals.[5] Granting that the boundary between what is external and what is internal is not entirely clear-cut, and granting that we are already habituated to reliance on a number of 'on-board' technological assistants, it is nonetheless a considerable step to the paradigm of internal regulation, namely a behavioural coding that is written into our genetics. However, is this kind of internal regulation even a remote possibility? Is this not a sci-fi dystopia? As ever, the future is not easy to predict.

First, those research projects that have sought to find a causal link between particular genetic markers and a disposition to commit crime have had only very limited success. These are challenging studies to undertake and their results are open to interpretation.[6] Summarising the state of the art, Erica Beecher-Monas and Edgar Garcia-Rill put it as follows:

> Genetic information, including behavioral genetics, has exploded under the influence of the Human Genome Project. Virtually everyone agrees that genes influence behavior. Scandinavian twin and adoption studies are widely touted as favouring a genetic role in crime. 'Everyone knows' that the cycle of violence is repeated across generations. Recently, alleles

---

4 Franklin Foer, *World Without Mind* (London: Jonathan Cape, 2017) 2.
5 For discussion, see e.g., Nuffield Council on Bioethics, *Novel Neurotechnologies: Intervening in the Brain* (London, 2013).
6 See, e.g., Nuffield Council on Bioethics, *Genetics and Human Behaviour: The Ethical Context* (London, 2002); and Debra Wilson, *Genetics, Crime and Justice* (Cheltenham: Edward Elgar, 2015).

of specific genes, such as those transcribing for monoamine oxidase A (MAOA) have been identified and linked with propensities to violence. Shouldn't adding genetic information to the mix produce more accurate predictions of future dangerousness?[7]

To their own question, Beecher-Monas and Garcia-Rill give a qualified answer. Better genetic information might lead to better decisions. However, poor science can be politically exploited; and, even in ideal polities, this science is complex—'while genes may constrain, influence, or impact behavior, they do so only in concert with each other and the environment, both internal and external to the organism carrying the genes.'[8] The fact of the matter is that behavioural genetics is not yet an easy answer to the problem of crime control; to assert that there is a demonstrated ' "genetic" basis for killing or locking a violent individual away and throwing away the key, is simply unsupported by the evidence'.[9] Rather, the authors suggest, we would do better to 'require that experts testifying about human behavior acknowledge the complexity of the environmental (nurture) and biological (nature) interactions, and ultimately recognize that human beings can and do change their behavior'.[10]

Nevertheless, with important developments in the tools and techniques that enable genetic and genomic research, it would be foolish to rule out the possibility of further headway being made. Genetic sequencing is many times cheaper and faster than it was twenty years ago. Researchers who now have access to large data-sets, including genetic profiles, which can be read very quickly by state-of-the-art artificial intelligence and machine learning might find correlations between particular markers and particular behavioural traits that are significant. As with any big data analytics, even if we are struggling to find a causal connection, the correlations might be enough. Moreover, the availability of the latest gene editing techniques, such as CRISPR/Cas9, might make it feasible not only to knock out undesirable markers but also to introduce more desirable regulatory markers. We cannot disregard the possibility that, by 2161, or before, we will have the technology and know-how to rely on coded genomes to do some of our regulatory work.

Secondly, however, if measures of external technological management work reasonably well in preventing crime, there might be little to be gained by committing resources to extensive research in behavioural genetics. To be sure, much of the research might, so to speak, piggy-back on research that

---

7   Erica Beecher-Monas and Edgar Garcia-Rill, 'Genetic Predictions of Future Dangerousness: Is There a Blueprint for Violence?' in Nita A. Farahany (ed), *The Impact of Behavioral Sciences on Criminal Law* (Oxford: Oxford University Press, 2009) 389, 391.
8   *Ibid.*, 392.
9   *Ibid.*, 436.
10 *Ibid.*, 437.

is targeted at health care; but, even so, a successful strategy of external technological management of crime is likely to militate against a serious investment in behavioural genetics research or, indeed, any other kind of internal regulation.

Thirdly, if behavioural geneticists do identify markers that are interpreted as indicating an elevated risk of anti-social or criminal conduct, a community might decide that the most acceptable option for managing the risk of criminality is to regulate the birth of individuals who test positive for such markers. As we noted in Chapter Twelve, there is now a simple blood test (NIPT) that enables pregnant women to check the genetic characteristics of their baby at a relatively early stage of their pregnancy. The test is intended to be used to identify in a non-invasive way babies that have the markers for one of the trisomies, in particular for Down syndrome.[11] In principle, though, the test could be used to interrogate any aspect of the baby's genetic profile including markers that are behaviourally relevant (if and when we know what we are looking for). If a community has to choose between regulating reproduction to eliminate the risk of future criminality and developing sophisticated forms of bioregulation by internal coding, it might well choose the former.[12]

If, in the event, regulatory signals are internally coded into regulatees, this will, as I have said, call for a further exercise of re-imagining the regulatory environment. However, long before such third-generation regulatory measures are adopted, there should have been a check on whether regulators are discharging their responsibilities. Are any red lines being transgressed? In particular, is such internal technological management compatible with respect for the commons' conditions? What is the significance of such management relative to the conditions for agency and moral development? Even if there is no compromising of the commons, regulators and regulatees in each community should ask themselves whether this is the kind of community that they want to be. What are the fundamental values of the community? Just as the draft Monitoring Bill in *Gnomon* invites readers to interrogate their deepest values, any proposed use of behavioural genetics to manage the risk of crime, whether by controlling reproduction or by applying a technological fix to modify the otherwise open futures of agents, should be very carefully scrutinised.

It is trite but true that technologies tend to be disruptive. To this, technological management, whether operationalised through external signals or through internal signals, whether coded into dry hardware and software or into wet biology, is no exception, representing a serious disruption to the

---

11 See Roger Brownsword and Jeff Wale, Testing Times Ahead: Non-Invasive Prenatal Testing and the Kind of Community that We Want to Be' (2018) 81 *Modern Law Review* 646.

12 Compare the vision of regulated reproduction in Henry T. Greely, *The End of Sex and the Future of Human Reproduction* (Cambridge, Mass.: Harvard University Press, 2016).

regulatory order. We need to be aware that technological management is happening; we need to try to understand why it is happening; and, above all, we need to debate (and respond to) the prudential and moral risks that it presents. But, how are these things to be done and by whom?

To do these things, I have suggested, we do not need to provoke objections by insisting that 'code' is 'law'; to bring technological management onto our radar, we need only frame our inquiries by employing an expansive notion of the 'regulatory environment' and our focus on the complexion of the regulatory environment will assist in drawing out the distinction between the normative and non-normative elements. As for the 'we' who will take the lead in doing these things, no doubt, some of the work will be done by politicians and their publics but, unless jurists are to stick their heads in the sand, they have a vanguard role to play. Importantly, in the law schools, it will not suffice to teach students how to 'think like lawyers' in the narrow sense of being able to advise on, and argue about, the legal position by reference to the standard rules and conventions. Beyond all of this, beyond the positive rules of law and traditional coherentist ways of thinking, students need to understand that there is a wider regulatory environment, that there is more than one way of thinking like a lawyer and that the technological management of our lives is a matter for constant scrutiny and debate.[13]

In sum, to repeat what I said at the beginning in the Prologue, the message of this book is that, for today's jurists, some of the issues can be glimpsed and put on the agenda; but it will fall to tomorrow's jurists to rise to the challenge by helping their communities to grapple with the many questions raised by the accelerating transition from law to technological management—and, possibly, by the year 2161, a transition from externally to internally signalled technological management.

---

13 For a glimpse of this future, see Lyria Bennett Moses, 'Artificial Intelligence in the Courts, Legal Academia and Legal Practice' (2017) 91 *Australian Law Journal* 561.

INDEX